8th Process Safety Management Mentoring (PSMM) Forum 2019

Topical Conference at the 2019 AIChE Spring Meeting
and 15th Global Congress on Process Safety

New Orleans, Louisiana, USA
31 March - 4 April 2019

ISBN: 978-1-5108-8689-6

Printed from e-media with permission by:

Curran Associates, Inc.
57 Morehouse Lane
Red Hook, NY 12571

Printed by Curran Associates, Inc. (2019)

AIChE
120 Wall Street, FL 23
New York, NY 10005-4020

Phone: (800) 242-4363
Fax: (203) 775-5177

www.aiche.org

Additional copies of this publication are available from:

Curran Associates, Inc.
57 Morehouse Lane
Red Hook, NY 12571 USA
Phone: 845-758-0400
Fax: 845-758-2633
Email: curran@proceedings.com
Web: www.proceedings.com

TABLE OF CONTENTS

The Effects of Oxygen Enrichment

on the

Explosibility Properties of Flammable Gases and Vapors

Richard W. Prugh
DEKRA Process Safety
113 Campus Drive
Princeton, NJ 08540
www.dekra-process-safety.com
prughrw@aol.com

Prepared for Presentation at
American Institute of Chemical Engineers
2019 Spring Meeting and 15th Global Congress on Process Safety
New Orleans, LA
March 31 – April 3, 2019

The Effects of Oxygen Enrichment on the Explosibility Properties of Flammable Gases and Vapors

Richard W. Prugh
DEKRA Process Safety
113 Campus Drive
Princeton, NJ 08540

Keywords: Flammables, Oxygen, Explosion

Abstract

In the chemical industry, oxygen-containing chemicals are used in the production of partial-oxidation products, including many in the pharmaceutical industry. The vapor spaces in oxidation reactors can contain high concentrations of oxygen gas, in combination with flammable vapors. Thus, it is important to determine the explosibility properties of organic-compound vapors in atmospheres containing high oxygen concentrations.

1. Introduction

In the chemical industry, "nascent" oxygen is involved in the production of partial-oxidation products, including many in pharmaceutical operations. The oxygen is initially in the form of peroxides, such as hydrogen and sodium peroxides, and also as chlorates and perchlorates, periodates, nitrates, and dichromates [1]. The vapor space in oxidation reactors can contain high concentrations of oxygen gas, in combination with the flammable vapors of organic compounds. To evaluate the possible explosion hazards of such mixtures of fuel and oxidizer, it is important to determine the explosibility properties of organic compounds in atmospheres containing high oxygen concentrations.

The basic flammability/explosibility data for this study are from the following references:

Table 1. References for Flammability Data

REFERENCE	FLAMMABILITY/EXPLOSIBILITY PROPERTY					
	Flammable Limits	Auto-Ignition Temperature	Minimum Ignition Energy	Maximum Explosion Pressure	Deflagration Index Kg	Oxygen Conc. Limit
NFPA 49; NFPA 325 [2]	√	√				
NFPA 68 [3]				√	√	
NFPA 69 [4]						√
NFPA 77 [5]	√ *		√			
NFPA 497 [6]	√	√	√			
USBM 503 [7]	√ *					√
USBM 627 [8]	√ *	√ *				
Babrauskas [9]	√	√ *	√ *			
Britton [10]	√ *		√ *			√
Crowl and Louvar [11]	√ *	√	√	√	√	√
Yaws [12]	√	√				

* Limited data for oxygen

Several methods are presented in this paper for estimating explosibility properties for preliminary process design. However, it is prudent to confirm the fuel/oxygen relationships with tests where process safety depends on the reliable control of oxygen concentration.

2. Flammability/Explosibility Properties

The explosibility properties of combustible materials can be divided into Ignition Sensitivity and Explosion Severity [13, 14]. The Lower Flammable Limits, the Autoignition Temperatures, and the Minimum Ignition Energies would be important in evaluating the Ignition Sensitivity. The Maximum Explosion Pressure and the Deflagration Index – a measure of burning rate – would be important in evaluating the Explosion Severity. The following sections describe the relationships of these properties to oxygen concentration.

2.1 Lower Flammable Limit

Data from many sources [15, 16, 17, 18] indicate that there is no significant effect of oxygen concentration on the Lower Flammable Limit [LFL] from the 21 volume-percent oxygen in air to 100% oxygen. At oxygen concentrations greater than the LFL in air, the specific heat of oxygen is about the same as the specific heat of nitrogen; that is, excess oxygen behaves as a diluent at the LFL.

For comparison with the Minimum Explosible Concentrations of combustible dusts and for evaluating the hazards of mixtures of flammable gases/vapors of combustible dusts, the values of LFL can be converted to mass densities by:

$$C = [\,LFL / (100 - LFL)\,]\,[\,MW / 0.024\,] \quad \text{grams per cubic meter}$$

where 0.024 is the number of cubic meters per gram-molecular weight [at 25°C and one atmosphere pressure (14.7 psia)].

For example, the LFL for propane in air is 2.1 volume-percent and, for the molecular weight of 44.1, the Minimum Explosible Concentration would be given by:

$$C = [\,2.1 / (100 - 2.1)\,]\,[\,44.1 / 0.024\,] = 39 \text{ grams per cubic meter}$$

Since the LFL is not significantly affected by oxygen concentration, oxygen enrichment would not affect the ventilation and dilution or inerting precautions that should normally be taken for a flammable vapor or gas.

2.2 Autoignition Temperature

The Autoignition Temperature [AIT] of a fuel is a kinetic phenomenon; that is, the AIT is low for very-long exposures to a given temperature, and it is quite high for short exposures to a given temperature. This relationship can be shown to follow the Arrhenius "law" [19], as:

$$r = A\,e^{-Ea/RT} \quad \text{fraction per second}$$

where A is the pre-exponential factor (in "per second");
Ea is the activation energy (in calories per gram-mole);
R is a constant (1.987 calories per gram-mole per °K); and
T is the temperature (in °K).

If equal reaction rates are assumed for combustion in air and oxygen, at two different temperatures, a value for the activation energy can be obtained, from:

$$Ea = 1.987\,[\,\ln(100/21)\,] / [\,(1/T_{AIR}) - (1/T_{O2})\,] = 3.1 / [\,(1/T_{AIR}) - (1/T_{O2})\,] \quad \text{cal./gmol}$$

The Autoignition Temperatures for 90 chemical compounds in air and in 100% oxygen are available [20]. The average AIT in air was calculated to be about 394°C [about 667°K], and the average AIT in oxygen was about 361°C [about 634°K]. Thus, the differences in AIT for air and oxygen is quite small, at about 5% lower.

The calculated activation energy Ea, as based on the referenced data, would be about 40,000 calories per gram mole, which is a "reasonable" value for an oxidation/ combustion reaction [21].

Thus, the AIT for a given oxygen concentration can be estimated from:

$$AIT_{O2} = AIT_{AIR} / [\ 1 + (\ 1.987\ AIT_{AIR} / 40{,}000\ )\ (\ln (C_{O2}/21)\ ] - 273 \qquad {}^{o}C$$

Oxygen enrichment would not affect the precautions that should normally be taken for a flammable or combustible vapor, such as avoiding exposure to hot surfaces in ovens, in dryers, and on electrical equipment such as motors.

2.3 Minimum Ignition Energy

Previous investigators had found that the Minimum Ignition Energy [MIE] was inversely related to oxygen concentration [22]. The Minimum Ignition Energies for 13 chemical compounds in air and in 100% oxygen are available [23]. The average MIE in air [excluding hydrogen (very low) and trichloroethylene (very high)] was calculated to be about 0.30 millijoule, and the average MIE in oxygen was about 0.0027 millijoule. Thus, oxygen concentration has a very strong effect on MIE, with about a factor of 100 reduction for 100% oxygen. The exponent for oxygen concentration for the 13 chemical compounds [including hydrogen and trichloroethylene] is about 2.9.

Thus, the MIE for a given oxygen concentration can be estimated [+/- 50%] from:

$$MIE_{O2} = MIE_{AIR} / [\ (\ C_{O2} / 21\ )^{2.9} = 6{,}200\ MIE_{AIR} / C_{O2}^{\ 2.9} \qquad \text{millijoule}$$

As indicated above, oxygen enrichment would have a significant effect on the precautions [such as inerting] that should be taken for flammable gases and liquid against very weak electrostatic [and other] ignition sources, including corona discharges [24].

2.4 Maximum Explosion Pressure

The Maximum Explosion Pressure [Pmax] – for combustion in air or in an oxygen-enriched atmosphere – can be calculated rather accurately if the heat of formation of the fuel is known and the degree of dissociation of the carbon dioxide and steam combustion products can be determined. For combustion in an oxygen-enriched atmosphere, the degrees of dissociation are very high; thus, the Pmax in 100% oxygen could be expected to show a rather small increase in pressure as compared to combustion in air.

There is a direct relationship between flame temperature and explosion pressure, as governed by the Ideal Gas Law, as:

$$P_{Max} = 14.7 (T_{FLAME} + 273) / (25 + 273) - 14.7 \qquad \text{pounds/sq.in. [psig]}$$

The available flame-temperature explosion-pressure data are as follows [25]:

Table 2. Flame Temperatures and Explosion Pressures for Fuel/Oxygen Mixtures

FUEL	FLAME TEMPERATURES (°C)			MAX. PRESSURES (barg)		RATIO (bara/ bara)	IDEAL-GAS-LAW RATIOS (°K/bara)	
	In Air	In Oxygen	Ratio [°K/°K]	In Air	In Oxygen		In Air	In Oxygen
Methane	1,950	2,705	1.37	7.1	10.4	1.41	274	261
Propane	1,988	2,822	1.37	7.9	10.2	1.26	254	276
Ethylene	2,088	2,902	1.34	8.0	10.7	1.30	262	271
Acetylene	2,262	3,069	1.32	10.6	11.2	1.05	219	274

Thus, approximate values for Pmax in mixtures of oxygen and in air would be given by:

$$P_{Max,O2} = [(T_{FLAME} +273) / 261] -1 \qquad \text{barg}$$

Thus, if the flame temperature can be calculated, from heats of formation and specific heats – with consideration of dissociations – the maximum explosion pressure for a hydrocarbon can then be calculated. It is likely that this relationship would also apply to other hydrocarbons and carbohydrates. Also, as indicated in the preceding Table 2, the values of P_{MAX} in 100% oxygen are about 30% greater than the value in air [excluding acetylene].

The data in Table 2 above can be used to estimate the Maximum Explosion Pressure [P_{MAX}] for a given concentration of oxygen as:

$$P_{MAX,C} = 0.035\, C_{O2} + 0.92\, P_{MAX,AIR} \qquad \text{bara}$$

The following Table presents the results of calculations, based on the above relationship, for two typical values for P_{MAX} in air [26]:

Table 3. Maximum Explosion Pressures for Fuel/Oxygen Mixtures

$P_{MAX,AIR}$	MAXIMUM EXPLOSION PRESSURES IN FUEL/OXYGEN MIXTURES (barg)							
	30 %O_2	40 %O_2	50 %O_2	60 %O_2	70 %O_2	80 %O_2	90 %O_2	100 %O_2
7 barg	7.4	7.7	8.1	8.4	8.8	9.1	9.5	9.8
8 barg	8.3	8.6	9.0	9.3	9.7	10.0	10.4	10.7
9 barg	9.2	9.6	9.9	10.3	10.6	11.0	11.3	11.7
10 barg	10.1	10.5	10.8	11.2	11.5	11.9	12.2	12.6

Thus, oxygen enrichment would have a significant effect on the extent of damage that could result from ignition of confined flammable gases and vapors, and could affect the design of explosion-protection systems and devices, such as explosion vents.

2.5 Deflagration Index, Kg

The Deflagration Index [Kg for flammable gases and vapors] is a measure of the speed of combustion and, thus, is directly related to the speed of flame through a mixture of fuel and oxidant. In the absence of strong turbulence, the flame speed is related to the fundamental burning velocity [FBV] by the ratio of the densities of the unburned mixture and the density of the combustion gases, by [27]:

$$S_{FLAME} = FBV\ (d_{UNBURNED}/d_{BURNED}) \quad \text{centimeters per second}$$

Typically, the ratio of the flame speed to FBV is about 6.

Previous investigators had observed that there is a relationship between the FBV and the Minimum Ignition Energy [MIE], as given by [28]:

$$FBV = K\,/MIE^{0.8} \text{ centimeters per second}$$

Some data are available for both the FBV and the MIE, as indicated in the following Table [29] :

Table 4. The Relationship Between FBV And Oxygen Concentration

GAS/VAPOR	FBV, Air (cm./sec.)	Kg, Air (bar-m/s)	MIE, Air (millijoule)	FBV, O_2 (cm./sec.)	MIE, O_2 (millijoule)	Kg, O_2 (bar-m/s)	Kg Ratio O_2/Air
Hydrogen	312	638	0.016	1400	0.0012	3,000	4.5
Methane	45	55	0.21	450	0.0027	450	4.6
Ethane	47.5	103	0.24	375	0.0019	670	6.5
Propane	46	100	0.25	370	0.0021	720	7.2
Butane	-	92	0.25	-	0.009	-	-
Hexane	-	-	0.24	-	0.006	-	-
Heptane	42.5	-	0.24	-	-	-	-
Averages	-	-	-	-	-	-	5.7

As indicated in the above Table, the Deflagration Index – and the speed of combustion – in oxygen is almost six times greater in 100% oxygen than it is in air [at 21% oxygen]. The data in the preceding Table can be combined to yield two approximate expressions for the Deflagration Index in given concentrations in oxygen, as:

$$Kg_{,O2} = 0.04 \, (K_{G,AIR}) \, (C^{1.2}) \qquad \text{bar-meters/second}$$

$$Kg_{,O2} = 1.85 \, C^{1.1} / MIE_{AIR}^{0.46} \qquad \text{bar-meters/second [excluding } H_2]$$

Thus, if the value of the Deflagration Index in air is known, or if the Minimum Ignition Energy in air is known, an approximate value for the Deflagration Index at a given oxygen concentration can be determined.

3. Combined Relationships

The explosibility/flammability properties of a given flammable gas or vapor can be compared with a standard or typical flammable gas, through the use of Ignition Sensitivity and Explosion Severity. These property combinations were first developed and utilized for the comparisons of the hazards of coal mines, starting in 1940, where combustible dusts presented an explosion hazard [30].

3.1 Ignition Sensitivity

The product of LFL [in terms of grams per cubic meter], AIT, and MIE form the basis for a calculation of Ignition Sensitivity. A high value of this product indicates a low likelihood of ignition by relatively-weak ignition sources, such as electrostatic discharges, friction, impact, and hot surfaces. For propane in a mixture with 40% oxygen, the product is only 39 x 445 x 0.035 = 600. For comparison, the similar product for typical combustible dusts is about 45,000.

To determine the Ignition Sensitivity of a combustible dust, the LFLxAITxMIE product would be divided into the corresponding product for Pittsburgh [bituminous] Coal Dust, at 2,000,000. Thus, the Ignition Sensitivity for typical combustible dusts is about 2,000,000/45,000 = 45. By comparison, the Ignition Sensitivity for propane in a 40% oxygen mixture would be about 2,000,000/600 = 3,300, indicating the much greater sensitivity of this propane mixture to ignition sources.

As another indication of the effects of oxygen enrichment, the Ignition Sensitivity for a stoichiometric [worst-case] mixture of propane and air would be about 450. Thus, the 40% oxygen mixture would be about 7 times more sensitive to ignition [3,300/450].

3.1 Explosion Severity

The product of Deflagration Index Kg and the Maximum Explosion Pressure Pmax [in terms of barg] forms the basis for a calculation of Explosion Severity. A low value of this product indicates a low likelihood of severe damage from explosion. For propane in a mixture with 40% oxygen, the product is about 275 x 9 = 2,500. For comparison, the similar product for typical combustible dusts is about 1,500.

For combustible dusts, the product is divided by the corresponding product for Pittsburgh [bituminous] Coal Dust, at 840. Thus, the Explosion Severity for typical combustible dusts is about 1,500/840 = 1.8. By comparison, the Ignition Sensitivity for propane in a 40% oxygen mixture would be about 2,500/840 = 3.0, indicating the greater likelihood of damage if a confined mixture of propane and 40% oxygen were to be ignited.

The effect of oxygen enrichment can be shown by the Explosion Severity for a stoichiometric [worst-case] mixture of propane and air, at about 0.9. Thus, the 40% oxygen mixture could be about 3 times more damaging than an air mixture [3/0.9].

3.1 Combination of Ignition Sensitivity and Explosion Severity

If the Ignition Sensitivities of fuel/oxidant mixtures are less than 0.2 AND the Explosion Severities are less than 0.5, then such mixtures "are generally not considered to be significant explosion hazards" [31].
For the mixture of propane and 40% oxygen – for which the properties are shown above – the combination of Ignition Sensitivity and Explosion Severity is about 5 orders of Magnitude [a factor of almost 100,000] above the "no significant explosion hazard" criterion. Thus, it is unlikely that a mixture of air with any flammable or combustible vapor could be considered to present "no significant explosion hazard".

3.1 Combination of Ignition Sensitivity and Explosion Severity as Explosibility Index

The product of Ignition Sensitivity and Explosion Severity can be considered to be an Explosibility Index [32 to 37]. This Index has been found to be useful in comparing the hazards of combustible dusts, but had not yet been applied to flammable gases and vapors.

4. <u>Graphical Presentation of Flammability/Explosibility Properties</u>

Figure 1 presents the flammability/explosibility properties of a mixture of flammable fuel, oxygen, and nitrogen, in terms of Ignition Sensitivity [IS], Explosion Severity [ES], and Explosibility Index [EI] at atmospheric pressure and at an initial temperature of 25°C, as compared to the similar explosibility properties of propane at room temperature and atmospheric pressure. The ratios are based on the available data and the results of the calculations that are described above. As could be expected, the explosion hazards increase as the stoichiometric line – where all carbon is oxidized to carbon dioxide, and all hydrogen is oxidized to steam – is approached.

5. Conclusions.

To evaluate the explosion hazards of mixtures of fuels with oxygen-enriched atmospheres, the available data were utilized to obtain methods for estimating the values of pertinent explosibility properties. The relationships are as follows:

Lower Flammable Limit	Equal to that of fuel/air mixtures	
Autoignition Temperature	$AIT_{O2}=AIT_{AIR}/[1+(2.0AIT_{AIR}/40,000)(\ln(C_{O2}/21)]-273$	°C
Minimum Ignition Energy	$MIE_{O2} = 6,200\ MIE_{AIR} / C_{O2}^{\ 2.9}$	millijoule
Max. Explosion Pressure	$P_{Max,O2} = [[0.035\ C_{O2}) + (0.92\ (P_{Max,AIR} + 1))] - 1$	barg
Deflagration Index	$Kg_{,O2} = 0.04\ (K_{G,AIR})\ (C^{\ 1.2})$	bar-meters/second
	$Kg_{,O2} = 1.85\ C^{\ 1.1} / MIE_{AIR}^{\ 0.46}$	bar-meters/second

In addition, a method is presented for evaluating the Ignition Sensitivity and the Explosion Severity of fuel/oxygen mixtures, using the above relationships, together with a method for graphically showing the relative hazards of such mixtures.

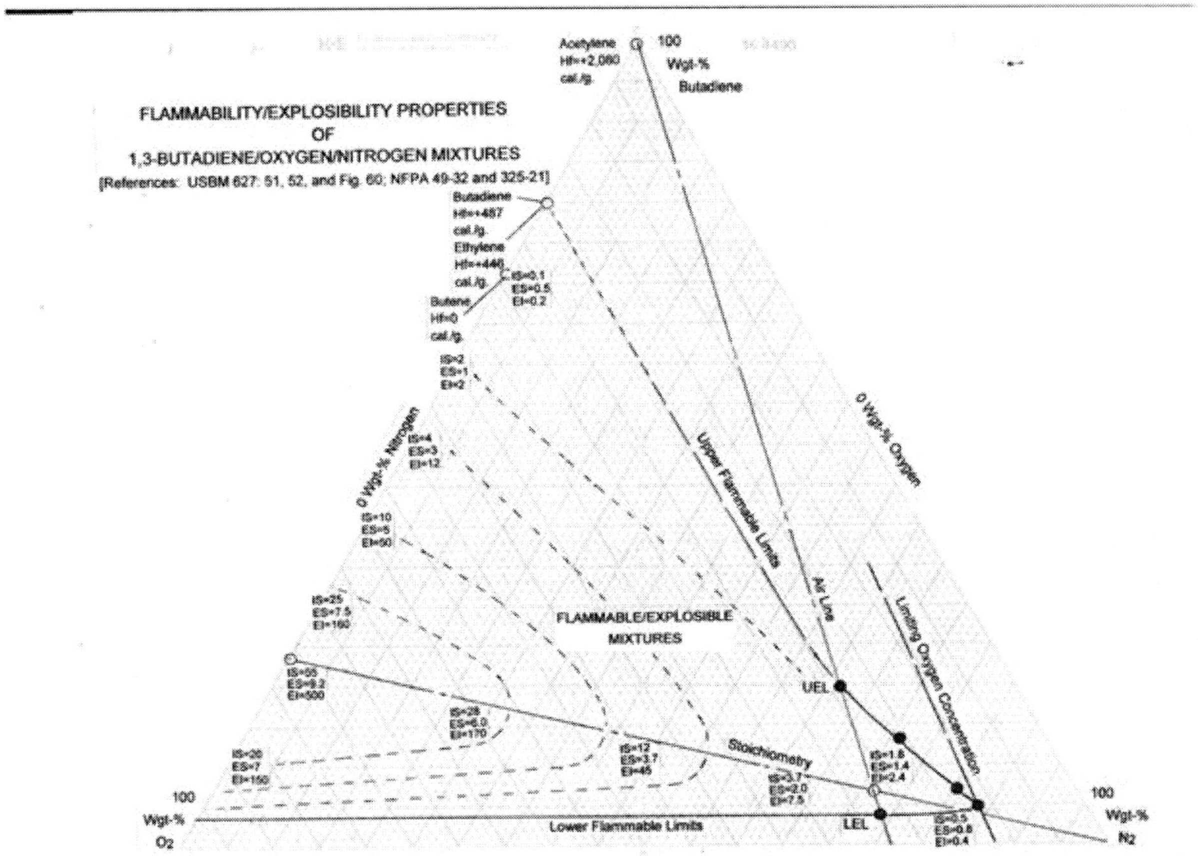

Figure 1 – Weight Concentrations of Gas Mixtures
with
Ignition Sensitivities, Explosion Severities, and Explosion Index
as compared to the properties of Propane

6. References.

1. Urben, P. G. [editor], "Bretherick's Handbook of Reactive Chemical Hazards", 6th Ed., Vol. 2, Page 287 (1999).

2. National Fire Protection Association, "Fire Protection Guide to Hazardous Materials", 13th Ed., including NFPA 49 and NFPA 325 (2002).

3. National Fire Protection Association, "Guide for Venting of Deflagrations", NFPA 68, Table D.1 (2002).

4. National Fire Protection Association, "Standard on Explosion Prevention Systems", NFPA 69, Table C.1(a) (2014).

5. National Fire Protection Association, "Recommended Practice on Static Electricity", NFPA 77, Table B.1 (2014).

6. National Fire Protection Association, "Recommended Practice for the Classification of Flammable Liquids, Gases, or Vapors and of Hazardous (Classified) Locations for Electrical Installations in Chemical Process Areas", NFPA 497, Table 4.4.2 (2017).

7. Coward, H. F., and G. W. Jones [U. S. Bureau of Mines], "Limits of Flammability of Gases and Vapors", Bulletin 503, Table 44 [and many "nose" Figures] (1952).

8. Zabetakis, M. G. [U. S. Bureau of Mines], "Flammability Characteristics of Combustible Gases and Vapors", Bulletin 627, Table A-1 [and many "nose" Figures] (1965).

9. Babrauskas, V., "Ignition Handbook", Table 1A (1992).

10. Britton, L. G., "Avoiding Static Ignition Hazards in Chemical Operations", Page 223 (1999).

11. Crowl, D. A., and J. F. Louvar, "Chemical Process Safety", 3rd Ed., Pages 251, 259, 261, 287, and 654 (2011).

12. Yaws, C. L., "Chemical Properties Handbook", Table 25-1 (1999).

13. National Fire Protection Association, “Recommended Practice for the Classification of Combustible Dusts and of Hazardous (Classified) Locations for Electrical Installations in Chemical Process Areas”, NFPA 499, Paragraph A.5.2.3 (2017).

14. Frank, W. L., and S. A. Rodgers, “NFPA Guide to Combustible Dusts”, Page 92 (2012).

15. Britton, L. G., “Avoiding Static Ignition Hazards in Chemical Operations”, Page 223 (1999).

16. Babrauskas, V., “Ignition Handbook”, Page 1023 (1992).

17. Zabetakis, M. G. [U. S. Bureau of Mines], “Flammability Characteristics of Combustible Gases and Vapors”, Bulletin 627, Page 9 (1965).

18. Crowl, D. A., and J. F. Louvar, “Chemical Process Safety”, 3rd Ed., Pages 263, 265 to 267, and 269 (2011).

19. Crowl, D. A., and J. F. Louvar, “Chemical Process Safety”, 3rd Ed., Pages 398 and 400 (2011).

20. Babrauskas, V., “Ignition Handbook”, Pages 1023 and 1024 to 1058 (1992).

21. Green, D. W., and M. Z. Southard, “Perry’s Chemical Engineers’ Handbook”, 9th Ed., Page 19-24 (2019).

22. Babrauskas, V., “Ignition Handbook”, Pages 75, 76, 203, and 625 (1992).

23. Babrauskas, V., “Ignition Handbook”, Pages 1024 to 1055 (1992).

24. Babrauskas, V., “Ignition Handbook”, Page 625 (1992).

25. Lewis, B., and G. von Elbe, “Combustion, Flames, and Explosions of Gases”, 3rd Ed., Pages 717 and 719 (1987).

26. National Fire Protection Association, “Guide for Venting of Deflagrations”, NFPA 68, Table D.1 (2002).

27. Babrauskas, V., “Ignition Handbook”, Page 35 (1992).

28. Babrauskas, V., “Ignition Handbook”, Page 74 (1992).

29. Zabetakis, M. G. [U. S. Bureau of Mines], “Flammability Characteristics of Combustible Gases and Vapors”, Bulletin 627, Page 43 and Figures 44 and 45 (1965).

30. Nagy, J., et al. [U. S. Bureau of Mines], “Explosibility of Carbonaceous Dusts”, Report of Investigations 6597, Page 2 (1965).

31. National Fire Protection Association, “Recommended Practice for the Classification of Combustible Dusts and of Hazardous (Classified) Locations for Electrical Installations in Chemical Process Areas”, NFPA 499, Paragraph A.5.2.3 (2017).

32. Dorsett, H. G., et al. [U. S. Bureau of Mines], Laboratory Equipment and Test Procedures for Evaluating Explosibility of Dusts”, Report of Investigations 5624, Page 2 (1960).

33. Jacobson, M., et al. [U. S. Bureau of Mines], “Explosibility of Dusts used in the Plastics Industry”, Report of Investigations 5971, Tables A-1, A-2, and A-3 (1962).

34. Jacobson, M., et al. [U. S. Bureau of Mines], “Explosibility of Metal Powders”, Report of Investigations 6516, Tables A-1, A-2, and A-3 (1964).

35. Nagy, J., et al. [U. S. Bureau of Mines], “Explosibility of Carbonaceous Dusts”, Report of Investigations 6597, Page 2 (1965).

36. Dorsett, H. G., and J. Nagy [U. S. Bureau of Mines], “Dust Explosibility of Chemicals, Drugs, Dyes, and Pesticides”, Report of Investigations 7132, Tables A-1, A-2, and A-3 (1968).

37. Grossel, S. S., “Safety Considerations in Conveying of Bulk Solids and Powders”, Page 62 in the “Journal of Loss Prevention in the Process Industries”, Butterworths, Vol. 1, No. 2 (April 1988).

7. Legal Disclaimer and Liability

(a) Limitation of Liability. The consulting services conducted by DEKRA Process Safety (the "Company") were performed using generally accepted guidelines, standards, and/or practices, which the Company considers reliable. Although the Company performed its consulting services pursuant to reliable and generally accepted practices in the industry, the Company does not guarantee or provide any representations or warranties with respect to Client's use, interpretation or application of the findings, conclusions, and/or suggestions of the consulting services provided by the Company. Moreover, the findings, conclusions, and the suggestions resulting from the consulting service are based upon certain assumptions, information, documents, and procedures provided by the Customer. AS SUCH, IN NO EVENT AND UNDER NO CIRCUMSTANCE SHALL THE COMPANY BE LIABLE FOR SPECIAL, INDIRECT, PUNITIVE OR CONSEQUENTIAL DAMAGES OF ANY NATURE WHATSOEVER, INCLUDING WITHOUT LIMITATION, ANY LOST REVENUE OR PROFITS OF THE CUSTOMER OR ITS CUSTOMERS, AGENTS AND DISTRIBUTORS, RESULTING FROM, ARISING OUT OF OR IN CONNECTION WITH, THE SERVICES PROVIDED BY THE COMPANY. The Customer agrees that the Company shall have no liability for damages, which may result from Client's use, interpretation or application of the consulting services provided by the Company.

(b) The Company's pricing of the consulting services provided does not contemplate that the Company shall have any liability resulting from its performance of the consulting services, except as otherwise set forth in the Quotation from the Company. Accordingly, the Customer shall indemnify and hold harmless the Company, its shareholders, directors, officers, employees and agents (the "Indemnified Parties") from and against any and all loss, cost, liability and expense, including reasonable attorney's fees and costs, which any of the Indemnified Parties may incur, sustain or be subject to, as a result of any claim, demand, action, investigation or proceeding arising out of or relating to either: (a) the consulting services provided by the Company; or (b) any material, equipment, specifications or safety information (or lack thereof) supplied to the Company (or which should have been supplied to the Company) by Customer and/or any failure of such materials, equipment, specifications and safety information to comply with any federal, state or local law or safety standard.

(c) For additional terms and conditions, which apply with respect to the provision of this report, see the Quotation provided by the Company and executed by Customer. If any terms set forth in the Quotation conflict with the terms set forth herein, the terms set forth herein shall apply.

Bridging Chemical Engineering (Risk Based Process Safety) to the Humanities – 7 Mega Elements' Approach (Protocol) of Accident Investigation in Light of Ethical Dimension, Architecture & Epistemology

Long(Steven) Zhang
No affiliation since Jul 2014 off-hiring from COP
11455 Bissonnet St Apt #514, Houston TX 77099
longz.houston@gmail.com

Prepared for Presentation at
American Institute of Chemical Engineers
2019 Spring Meeting and 15th Global Congress on Process Safety
New Orleans, LA
March 31 – April 3, 2019

Bridging Chemical Engineering (Risk Based Process safety) to the Humanities – 7 Mega Elements' Approach (Protocol) of Accident Investigation in Light of Ethical Dimension, Architecture & Epistemology

Long(Steven) Zhang
No affiliation since Jul 2014 off-hiring from COP
11455 Bissonnet St Apt #514, Houston TX 77099
longz.houston@gmail.com

Keywords: Investigation, Diagnosis, Humanities, *the Art of War, Narrative, Epistemology*

Abstract

In this paper, facing the challenges of Incident Investigation, actually whole CCPS 4th Pillar "Learn from Experience", chemical engineers' (process safety practitioners') brand new vocabulary (7 Mega Elements' narrative on failure diagnosis) would be presented in context of bridging chemical engineering (RBPS) to the Humanities, of which 3 core topics i.e. ethical dimension, architecture and epistemology would be explored and reflected upon so that a new era of RBPS diagnosis (Learn from Experience) predicted.

1 Introduction

In 2007 Center for Chemical Process Safety (CCPS) unprecedentedly presented to process industry Risk Based Process Safety (RBPS) 4 Pillars (and relevant 20 Elements), among which 4th Pillar "Learn from Experience" consists of 4 Elements: Incident Investigation, Measurement & Metrics, Auditing, and Management Review & Continuous Improvement[1].

In latest decades chemical engineers (process safety practitioners) have been facing some big challenges from accident/incident investigations, actually from whole CCPS 4th Pillar (Learn from Experience), for which there might have been many efforts, discussions, and even some 'solutions'. The author argues, however, the challenges comes not only from the needs of high quality and consistency of accident investigation, but, more fundamentally, from building up a brand new vocabulary (protocol) for chemical engineers to competently communicate of, and negotiate on, accident investigations or "why they occur and how ultimately to prevent".

Toward building a new vocabulary/protocol/narrative on accident investigation or, more broadly, on how Experience could be truly Learned (CCPS 4th Pillar), since 9th Global Congress, the author has had a series of efforts for exploration and reflections on the possible way forward, e.g. RBPS Core value (landlord for temple), Process Safety Cultural Communication (landlord's livelihood) & Egg model[2]; 4th Pillar as "espionage" not to enemies but ourselves[3][4]; 7 Mega Elements & 28 infra cultural activities (practices); RBPS Socio-Technical systems[5]; an ethical dimension of Vision 20/20[6] etc.. And in broader context of bridging chemical engineering (Risk Based Process Safety) to the Humanities, the author finally arrived at 7 Mega Elements' vocabulary/ protocol/ narrative/ "cosmology" on RBPS diagnosis for chemical engineers (process safety practitioners)[7].

2 7 Mega Elements' Protocol – Chemical Engineers' Vocabulary

Louis L. Bucciarelli claimed that he had taken seriously the "two cultures" challenge set by C. P. Snow and been working at bridging engineering to the Humanities for 35 years[8]; instrumental rationality does not suffice itself in engineering design, diagnosis and education[9]. And the author's experiences during the humble quest (through relevance of anthropology, intercultural communication, *the Art of War* etc.) for an ultimate solution to catastrophic (process safety) accidents (chemical engineering failure) seems not only resonating with Bucciarelli's bridging journey but a little bit unique, not only toward chemical engineering (Risk Based Process Safety) but also to engineering discipline at large, i.e., through benchmarking RBPS approach against Sun Tzu's *the Art of War,* 7 Mega Elements (vocabulary/ protocol/ narrative) not only were discovered for (chemical) engineers but function as 7 fundamental (chemical) engineering thoughts, which eventually sheds light on those entailed 28 practices of (chemical) engineering discipline.

Unlike Bucciarelli who tried to bridge philosophy and engineering, the author proceeded another way, i.e. first making relevance of anthropology & intercultural communication to chemical engineering (Risk Based Process Safety), then benchmarking Egg model against *the Art of War*, what the author calls 1st main ancient source for bridging (chemical) engineering to the Humanities, compared to 1st main modern one - Bucciarelli's *Engineering Philosophy.*

When guiding all visitors to some biggest lessons ("more original" two cultures) of ancient cultural traditions or civilizations of east and west, the author claimed that there should be an landlord (core value) to live within the temple building (CCPS 4 Pillars) and, furthermore, the landlord should have livelihood (Process Safety Cultural Communication (PSCC)) as well, and that should the landlord died the building seems still beautiful outwardly, however, with a purpose of tomb. In another words, when (chemical) engineers' vocabulary/ protocol/ narrative becomes deficient, or landlord and his livelihood are absent in the vocabulary, catastrophic (process safety) accidents are bound to happen, no matter how technically over competent relevant chemical engineers (process safety practitioners) might be, e.g., the lesson from the case of Bian Que–Duke Cai.

Through anthropology and intercultural communication, **RBPS Core Value** was defined in two ways, an unfolded one: *RBPS Core Value functions as the root of the tree of Process Safety*

Culture. Deficiencies are symptoms or signals for a dying or dead state of a core value, which means that the organization has strategically failed at the highest levels in managing process safety through a risk based approach – even if there has been no occurrence of process safety accidents till that moment. In another words, the RBPS temple is bound to collapse, sooner or later, due to the change or drift from the original design intent of the temple. And RBPS Core Value's simpler definition, through the concept of Power Distance in value studies: *there is no power, authority or governance on process safety except on behalf of process safety expertise (PSE)*. (PSE means the divine essence of the highest strategic insights and vision of process safety management; the Tao for process safety culture; and the soul of *The Art of War* against catastrophic process safety accidents)

And process safety culture does not exist except through ***PSCC****, whose ultimate purpose is to convey the message of RBPS Core Value. Without nourishment from RBPS Core Value, the tree of Process Safety Culture or PSCC would get dry and dead.* So the temple building was eventually upgraded to Egg model, as showed within the scheme 1st. And these 2 new universes (or 2 Mega Elements) might be deemed what Bucciarelli called the rhetoric of engineering as the parcel of instrumental explanation[10], and probably 2 most fundamental (chemical) engineering thoughts as well.

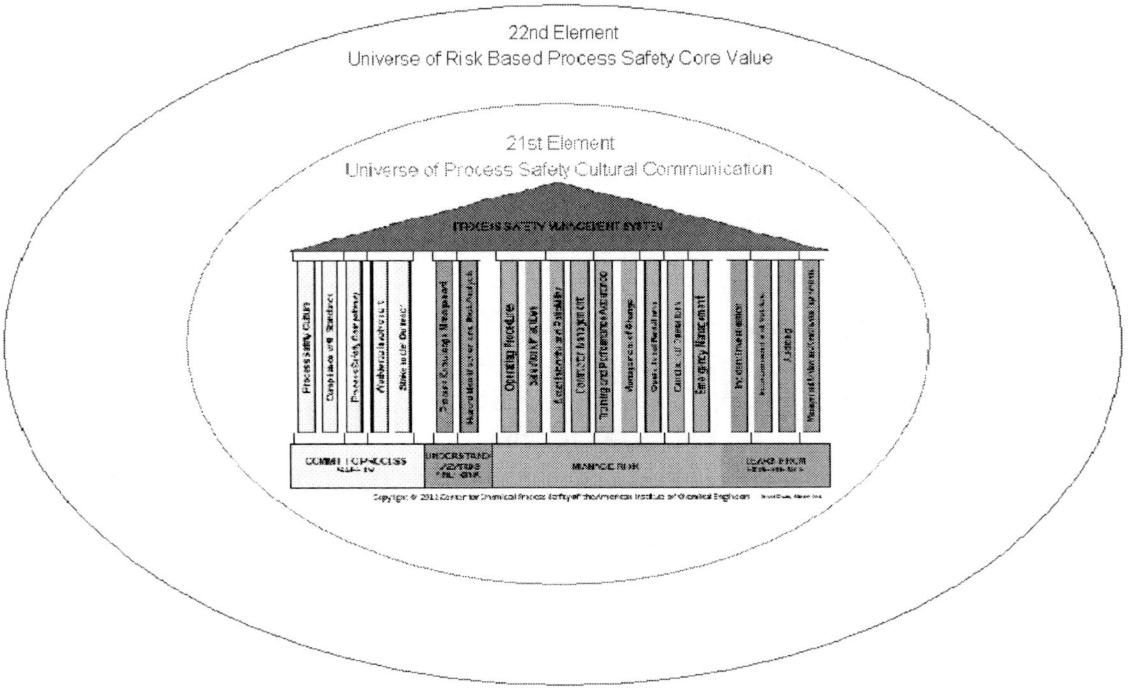

Egg Model (CCPS Temple + 2 universes) - Scheme 1st

The Temple (4 pillars) Figure Source: CCPS, Process Safety Boot Camp, contributed by David Guss, Nexen Inc. Copyright 2010 AIChE and used with permission

Then, in spite of the formidable strengths of most, actually 17, elements of CCPS 4 Pillars, except Process Safety Culture, Process Safety competency & Stakeholder Outreach, the outcome of benchmarking Egg model (4 Pillars + 2 Universes) against *The Art of War,* what the author calls a matured ancient vocabulary on engineering (military) diagnosis, not only shows that those 4 Pillars' (20 elements') narrative, even together with the new 2 Universes, would probably be not so architecturally organized at negotiation and communication on why process safety (chemical engineering) accidents (failure) occur & how to prevent, but makes the author further believe that Sun Tzu's vocabulary (6 Mega Elements, refer to Table 1st) might potentially become a narrative/ protocol paradigm for future (chemical) engineering diagnosis and *The Art of War* an main ancient source of bridging engineering to the Humanities.

For the fact that, however, we could not neglect the historical context of narrative/ protocol of *The Art of War*, the author, in light of modern globalization and global operations, fundamentally redefined Process safety Culture into Process Safety Cultural Communication (PSCC) as the 7th Mega Element, and upgraded another 3 ones i.e. Leadership, Weather/Climate & Espionage as below:

Leadership: a new chain of command, inherently from RBPS Core Value, rather than blindly from superiors;

Weather/Climate: RBPS Core Value enlightened outreach to the stakeholders & community, rather than only tactically maintaining seemingly good relationship with community; &

RBPS Egg		The Art of War	Philosophy of Process Safety	
RBPS Core Value - Tao - **RBPS Worldview**				Worldview
Commit to Process Safety	Process Safety Competency	Leadership	People	Process Safety Cultural Communication (Process Safety Culture) **RBPS Methodology**
	Workforce Involvement			
	Compliance with Standards	Weather/Climate	Plant	
	Stakeholder Outreach			
Understand Hazards & Risk		Terrain		
Manage Risk		Discipline	Procedure	
Learn from Experience		Espionage	Engineering & Humanities	

Table 1st

Espionage: 4th Pillar *Learn from Experience* (4 elements: *investigation, metrics, audit & review*) plays role "espionage", not to the enemies but to ourselves our organizational "national security"

And 7 Mega Elements might also be deemed as 7 (chemical) engineering thoughts.

3 Ethical Dimension (2 Universes) - What Chemical Engineers Count & How

While declared *there is no power, authority or governance on process safety except on behalf of Process Safety Expertise (PSE)* the author claimed that PSE shall eternally and exclusively chair at the highest strategic levels throughout all the processes of decision making and behavior determining, i.e. chemical engineers (process safety practitioners) need getting inspired from RBPS Core Value and persistently conveying the message of RBPS Core Value, both of which make up the ethical dimension (of 7 Mega elements' vocabulary/ protocol/ narrative), probably what Bucciarelli calls "best left sacred"[11], which ensures, or anthropologically designs, all relevant decision making and behavior determining correct. In this sense, at highest strategical levels, chemical engineers do not design at all but RBPS Core Value does.

From philosophical perspective, Bucciarelli claimed that the tensions at the interfaces between different design participants could not be solved by instrumental means alone [12]. And imitating American anthropologist Edward T. Hall's saying the author ever described process safety culture as a prison, unless we have RBPS Core Value as the key to unlock it and PSCC the brightness of the key. And Bucciarelli pointed out that *designing is a social process of negotiation, of iteration, of rectifying mis-steps, even misconceptions – a process rich in ambiguity and uncertainty* [13]. Similarly, the author in metaphor described PSCC as the photosynthesis of tree of process safety culture, the impulses from the heart (RBPS Core Value) and nourishment (highest designing principles) transportation from the root (RBPS Core Value). What does Bucciarelli mean by that 'social process'? What does he mean by "contribute to a sense of community among the citizenry" [14]? And what does the author mean by "photosynthesis" and "nourishment"? With hesitation & obligations, the author claims that the said social process, what the author puts PSCC, traditionally should be called Constitutional Democracy; and RBPS Core Value a nickname for Universal Value in chemical engineering.

As Universal Value & Constitutional Democracy in (chemical) engineering seem more subtle than those traditional ones the author here might call them Universal Value & Constitutional Democracy 2.0, what Bucciarelli described "old themes but in a new light" [15]. And this 2.0 might be acknowledged as two fundamental (chemical) engineering thoughts and the ethical dimension of (chemical) engineers' vocabulary/ protocol/ narrative on accident investigation and whole (chemical) engineering diagnosis (4th Pillars) as well. And probably only in this way could (chemical) engineers truly acquire what Bucciarelli called "new ways of perceiving and reading the world" [16], eventually leading the industries and communities into a new enlightened era of accident investigation and (chemical) engineering diagnosis. And the author might further states that, as the ethical dimension of (chemical) engineers' vocabulary (protocol), Universal Value & Constitutional Democracy 2.0 should function as the highest communication (designing) principles on accident investigation and (chemical) engineering diagnosis, and on accident prevention (foreknowledge) as well.

4 7 Mega Elements' Architecture – 28 Practices' Diagnosis Map

Inspired from Edward T. Hall's map of culture, the author developed a map of 28 infra cultural activities of Process Safety Culture (refer to 7 ones in the diagonal from upper left to lower right and 21 ones above), what Bucciarelli might like to call another, slightly different, "more complex vision of engineering thoughts and practices" [17]. And each infra cultural activity was defined as the critical statements & behaviors related to the interaction between Mega Element A and Mega Element B, e.g. infra activity 04 means: verbally & behaviorally communicate that the actual death of RBPS Core Value might be one of the root causes of accidents. And those 7 ones, extremely, are in fact of the critical verbal statements & behaviors related to the definition of the Mega Element (or engineering thought) concerned, e.g. infra activity 22: verbally and behaviorally embrace and reflect on Leadership - the engineering thought of new chain of command inherently from RBPS Core Value rather than blindly from superiors.

This matching between verbal statements & behaviors, or the inherent relationship between the message of RBPS Core Value and all relevant decision making & behavior determining, seems some resonance with Wittgenstein's recognition that rules and meanings (what the author calls ***tone***) in talk and texts (what the author calls vocabulary/ protocol/ narrative) were mixed with practice [18], or with what Bucciarelli put another way "the meaning of statements is embedded in practice" [19]. So these 28 infra cultural activities' map, an unfolded version of what & how chemical engineers (process safety practitioners) count, might accommodate as highest strategic encompassing diagnosis protocol (map).

	RBPS C.V. 0	P.S.C.C. 1	Leadership 2	Weather/C. 3	**Espionage (Diagnosis) 4**	**Terrain 5**	**Discipline 6**
RBPS Core Value 0	**00**	**01**	**02**	**03**	***04***	**05**	**06**
Process Safety Cultural Communication (PSCC) 1	10	**11**	**12**	**13**	***14***	**15**	**16**
Leadership 2	20	21	**22** 4 dimensions	**23**	***24***	**25**	**26**
Weather/ Climate 3	30	31	32	**33**	***34***	**35**	**36**
Espionage (Diagnosis) 4	40	41	42	43	***44***	***45***	***46***
Terrain 5	50	51	52	53	54	**55**	**56**
Discipline 6	60	61	62	63	64	65	**66**

28 Infra Cultural Activities' Diagnosis Map (Table 2nd)

As the Humanities e.g. philosophy, anthropology & intercultural communication, Sun Tzu's *The Art of War* etc. sheds more lights on the old themes (i.e. RBPS Core value, PSCC, Leadership and Weather/Climate), upon this Diagnosis Map, "another" complex vision of engineering thoughts and practices, the author claims that *the yellow grounded, or more exactly red number lettered area, are all about organizational "national security", and that the infra activity 44 (Espionage)'s 2nd rationality, next to the instrumental one, whose protocol area is painted green, SHALL be reactivated so that "espionage" to our organization would shift to a more proper name - "counter-intelligence", on accident investigation & prevention and on engineering diagnosis at large.*

Through maintaining the architecture/protocol of 7 Mega Elements could all these 22.5 infra activities of "national security" area be reliably visible, open and "problematic" over there, together with "traditional" green area (5.5 infra activities), to critique and diagnosis among board of directors, executive management, operation managers, strategy, legal, HR, chemical engineers (process safety practitioners), and even beyond. Pragmatically, from designing perspective, chemical engineers should pursue this more matured architecture/protocol/vocabulary, getting inspired from RBPS Core Value, persistently conveying its message and maintaining alive this complex vision of engineering thoughts & practices i.e. this 28 Practices (infra activities)' Diagnosis Map, otherwise the insightful topic "social construction of failure/ accident" [20] would turn out to be a truism. In another words, either RBPS Core Value highest strategically designs safety (and tranquility) through PSCC, or otherwise (chemical engineering/ process safety) failures/ accidents/ catastrophes are socially constructed.

5 Epistemology – Future (Chemical) Engineers' Knowledge Hierarchy

Walter Vincenti's understanding of *engineering knowledge,* as J. C. Pitt ever explained, *is task-specific and aims at the production of an artifact to serve a predetermined purpose* [21], of which, the author believes, CCPS 4 Pillars (20 Elements), as of (chemical) engineering/RBPS knowledge, might have been satisfactorily reflecting the main characters. Bucciarelli discovered, however, that engineering knowledge, of which he described *what engineers do, the reasons they do what they do, and their justification for believing the latter are good and true,* is *historically and culturally contingent, not fixed by the dictates of science and empirical fact alone.* [22]. The author might explain here, echoing the former statement "the landlord … even to buy a new house with new style of design to live within", that the (chemical) engineering knowledge we have here with CCPS 4 Pillars might not be the single unique design/house for the landlord [23], but the landscape (landlord, livelihood & house) of the contingencies, what Bucciarelli calls "a picture of both possibilities and limitations of engineering thinking" [24], is unique. In another words, both the landlord (RBPS Core Value), deeply rooted in most cultural traditions and civilizations, and his livelihood (PSCC), what the author likes to call Constitutional Democracy, should preside the landscape of the contingencies and chair and guide at the highest strategic levels the whole engineering design process, for surety including on accident investigation and diagnosis, so that the objects of design, which "traditionally" have been confined to artifacts e.g. a mega petrochemical project/ operations, might also need to include all processes of decision making and behavior determining (refer to Table 2nd – 28 Infra Activities' Diagnosis Map).

The author might state that *a complete (chemical) engineering design process, including on accident investigation, diagnosis & prevention, shall consist of 3 components: RBPS Core Value (Universal Value), PSCC (Constitutional Democracy) and the application of (chemical) engineering knowledge (e.g. one with CCPS 4 Pillars).* And at same time, we recognize that *future (chemical) engineers need a new epistemology - a new hierarchy of knowledge which consists of Universal Value Knowledge, Constitutional Democratic Knowledge & the relevant (chemical) engineering knowledge.* Edward T. Hall had an earth-breaking saying: *culture is communication and communication is culture* [25], from which the author got inspired and developed the concept of PSCC. When (chemical) engineers verbally (of text) and behaviorally (of practice) abandon the ethical dimension of 7 Mega Elements' vocabulary/ protocol/ narrative, they lose the whole process safety culture or, we could say, both their knowledge hierarchy is deficient, akin to a genocide, and their engineering design process seriously flawed, like building concentration camps.

6 Conclusion

In context of bridging chemical engineering to the Humanities, on accident investigation, diagnosis and prevention, and on engineering design process at large, the author presents 7 Mega Elements' Vocabulary/ Protocol/ Narrative, 28 Infra Cultural Activities' Diagnosis Map and a newly galvanized old Hierarchy of Knowledge, so that some outlooks might be made: a) as one of the largest communities of engineering professions, AIChE would pursue leading roles at dialogues between engineering & the Humanities; b) some small groups of chemical philosopher/anthropologist-engineers would emerge; c) an enlightened era of (chemical/process safety) accident investigation, diagnosis & learning, and engineering design process at large, would arrive; & d) engineering education's future would get brighter than ever.

7 References

[1] Center for Chemical Process Safety, *Guidelines for Risk Based Process Safety*, John Wiley & Sons, Inc., 2007

[2] Zhang, L., An Upgraded Definition of Risk Based Process Safety Core Value and Its Impact in Light of Anthropology & Intercultural Communication on the Future Evolution of Process Safety Culture & Management, 2013 AIChE Spring Meeting and 9th Global Congress on Process Safety, San Antonio, April 2013

[3] Sun-Tzu and Ames, Roger T., Sun-tzu: *The Art of Warfare*, Ballatime Books, 1993

[4] Zhang, L., Benchmarking Risk Based Process Safety Egg (RBPS Temple+2 Universes) against The Art of War, 1st CCPS Asia-Pacific Conference on Process Safety, Qingdao, Sept 2013

[5] Zhang, L., People Means Leadership, NOT Simply mean People – 4 New Dimensions of Process Safety Competency, 2014 AIChE Spring Meeting and 10th Global Congress on Process Safety, New Orleans, April 2014

[6] Zhang, L., How Committed to Process Safety is Committed Enough? – Doing Process Safety Cultural Communication (PSCC) as the Inevitable Ethical Dimension for Vision 20/20, 2016 AIChE Spring Meeting and 12th Global Congress on Process Safety, Houston, April 2016

[7] Zhang, L. and Wasel, W., Corporate Memory Loss Ongoing – Except CCPS 4th Pillar (Learn from Experience) as Espionage, 2018 AIChE Spring Meeting and 14th Global Congress on Process Safety, Orlando, April 2018

[8] Bucciarelli, L. L., *Engineering Philosophy*, Delft: DUP Satellite, 2003, p.2

[9] Ibid. p.99

[10] Ibid. p.6

[11] Ibid. p.4

[12] Ibid. p.4-5

[13] Ibid. p.7

[14] Ibid. p.5

[15] Ibid. p.2

[16] Ibid. p.7

[17] Ibid. p.101

[18] Ibid. p.6

[19] Ibid. p.6

[20] Ibid. p. 101

[21] Pitt, J. C., "What Engineers Know", *Techne* 5:3, Spring, 2001, p.25

[22] Bucciarelli, *Engineering Philosophy*, p.99

[23] Pitt, "What Engineers Know", p.26

[24] Bucciarelli, *Engineering Philosophy*, p.99

[25] Hall, E. T., *The Silent Language*, Anchor Books, 1990, p.186

548354 Applying PSM in Vaca Muerta Shale Oil & Gas Operations: The Experience

How to develop, implement and sustain Process Safety in greenfield shale oil and gas projects? This has been a challenge to many new ventures.

Although conventional oil operations have been in place for decades, the exploration, drilling and fracking operations in unconventional shale fields are relatively new in the World. Most of the projects have kicked off less than eight years ago and there are some aspects of the operations that are unique to unconventional fields.

The objective of this paper is to share with the Process Safety global community, how a fit for purpose Process Safety program was developed and is being implemented in the Upstream Unconventional shale oil and gas YPF & Chevron joint venture in Vaca Muerta / Argentina. The project started in 2013, has currently 14 D&C rigs, 5 fracking sets, several separation and treatment production facilities, 3,500 employees and is expected to grow even more in the next 5 years. The paper describes how the project team developed the idea, planned the steps, engaged all stakeholders and selected the key elements of the Process Safety Management System that would add more value in each stage of the project, including drilling, fracking and oil and gas production. The paper also explains the key challenges, such as the difficulties faced to instill Operational Discipline and other key cultural and technical aspects related to Contractor field Operation crews. The journey to implement Process Safety in the Unconventional shale oil & gas operations is very dynamic and is in progress, however, there are already several positive aspects as well as pitfalls to take away as a lesson learned.

548596 A Survey of Process Safety Workers' Attitudes, Behaviors, and Outcomes: Moving Beyond Human Error and Toward a Systems Model

In an effort to move beyond the "human error" explanation for safety incidents, we explored issues with procedures from the worker's point of view by using an anonymous, holistic survey developed from interviews with then currently employed operators. The current sample ($N = 174$) included individuals employed in the process safety industry (MKO/Advanced Procedures Initiative) and were primarily from the Oil & Gas and Chemical industry. The survey was deployed over the course of approximately four weeks. Twelve distinct constructs emerged from the survey (e.g., perceptions of procedure quality, procedure deviation, attitudes toward the procedure change process, etc.). Results indicated that perceptions of procedure quality was the focal variable in all analyses including positive relationships with attitudes toward the procedure change process and negative relationships with procedure deviations, and safety incidents and near-misses. We conclude that these elements are important factors to consider when evaluating and developing procedure systems and provide additional information beyond what the industry considered paramount.

546680 A Method of the Evacuation Pathway Optimization Under Toxic Gas Releases in Chemical Plants

Toxic gases release in chemical plant is a key factor contributing to major accidents and may cause tremendous human life and economic losses. A scientific planning evacuation pathway is an important guarantee of reducing the losses of accidents. Many studies regarding evacuation planning in chemical plant have been carried out by simulating the accident scenarios. However, at present, a systemic method of evacuation pathway optimization has not been developed, and the effectiveness of emergency preparedness has been compromised. To address this problem, an approach based on the CFD modelling of hazardous gas release consequences to plan evacuation pathways and appropriate shelters is proposed in this paper. The methods stochastic programming and dose-response model are mainly employed in this paper. The toxic gas concentration can be converted into poisoning fatality probabilities by dose-response model. Then, stochastic programming method is used to select all the optimal pathways from the evacuation point to the safe area. Setting the minimal fatality probability as the optimization goal, the optimized evacuation pathways under all the leakage scenarios are finally determined through solving the stochastic programming model. As for places with limited evacuation opportunities, an approach of setting up a safe room is proposed in this paper and relevant design guidance are put forward in this paper considering the detailed situation of the chemical plant. A case study of the optimization of evacuation pathways and safe room setup in a salt chemical plant that contains 88 leakage scenarios of chlorine leakage is investigated. With the proposed methodology, the optimized evacuation pathway and the appropriate safe room can provide scientific references for decision makers and evacuees on reducing losses when toxic gases release in chemical plants.

Four Major Gaps That Are Preventing Most Companies Worldwide from Achieving Excellent Process Safety Performance

William G. Bridges
Process Improvement Institute Inc. (PII)
1321 Waterside Lane, Knoxville, TN 37922 USA
wbridges@piii.com

Jeff Thomas
Process Improvement Institute Inc. (PII)
jthomas@piii.com

Prepared for Presentation at
American Institute of Chemical Engineers
2019 Spring Meeting and 15th Global Congress on Process Safety
New Orleans, LA
March 31 – April 3, 2019

FOUR MAJOR GAPS THAT ARE PREVENTING MOST COMPANIES WORLDWIDE FROM ACHIEVING EXCELLENT PROCESS SAFETY PERFORMANCE

William G. Bridges
Process Improvement Institute Inc. (PII)
wbridges@piii.com

Jeff Thomas
Process Improvement Institute Inc. (PII)
jthomas@piii.com

Keywords: process safety, PSM, human factors, near misses, PHA, HAZOP, procedures

Abstract

Process safety requires implementing many management systems, specific engineered features, and operating and maintenance practices effectively. Most companies believe they have done just that, and yet major accidents continue to occur. Why is that? What is missing?

This paper looks at the statistics of major accidents, combined with results from audits and assessments from more than 50 chemical, petrochemical, oil/gas, and related processing companies world-wide. The paper illustrates the four major gaps that are common to the companies/sites that keep having major accidents, compared to those companies/sites that do not have such accidents:

- Accurate and clear operating and maintenance work instructions (procedures)
- PHA of all modes of operation, especially startup, shutdown, abnormal operations, and online maintenance modes
- Getting Near Misses reported and investigated
- Addressing human factors that are missing from most management systems

The paper also provides guidance for closing each of these gaps.

Introduction

"Process safety management (PSM) is the application of management principles to the identification, understanding, and control of process hazards to prevent process-related incidents"[1]. PSM entails development and implementation of programs or systems to ensure that the practices and equipment used in hazardous processes are adequate and are maintained appropriately. The primary categories of programs or systems have come to be called elements of PSM. However, the basic elements of PSM have been defined by many groups in a number of ways. Table 1 lists the elements of PSM systems from various industry and government groups. Many of the elements with different names have essentially the same meaning. For instance, "maintenance and inspection of facilities," together with some aspects of "personnel" practices, both under American Chemistry Council's (ACC's, formerly CMA's) Process Safety Code of Responsible Care™[2] are essentially the same as the single element, "mechanical integrity," under 29 CFR 1910.1190)[3].

Table 1. Comparison of PSM systems

OSHA 29 CFR 1910.119 EPA 40 CFR 68	AIChE/CCPS Risk-Based Process Safety (RBPS) Standard	Responsible Care© Process Safety Code
Management System	**Commitment to Process Safety**	**Management Leadership**
Employee Participation	Process Safety Culture*	Commitment*
Process Safety Information	Compliance with Standards	Accountability*
Process Hazard Analysis*	Process Safety Competency*	Performance Measurement
Operating Procedures*	Workforce Involvement*	Incident Investigation
Training*	Stakeholder Outreach	Information Sharing
Contractors	**Understand Hazards and Evaluate Risk**	CAER Integration
Pre-Startup Safety Review	Process Knowledge Management	**Technology**
Mechanical Integrity	Hazard Identification and Risk Analysis	Design Documentation
Hot Work Permit	**Manage Risk**	Process Hazard Information
Management of Change	Operating Procedures*	Process Hazard Analysis
Incident Investigation	Training and Performance*	Management of Change
Emergency Planning and Response	Safe Work Practices *	Facilities
Compliance Audits	Asset Integrity and Reliability	Siting
Trade Secrets	Contractor Management	**Codes and Standards**
	Management of Change	Safety Reviews
	Operational Readiness*	Maintenance and Inspection
	Conduct of Operations*	Multiple Safeguards*
	Emergency Management	Emergency Management
	Learn from Experience	**Personnel**
	Incident Investigation	Job Skills*
	Measures and Metrics	Safe Work Practices*
	Auditing	Initial Training*
	Management Review and Continuous Improvement	Employee Proficiency*
		Fitness for Duty*
		Contractors

* Contains some of the 6 elements of a Human Factors element, but all missing elements not covered completely. Mostly covered in CoC/OD element

Note that the newest definition of process safety is CCPS's *Risk Based Process Safety (RBPS)*[4] replaces their earlier process safety definition. In the older definition from CCPS [1], there was an element on Human Factors, which brought strong focus to this necessary element; in RBPS, the human factors subelements are now spread across 6 different elements, though some human factors were inadvertently weakened in the transition to *RBPS*.

Although nearly the entire industry agrees that implementing PSM is the right thing to do, interpreting and converting the PSM requirements into practices is unique to each company, and even unique to each plant site. Not only can the requirements be interpreted differently for each site based on local needs, but each site also starts from a different point when they begin to implement a system that is consistent with this regulation.

This paper looks at the statistics of major accidents, combined with results from audits and assessments from more than 50 chemical, petrochemical, oil/gas, and related processing companies world-wide. The paper illustrates the four major gaps that are common to the companies/sites that keep having major accidents, compared to those companies/sites that do not have such accidents:

- Accurate and clear operating and maintenance work instructions (procedures)
- PHA of all modes of operation, especially startup, shutdown, abnormal operations, and online maintenance modes
- Getting Near Misses reported and investigated
- Addressing human factors that are missing from most management systems

The paper also provides guidance for closing each of these gaps.

1. Procedures Need to be More Accurate and Clear

Operating and maintenance procedures have always been crucial to the safety, quality, and productivity of process systems. With the advent of new safety and quality standards such as OSHA's process safety management (PSM) regulation, OSHA's personal protective equipment (PPE) regulation, EPA's risk management program (RMP) regulation, and the ISO 9000 quality standard, many companies are facing the daunting task of developing or upgrading their procedures to satisfy varied and sometimes complex and conflicting requirements.

Although there are overlapping characteristics that the various regulations and standards share, each approach the procedure-writing process from a somewhat different perspective (e.g., quality or safety, protection of workers or protection of the public/environment), and the required level of detail for documenting the procedures differs greatly between these regulations and standards. For instance, the ISO 9000 quality standard advocates documenting procedures and work instructions that impact quality and suggests keeping the procedures as simple as possible, while the OSHA PSM regulation requires detailed procedures that address: all operating modes (startup, shutdown, etc.), operating limits, consequences of deviations, means to avoid hazards, safety and health considerations, and safety systems and their functions. How can companies develop and maintain procedures that ensure productivity and simultaneously satisfy the different regulations and standards? The key is to remember the ultimate goal of the regulations and standards: to reduce human errors that can impact quality, productivity, and/or safety.

Procedure-related errors are errors that occur because some characteristic of the procedure caused task performance to fail. **This is currently the most critical human factor at most sites since 90% of accidents have at least one root cause related to mistakes within procedures**. Reducing these procedure deficiencies can reduce human error rates by a factor of 2 to 20 [5].

1.1. Identified problems

Deficient Procedures are the most prevalent problem in process industries since procedures have not traditionally been developed from the perspective of optimizing human factors; instead procedures have been traditionally developed to meet a compliance requirement. Examples of procedure deficiencies (inaccuracies) include:

- Incorrect/incomplete/nonexistent (most procedures we have audited have been only 70-75% accurate – the inaccuracies include missing critical steps, steps as written are not what needs to be done, or the steps are out of sequence) [5]
- No/misplaced/incorrect information in warnings (for example, a warning should **never contain** the action to take; it should instead **emphasize** the action to take) and warnings should be located BEFORE the step applicable to the warning
- Poor format and presentation rules

1.2. Objective

The flowchart (Figure 1) nd the procedure quality checklist (Table 2) describes the basic sequence of steps for writing effective procedures. Note that some basic rules are contained in the flowchart, such as "operators write operating procedures," "maintenance craft-persons write maintenance procedures," and "Lab technicians write lab procedures." Procedures are NOT written by engineers or department superintendents. This is a critical to ensure procedures are written in the user's common language and are written at the right level of detail.

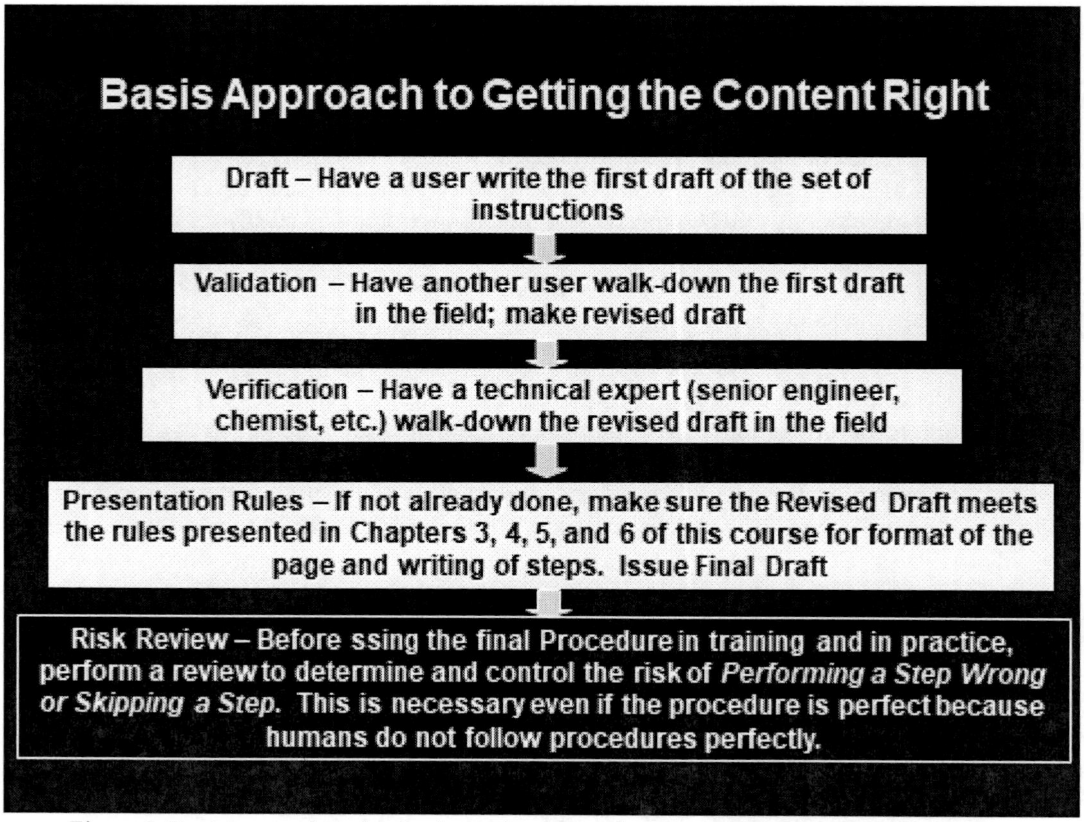

Figure 1. Best Approach to developing written procedures, once the unit has decided which procedures are needed (Extracted from PII Course, Writing Effective Operating and Maintenance Procedures © PII 2003-2017) [6]

The procedure content controls the human error related to "procedure-based" error – if content is wrong, the trainee is very likely to learn the step wrong (trainer uses the procedure as a basis for training or the trainee refers to the procedure step that is wrong when executing the procedure).

Even if the content is correct, not following page formatting best practices can still increase human error rates. If the content of the step is wrong it does not matter if the presentation is clear. If the content is perfect, you can further reduce error rates by 50% to 70% by presenting the information properly.

Rules for developing content[7]

- The procedures should be written at the level of someone who has just completed the basic training for that task. Do not write the procedure for someone just hired or for the 10+ veteran.
- Clearly identify ahead of time what activities need procedures to help ensure error rates are controlled low enough, and identify which activities are considered common "skills" of all of the staff.

> *Example: Starting a pump is typically considered a "skill." This means a procedure step can simply be "Start the benzene recycle pump (P-119)." The procedure does not need to explain how to start the pump. But, starting a pump is a skill that nonetheless must be learned; it requires doing several sub-steps related to positioning of intake and discharge valves, how the pump is throttled at startup, checking local pressure gauge, etc. So, a training module is needed for starting pumps of various types, and the operator needs to learn this necessary skill. Once the skill is learned, it will be applied to so many various cases and so often, that a procedure is no longer needed as a "refresher" on how to "start the pump."*

- The first draft of the procedure must be walked-down in the field by another user. Simply reading through the procedure does not catch enough of the mistakes.
- The revised draft should be reviewed in the field by a technical staff person, such as a process or operations engineer (for operating procedures).
- The procedure needs to be checked to ensure it follow the page format and step writing rules described in the next section. This ensures the accurate steps are clearly presented.

The target is to reach an accuracy of 95% or better[7] (so, no more than one wrong or missing step out of 20 steps). The following have been observed in the field by PII staff by direct data collection at more than 110 sites/plants around the world, based on walk-down review of several to dozens of procedures at each unit/plant:

- At 95% accuracy or better and when the same procedures follow 80% of the rules for procedure clarity (presented next), then most users will follow the written procedures and will try to keep the procedures up-to-date
- At about 85% accuracy or less, about half of the users stop using the procedures
- At about 75% accuracy or less, less than 10% of the users will refer to the procedure or will try to keep it up-to-date. **So, the written procedures are not very effective.**
- Unfortunately, the typical operating procedure walked down by PII staff (accompanied by senior operators, a process engineer, and a shift supervisor) is about 75% accurate (so one step in four is missing or wrong). Usually these inaccurate procedures also follow less than half of the best practices for procedure format, presented next[7].

Procedure Clarity

The importance of the accuracy of the procedure steps were discussed above. The clarity of the procedure steps (how they are written and how the page is formatted) is also important. Following best practices for step and page format reduces human errors by a multiplying factor of 3 to 5.[5],[7]

The best practice rules for writing and validating procedures have been published for many years *(see Bridges & Williams, 1997; Madden & Bridges, 2016, 2017)* [7],[8],[9]. These rules include best practices for formatting of the pages and steps. These have been gradually improved over the past decades and now are incorporated into PII's training materials.[6]

Table 2. Procedure Quality Checklist (courtesy PII, copyright 2003-2017)[6]

#	Issue	Response
	Procedure Content Checklist	
1	Is the procedure drafted by a future user of the written procedure? (Engineers should not author procedures to be used by operators or maintenance staff.)	
2	Is the procedure validated by a walk-down in the field by another future user of the procedures?	
3	Is the procedure reviewed and commented on by technical staff (engineers or vendors)?	
4	Is the procedure checked versus the Page and Step format rules below?	
5	Is a hazard review of step-by-step procedures performed to make sure there are sufficient safeguards (IPLs) against the errors that will occur eventually (when a step is skipped or performed wrong)?	
6	Is the content measured using "newly trained operators" to judge the % of steps that are missing, steps that are confusing or wrong, and steps that are out-of-sequence? (A score of 95% accuracy of content is good; 98% should be the targeted average.)	
	Page Format Checklist	
1	Is the title of the procedure the largest item on the page?	
2	Is the procedure title clear and consistent with other titles, and does it uniquely describes the topic?	
3	Are the document control features the smallest items on the page?	
4	Are temporary procedures clearly identified?	
5	Is white space used effectively? • Is there one blank line between each step? • Does indentation help the user keep their place? • Are the margins large enough to reduce page congestion?	
6	Is type size is 12 pt font or larger?	
7	Is mixed case used for words of steps, with ALL CAPS used only for special cases (such as IF, THEN, AUTO, and WARNING)?	
8	Is the step number very simple (single level of number used)? Only an integer?	
9	Have sections or information not necessary to performing the steps been moved to the back or to another part of the manual or training guide?	
10	Are section titles bold or larger than the text font? Do sections have clear endings?	
11	Is the decision on electronic presentation versus hard copy correct? Are electronic linkages to procedures clear and accurate and easy to use? If paper is chosen, is the color of the paper appropriate?	
12	Is the overall page format (such as Outline format or Two-Column [T-Bar] format) appropriate to the use of the procedure?	
13	Are play script features added for tasks that must be coordinated between two or more users? • Play script is normally used when there are two or more hand-offs of responsibility for steps.	
14	Are rules followed for formatting of Warnings, Cautions, and Notes? (See annotated rules, such as Warnings are for worker safety and Warnings must clearly standout from rest of page.)	
	Step Writing Checklist	
1	Is each step written as a command?	
2	Is the proper level of detail used throughout? This is judged based on: • *Who* will use the procedures • Same level of detail used in similar procedure steps	
3	On average, is there only one implied action per instruction? Best practice is to average 1.2.	
4	Does the procedure indicate when sequence is important? • If sequence matters, each step should be numbered (with an integer or letter)	

	• If sequence does not matter, bullet lists should be used	
5	Are only common words used? Apply "education" level test (5 grade reading level is best)	
6	Do all acronyms, abbreviations, and jargon aid understanding? • Develop a list of such terms for use in procedures *and* communication. • Use terms that users use (within reason)	
7	Is each step specific enough? No room left to guess/interpret: • The meaning of a word or phrase (Check vs. Make sure) • The intent of a step or series of steps • A desired quantity or value • To what equipment the step applies	
8	Is the procedure free of steps that require in-your-head calculations? • Values expressed as ranges rather than targets with error bands • Conversion tables, worksheets, or graphs provided where needed	
9	Are graphics to the user's advantage? • No explanatory paragraphs or lengthy instructions that could be replaced by a picture • No impressive graphics that provide no real advantage	
10	Are references to the user's advantage? • No lengthy explanations or instructions that could be replaced by branching to a reference • No references to a procedure that references still another • No gaps or overlaps between this procedure and a referenced document • If branching, must branch to a procedure, not to a specific step in a procedure	
11	Are rules followed for writing warnings, cautions, and conditional steps? • No more than 2 per page • No actions within a warning or caution (actions must always be numbered steps) • Warnings and Cautions contain descriptions of potential consequences	

For a complete description of best practices, refer to PII's paper *"Best practices for writing operating procedures and trouble-shooting guides"*. [7]

2. PHAs (Process Hazards Analyses) Need to Close Two Major Gaps

Process Hazards is a broad term. The relationship between many components define process hazards. For a PHA team, Process Safety Information (information on the chemicals, technology and equipment) and operating procedures are necessary to identify specific components of and analyze a hazard scenario. These components include: process deviations (parametric and procedural), their causes, the process safety consequence and various safeguards for detecting deviations and causes, preventing causes, detecting and mitigating consequences and intervening safeguards. A thorough review of incident investigations also offers an opportunity to further analyze process hazards.

2.1. Identified problems

Over the last twenty years the CSB has brought needed attention, through its investigations of industry incidences, to two significant gaps in many current PHAs; where the teams did not identify and analyze:

- Process hazards unique to non-routine modes of operations resulting in catastrophic consequences, and
- Process hazards related to Damage Mechanisms.

Though the intent of the US OSHA PSM regulation has always been for PHA teams to analyze the hazards for all modes of operation, OSHA did not give specific requirements for addressing damage mechanisms and only recently have they increased their enforcement attention on PHA of non-routine modes of operation.

Industry has responded by modifying the *Guidelines for Hazard Evaluation Procedures*, 3rd Edition, 2008, CCPS/AIChE[10] to improve the coverage of these two main issues. API in turn issues guidance on addressing damage mechanisms[11]. Industry best practices now exist for addressing both of these weaknesses. Now, with proper approaches, a team with process safety competencies and process knowledge can thoroughly address the issues during a PHA.

Another less significant gap noted by regulators and by the US CSB has been an incomplete consideration, within the PHA, of previous incidents.

This section of the paper lays out the case from the US CSB and US OSHA on these PHA gaps:

- Poor or no coverage of hazards during non-routine modes of operations
- Poor or no coverage of damage mechanisms

2.2. Objectives

2.2.1. Hazard Evaluation of Non-Routine Modes of Operation

Non Routine modes of operation include start-up, shut-down, online maintenance and other abnormal operations. The hazard evaluation of non-routine modes of operation involves reviewing procedures using a HAZOP, simplified HAZOP, or What-if analysis to uncover potential accident scenarios associated with non-routine operations, for continuous or batch operations. Human error is more likely and more critical during non-routine operations. By analyzing procedural steps where human error is more likely, and where human error or component failure could lead to a consequence of interest, risk can be reduced. The hazard evaluation team's objective is to evaluate the risk associated with skipping steps and performing steps wrong.

The purpose of a hazard evaluation of non-routine modes of operation (governed by written procedures) is to make sure an organization has enough safeguards for the inevitable instance when a step is either performed wrong or skipped (inadvertently or due to shortcutting or other reasons).

Industry has found that a HAZOP or what-if analysis, structured to address procedures, can be used effectively for finding the great majority of accident scenarios that can occur during non-routine modes of operation[10],[12],[13],[14]. Experience shows that reviews of non-routine procedures have revealed many more hazards than merely trying to address these modes of operation during the P&ID driven hazard evaluations.

General guidelines for analyzing non-routine modes of operation or batch processes

- Define the assumptions about the system's initial status. "What is assumed to be the starting conditions when the user of the procedure begins with Step 1?"
- Define the complete design intention for each step. "Is the step actually 3 or 5 actions instead of one action? If so, what are the individual actions to accomplish this task?"
- Don't analyze safeguard steps that start with ensure, check, verify, inspect, etc., or where the consequence of skip is "loss of one level of safeguard/protection against…" There is no reason to analyze these steps since they will show up as safeguards of deviations of other steps. This approach is similar to not analyzing a PSV during a HAZOP of continuous mode (i.e., during a parametric deviation analysis); instead the PSV is shown as a safeguard against loss of containment.
- Together with an operator before the review, identify the sections of the procedures that warrant use of:
  - 7-8 Guide Words (extremely large consequences can happen if deviations occur)
  - 2 Guide Words (the system is complex, mistakes are costly, or several consequences could occur)

  - On others, use What-If (no guide words or guide phrases; for use on simpler or lower hazard systems)
- Decompose each written step into a sequence of actions (verbs)
- Apply guide words directly to the intentions of each action

The following preparation steps may also be needed prior to the procedure PHA:
- Walk through procedure in the plant with one or more operators to see the work situation and verify the accuracy of the written procedure.
- Determine if the procedure follows the best practices for "presentation" of the content; the best practices will limit the probability of human error
- Discuss generic issues related to operating procedures, such as:
  - staffing (normal and temporary)
  - human-machine interface
  - worker training, certification, etc.
  - management of change
  - policy enforcement
- Review other related procedures such as lock out/tag out and hot work

- **If the procedures are NOT >90% accurate, then redo procedures before the PHA.**

2.2.2. Damage mechanisms

Recently, more and more regulators, including the US CSB, are making analysis of Damage Mechanisms (DMs) a PHA requirement. This requires that potential process damage mechanisms and their potential consequences of interest be identified. Damage mechanisms can be broken down into three main categories. These categories and examples of each are shown in the following table (Table 3).

Table 3. Types of Damage Mechanisms [15]

Category	Examples
Mechanical	• Mechanical loading failure • Mechanical fatigue • Buckling • Cracking • Embrittlement • Ductile fracture • Brittle fracture

Chemical	• Corrosion
	  - Uniform
	  - Localized
	  - Pitting

Physical	• Thermal failures
	  - Creep
	  - Thermal fatigue
	  - Transformation

The team may use aids such as piping specs, established literature and standards, any applicable MOC documents, etc. to help identify DMs. The PHA leader has several options of how to address damage mechanisms, but PII preferred approach is to discuss during HAZOP under "Loss of Containment" deviation in each node/section.

Damage mechanisms can (and sometimes ***should***) be reviewed prior to the PHA. For example, all MOCs should consider damage mechanisms and their potential consequences of interest before the change is approved. The MOC should also require updates to process safety information which might be relevant to protect against DMs.

Coverage of Damage Mechanism (DM) for each major section [14]

As mentioned above, this is PII's preferred method for thoroughly covering all DM within a PHA/HAZOP of an entire unit. The benefits are that the team can more easily catch changes in DM from section to section and they can more easily identify when unique safeguards, such as Remote Isolation Valves, segregated containments (dike), and unique materials of construction are needed. To facilitate this, a list of generic causes of loss of containment and a generic list of safeguards against loss of containment are covered in the Loss of Containment (LOC) deviation of each section.

For each node (each line, each vessel, each column, etc.), the PHA team should discuss and document each damage mechanism listed in Table 3 (as a cause of loss of containment), consequence of the failure if the damage mechanism occurs, and the safeguards in place to prevent the damage mechanism, detect the mechanism before failure, prevent the release, detect and response to the release, and mitigate or contain the release.

The documentation style varies between PHAs. There are two different styles that have been acceptable in the past.

- One style (Example A, Table 4) uses a reference to a summary table of typical causes (this summary table is not shown but is similar to Table 5 of this paper) instead of listing each individual damage mechanism. In Example A, the same approach is used for safeguards. This approach saves redundant text and some time, but it requires the leader and scribe to be diligent to cover everything in the Typical tables that were referenced in the Cause and Safeguard columns. A modified approach is to relegate some generic causes and safeguards to Typical tables for reference and then to carefully list the specific cause and/or safeguards of interest for LOC for each specific node.

Table 4. Example A – Referencing Generic Tables for Typical Causes and Typical Safeguards [14]

Dev	Deviation	Causes	Consequences	Existing Safeguards	Recommendations
9.10	Loss of containment	Accelerate corrosion External fire High pressure (linked from 9.7) Typical causes of loss of containment (see Table A.1)	Release to atmosphere leading to potential injury of workers and/or community	Pressure Relief Valve's 231A, B, C on Reactor and no valves in line to reactor Most lines / connections are welded construction; only a few flanges Drills conducted each year on evacuation, rescue and isolation Emergency Response personnel are trained at SABIC FTC Generic safeguards protecting against or mitigating process material releases (see Table A.2)	Safety 7. Consider changing the ITPM schedule for managing most PSVs whose inspections are too infrequent based on industry standards and best practice. For instance, the current inspection frequency for PSV-8220 on the Ammonia Receiver (I-2005G) is 9 years, whereas, consensus codes typically recommend testing/inspection every 1-4 years for PSVs in highly toxic services.

- Example B (**Table 5**) does not use or reference a generic or Typical list of causes (damage mechanism) or Typical safeguards, but instead develops a specific listing for the LOC deviation of each node. This style has proven easier to justify to regulators and other outside reviewers but takes more documentation effort.

Table 5. Example B – Specific Listing of Causes and Safeguards [14]

Dev	Deviation	Causes	Consequences	Existing Safeguards	Recommendations
1.9	Loss of containment	Corrosion/erosion External fire and/or flame impingement Gasket, packing, or seal failure Improper maintenance Material defect Operator failing to close or inadvertently opening a valve to the atmosphere (e.g., a valve at a hose connection) Railcar inadvertently derailed Valve leaking to the atmosphere High pressure (linked from 1.5) Acid corrosion caused by high concentration of water (linked from 1.8) High ambient temperature External impact (such as from a mini-engine or another railcar)	Catastrophic release of chlorine from a ruptured railcar Steady release of chlorine from a ruptured connection Steady release of chlorine from a leaky connection High pressure caused by thermal expansion of liquid chlorine if railcar is also over-full	Chlorine repair kit Derailer and warning flag to prevent impact by a mini-engine or another railcar Limited vehicular access to area Maintenance/operator response as required, including isolation if needed Operator periodically monitoring the railcar valves while unloading Personal protective equipment in the area Plugs installed in all chlorine valves to the atmosphere when the valves are not in use Relief valve on each railcar for mitigating releases caused by overpressure Supplier maintenance of railcars (per strictly enforced US DOT requirements) Video monitoring of the unloading area Concrete crossties on rail spur Dike preventing any combustibles spilled nearby from reaching the unloading rack area Concrete railroad ties in chlorine unloading area to prevent fires near railcar	10. Consider installing a chlorine detection system in the unloading and vaporizing area to help detect chlorine releases (especially at likely release points) 11. Verify that periodic maintenance and inspections are being performed in accordance with Chlorine Institute recommendations 12. Review the drainage system for the unloading are, and identify the areas that may be affected by a large chlorine release 13. Consider prohibiting the use of heavy equipment (e.g., cranes) in the unloading and vaporizing area unless special precautions to prevent equipment damage are enacted 33. Consider providing a high pressure alarm for each vaporizer 40. Consider providing a water deluge system in the unloading area to help mitigate chlorine releases from the railcar

Regardless of method, the team leader and scribe must ensure the team rigorously discusses all of the damage mechanisms in Table 5 and adequately documents the results of the damage mechanism review in the PHA/HAZOP analysis tables.

For a complete description of PHA best practices, refer to PII's paper *"Recipe for a Complete Process Hazard Analysis"*. [14]

3. Near Misses Need to be Reported and Enough of Them Investigated

We must learn from accidents and near misses to prevent recurrence. The first step in the learning process is investigation to determine the causes and underlying reasons why accidents and near misses occur. A thorough investigation of root causes will identify the management system weaknesses. Learning which management system weaknesses are leading to near misses and accidents is one of the highest value activities in which a company can invest, and learning from near misses is much cheaper than learning from accidents. Many chemical companies have implemented process safety management systems, and now they are beginning to focus on getting near misses reported and on root cause analysis. This is a very exciting trend. Unfortunately, the chemical industry gets very few near misses reported (the chemical industry is certainly not the only industry with this problem).

To understand more about near misses and getting them reported, it is best to first review the basic definitions.

An incident is either an Accident or a Near Miss.

An Accident is a sequence of unplanned events and conditions that result in harm to people, environment, process, product or image.

A Near Miss is an unplanned sequence of events that could have caused harm if conditions were different or is allowed to progress, but did not in this instance.

Given a consistent understanding of the definition of a near miss, it is possible to estimate how many near misses should be reported for every accident. Studies in several industries indicate that there are between 50 and 100 near misses for every accident. Also, data indicates that there are perhaps 100 erroneous acts or conditions for every near miss. This gives a total population of roughly 10,000 errors for every accident. **Figure 2** illustrates the relationships between accidents, near misses and non-incidents.

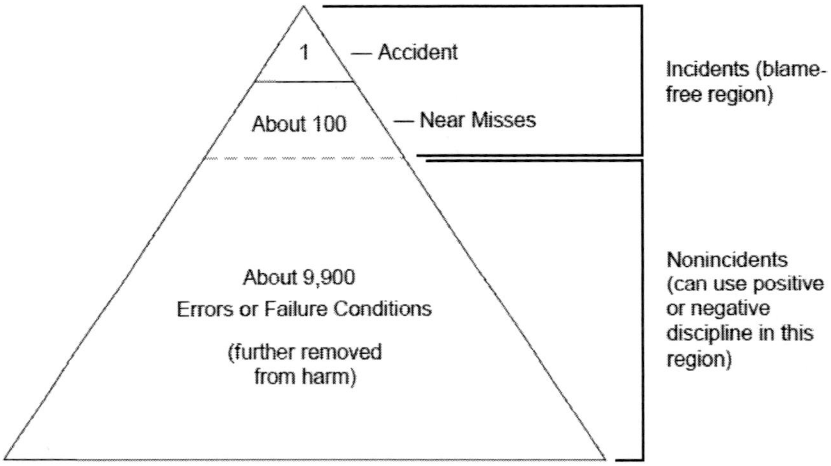

Figure 2. Relationship between Errors and Potential or Actual Impact [16]

Investigating near misses is critical to preventing accidents, because near misses share the causes and root causes of accidents; they are one or two barriers away from the loss/accident. We are very likely preventing many apparently unrelated accidents when we prevent the ones that are obviously related.

3.1. Identified problems

During the past 15 years, we have asked more than 5,000 students of our process safety management (PSM) courses and 3,500 students from our investigator leadership training courses how many near misses they get reported for every accident. The students represented about 400 companies, predominantly in chemical-related process industries (chemical, polymer/plastic, petrochemical, refining, oil and gas exploration, pharmaceutical, pulp and paper, etc.). The answers are quite disturbing. More than 95% said that their ratio of near misses reported to accidents reported falls in the range of 0 to 20. Less than 5% of the individuals said the ratio was greater than 5, and less than 2% said the ratio was higher than 10. Students noted that fear of disciplinary action, lack of management commitment, and lack of understanding of the difference between a near miss and a non-incident were the main reasons why near misses do not get reported. [16]

In conducting about 150 PSM audits and 7000 process hazard analyses (PHAs) during the past 22 years, we have found ratios from 0 to 105. For the first half of the 1990s, more than 90% of the facilities we talked to had ratios in the range of 0 to 0.5, and more than 95% had ratios in the range of 0 to 1. In the last half of the 1990s, more than 90% of the facilities had ratios in the 0 to1 range, and more than 95% had ratios in the 0 to 2 range. In addition, only a few (less than 2% of the facilities) had a ratio higher than 5. By 2005, many facilities had ratios in the 20 to 100 range and each of these had seen great gains (more on this in future articles), but the average across all companies we talked to increased only slightly. By 2011, the ratio of near misses to accidents (any loss event) was about 2 for companies we deal with. But by the end of 2011, many major companies had ratios well above 50! [16]

Our auditors and PHA leaders commented that the primary reasons for lack of reporting were about the same as those found in the survey. The barriers to getting near misses reported, discussed in detail later, are:

- Fear of disciplinary action.
- Fear of teasing by peers (embarrassment).
- Lack of understanding of what constitutes a near miss versus a non-incident.
- Lack of management commitment and lack of follow-through once a near miss is reported.
- An apparently high level of effort is required to report and to investigate near misses compared to low return on this investment.
- There is No Way to investigate the thousands of near misses per month or year!
- Disincentives for reporting near misses (e.g., reporting near misses hurts the department's safety performance).
- Not knowing which accident investigation system to use (or confusing reporting system).
- Company discourages near-miss reporting due to fear of legal liability if these are misused by outsiders.

The good news is that near-miss reporting appears to be improving, and the industry appears to recognize most of the barriers to near-miss reporting. The bad news is that the ratio is still very poor and improvement appears slow!

The number of companies that participated in the informal surveys suggests that the results are a statistically significant representation of the chemical industry.

A formal (written) survey was developed and e-mailed, faxed, and/or mailed to more than 100 companies. Of these, more than 12 replied originally in 1997. Since then, we have reviewed data in detail from more than a dozen major companies. These 25+ companies or affiliates provided data from more than 400 facilities, including more than 150,000 employees in manufacturing and operations. The data were from the chemical industry, polymer industry, refineries, drug/pharmaceutical companies, pulp and paper mills, petrochemical companies, and oil exploration/production.

Table 6 summarizes the results of both the informal and formal surveys.

Table 6. Summary of survey results [16]

Survey Items	Informal surveys		Formal Survey (or actual organization data)
	Classroom polls	Found during Audits and PHAs	
Current Near-miss Reporting Ratio Range (0-5 years prior)	1 to 105	2 to 20	0 to 105
Previous Near-miss Reporting Ratio Range (5-10 years prior)		1 to 5	NA
Current Near-miss Reporting Ratio Average (0-5 years prior)	2-3	3	5
Previous Near-miss Reporting Ratio Average (5-10 years prior)		1	NA
Goal for Near-miss Reporting Ratio (average)	20+	NA	20
Theoretical Upper Limit on Near-miss Reporting Ratio (average)	About 100	NA	100
Number of Participating Companies	400	300+	25+
Number of Participating Facilities	8500+	500+	400+

3.2. Objective

A company should strive to reach a ratio of 50-100 and investigate about 20 near misses per accident. This will provide a statistically significant sample of all incidents (and all important errors) and provide a company with sufficient feedback on which management system weaknesses are causing the errors and component failures. Various companies with different cultures have achieved high ratios with great return on investment.

Table 7 shows a brief description of possible solutions to overcome each one of the barriers identified in the survey.

Table 7. Summary of Survey Results[16]

Barriers	Solutions
Fear of disciplinary action	Implement a policy to NOT punish individuals when their errors lead to accidents and Near Misses.
Fear of teasing by peers (embarrassment)	Ensure that all employees understand the importance of near-miss reporting; demonstrate, through feedback of lessons learned.
Lack of understanding: Near miss vs Non-incident	Develop a list of "in-context" examples that illustrate what you consider to be Near Misses and what you consider to be non-incident
Lack of management commitment and lack of follow-through on reported near misses	Hold management accountable for achieving a Near-Miss reporting ratio
Apparently high level of effort is required to report/investigate Near Misses	Ensure that the data are entered in a database and queried regularly. Share the results with employees so they can see the value of the reported near misses
There is no way to investigate the thousands of Near Misses per month	Let front-line foremen or supervisors decide if a Near Miss or accident needs to be investigated
Disincentives for reporting Near Misses	Ensure that goals and incentives are not tied to lower incident rates (since this discourages reporting), but instead provide incentives for high Near-Miss reporting ratios
Not knowing which accident investigation system to use	Have ONE incident reporting system with ONE approach
Company discourages Near Miss reporting due to fear of legal liability	Involve legal on major Near Misses and accidents to ensure the results are protected as much as possible under attorney/client privilege.

For a complete description of the possible solutions for each barrier, refer to PII's paper *"Gains from Getting Near Misses Reported"*. [16]

4. Key Human Factors are Not being addressed

All accidents (or nearly all, if you consider that there are some natural phenomena that we either cannot guard against or choose not to guard against) **result from human error**. This is because humans govern and accomplish all of the activities necessary to control the risk of accidents. Humans influence other humans in the process – not only do humans cause accidents (unintentionally) by making errors directly related to the process itself, but they also cause errors by creating deficiencies in the design and the implementation of management systems (i.e., we make errors in authorities, accountabilities, procedures, feedback, proof documents and continual improvement provisions). Ultimately, these management systems govern the human error rate by directly causing or indirectly influencing the process. The process-related activities where errors have the most influence include:

- Designing a process
- Engineering a process
- Specifying components

- Receiving and installing equipment
- Commissioning
- Operating a process
- Maintaining, inspecting and repairing a process
- Troubleshooting and shutting down the process
- Managing process, procedure, materials, facility and personnel changes

The 10 human factors categories to be controlled by companies are:
- Available Time (includes staffing Issues) – for responses only
- Stress/Stressors (includes staffing issues)
- Complexity & task design
- Experience/Training
- Procedures
- Human-Machine Interface (includes tools)
- Fitness for Duty
- Work processes & supervision
- Work Environment
- Communication

4.1. Identified problems

Recent major accidents have highlighted the need for increased focus on human factors. The U.S. Chemical Safety Board (CSB) cited (US CSB Video, 2006) human factor deficiencies as one of the main contributors of the catastrophic accident at the BP Texas City Refinery in March 2005[17]. The human factor deficiencies included lack of control of worker fatigue, poor human-system-interface design, poor communication by radio/phone, out-of-date and inaccurate operating procedures, and poor (no) communication between workers at shift handover. The CSB cited similar issues from many other accidents[18] and has urged industry and the U.S. OSHA (the regulator) to pay much more attention to human factors. As a result, the recent U.S. OSHA National Emphasis Program for Refineries included human factors as one of the 12 core elements.

Implementing human factor engineering and policies to prevent accidents is not a new concept. Nearly all (or all, from a more complete perspective) of the causes and root causes of major accidents in the past 30 years have been the result of poor control of human factors. This has been cited in many root causes analysis reports and papers concerning these major accidents.

Process Safety Management (PSM) systems based on OSHA's PSM standard are likely lacking the fundamental human factor elements and implementation guides that, if applied across the applicable PSM elements, would work together to reduce human error. The *Risk Based Process Safety* guideline from the CCPS/AIChE (2007) does contain these human factor requirements.

Many site and company staff do not know there are standards for the control of human factors. However, at last count there were more than 300 non-governmental standards (such as from ANSI, ISO, IEEE, etc.) for the control of human factors, and there are more than 100 government regulations and standards and guidelines. Some of the better government regulations cover aviation, marine operations (shipping) and aerospace. An organization should devote resources to finding and implementing such best practices.

The following is a description of each of the human factors mentioned above and specific ways to control each. Our data from review of more than 2000 incidents reveal that:
- 90% of accidents had at least one root cause related to deficiencies in written work instructions.
- 70% of accidents had at least one root cause related to miscommunication between workers or between workers and supervisors.

- 40% of accidents had at least one root cause related to excessive fatigue of the worker. [5]

4.2. Objective

Not all errors can be prevented. Since the beginning of time, humans have tried to control error rate with more or less success. Human errors have been measured for hundreds of years. Psychologists have studied why humans make mistakes and have gradually put a science around human error probability. Today, the best models for control of human error in the workplace are generally agreed to be related to control of human factors; in turn, these have grown out of what was previously denoted performance shaping factors. What can be done to control each Human factor is discussed in greater detail next.

For a complete description of each human factor, its importance and how to control them; refer to PII's paper *"Human factors and their optimization"* [5].

Time available for the task

Task design is directly linked with this category. Process design, procedure development and worker scheduling need to take into consideration not only the time it takes to complete each task but also the time it takes to manage usual as well as potentially unusual or unexpected situations. True; we do not create norms for exceptions, but exceptions sometimes are more usual than realized.

Stress/Stressors

A well-designed, organized and managed work process helps maintain and promote individual health and well-being, but stresses will arise through no fault of the organization. Supervisors should be trained to recognize stress in workers and, in severe cases, reassign them and help them get assistance to deal with the stress and in other cases, reduce their job pressures for a day by shifting assignments. A wellness program may also be necessary to help cope with stress. Simple approaches, such as training workers to recognize and deal with stress, can help most individuals. Approaches to deal with stress include positive self-talk and practicing relaxation techniques (i.e., breathing control).

Complexity & task design

If roles and responsibilities for accomplishing tasks are not clearly defined, then there will be a risk of serious errors. It is also important that employees and engineers/designers understand the characteristics of the work elements involved, how each element passes information to the other, how each person involved communicates, and that each person has guidelines to adhere to during task design and procedure development.

For optimal performance, task designers often must integrate human and automated equipment in their task analysis.

It is crucial for task designers also to have an understanding of the appropriate allocation of roles and responsibilities to the various participants in a task. Understanding the characteristics and limitations of all of the humans and automated systems involved in the task (particularly critical tasks) will allow for establishing additional controls where needed.

The task analysis process includes *Task definition* which is the initial operational description of the operator tasks required for execution of a given system function; *Performance analysis* of all tasks with respect to human performance requirements; *Error and workload analysis* to be included in the design process, particularly when the likelihood of unacceptably operator error is high; and *Evaluation* since evaluating the design of a new or complex task is important to ensure that the task can be effectively performed as designed. For very high consequence scenarios involved complex human interaction, it is likely prudent to invest in a human reliability analysis (HRA).

Experience/Training

Training to master the Knowledge, Skills, and Abilities (KSAs) required for a job may be obtained from a variety of sources. Companies provide the training or it is obtained from other sources (e.g., trade schools, universities, contractor organizations, etc.) to reach the skills and knowledge established in job descriptions. There are several factors that may result in personnel not mastering the required KSAs for a job. These include course design and delivery methods, course completion, practical skills demonstrations or simulation, and training frequency. Course design begins when the learning objectives have been identified. The design process consists of determining the delivery methods (simulator, on-the-job training, etc.), number of hours required to cover the materials, instructor qualifications, etc. Although some methods, materials and instructors may be more effective or efficient than others, the important issue is that the course content is complete and addresses all of the relevant KSAs, so that the learning objectives are met and the KSAs mastered.

Another determinant of KSA mastery is course completion. Although this factor appears obvious, there are often competing demands on personnel that may pull them out of training at times. As a result, they may miss the instruction related to specific KSAs. Testing may not identify the KSA deficiency because it is impossible to test mastery of all KSAs. Sampling techniques are used to generate examinations. If attendance and participation are not controlled, some personnel may miss training on specific KSAs and testing may not identify the deficiencies before an error is committed.

Another factor affecting KSA mastery is forgetting. An individual's ability to perform a task will degrade over time unless the relevant KSAs are refreshed. Proficiency training (i.e., refresher) will be required for some tasks to maintain the level of mastery that was demonstrated following initial training. One function of training programs is to identify those tasks that require proficiency training.

If certain tasks are performed frequently, proficiency training may be unnecessary. By performing a task, personnel practice the task and obtain feedback on where they have weaknesses. Task performance refreshes the KSAs and successful task performance verifies that proficiency has been maintained. Furthermore, since humans learn by comparison to similar activities, as a worker practices one task, they are learning about similar tasks. Care must be taken here to avoid the practice drifting away from what the procedure requires. Also, if new operators are trained by experienced operators, you must ensure that the new employee is trained according to the procedure, not whatever practice is actually being used. (Of course, ideally, the practice and procedure will be identical.)

Operating and maintenance procedures

Many procedures do not follow best practices for controlling human error, and so the written procedure actually "contributes" to increased error rates. Additionally, many organizations do not have guides on how to troubleshoot (what to do when process deviations occur). The best practice rules for writing and validating procedures have been published for many years. [6], [7], [8], [9]

Procedures have not traditionally been developed from the perspective of optimizing human factors; instead, procedures have been traditionally developed to meet a compliance requirement to have written procedures. Examples of procedure deficiencies (inaccuracies) include:

- Incorrect/incomplete/nonexistent (most procedures we have audited have been only 70-85% accurate – the inaccuracies include missing critical steps, steps as written are not what needs to be done, or the steps are out of sequence)
- No/misplaced warnings or misuse of warnings (for example, a warning should never **state** the action to take; it should instead **emphasize** the action to take an the statement of the action should be in a numbered step)
- Poor format and presentation rules

Human-machine interface

Design for operability refers to designing the Human-System Interface (HSI) to be consistent with the abilities and limitations of the personnel who will be operating it. Weaknesses in the design processes can result in an HSI that is not well suited to the tasks that personnel must perform to ensure plant safety, resulting in increased workload, decreased performance by personnel, and an increased likelihood of errors.

Design for maintainability refers to designing the HSI and associated plant equipment to ensure that personnel are able to perform necessary maintenance activities efficiently. Weaknesses in the design process can result in systems that impose excessive demands on personnel for maintenance and, therefore, are prone to maintenance errors or problems with reliability and availability. *If it is hard to reach, workers will make more errors, including the error of deciding it is not worth it.*

Design for flexibility refers to the way that changes, such as upgrades to the HSI, are planned and put into use. A new HSI component may require the user to perform functions and tasks in new ways. Skills that the user developed for managing workload when using the former design, such as ways for scanning information or executing control actions may no longer be compatible with the new design. The new HSIs must also be compatible with the remaining HSIs so that operators can use them together with limited possibilities for human error. Also, HSI modifications may not be installed or put into service all at one time, causing the user to adapt to temporary configurations that are different from both the original and final configurations. Weaknesses in HSI implementation can increase operator workload and the likelihood of errors.

Fitness for duty

Implement a company fitness-for-duty program with a primary responsibility for detecting and preventing impaired personnel from performing tasks that may affect productivity and safety. Medical evaluations of personnel, behavioral observation programs, employee assistance programs, and drug and alcohol testing are used to detect impairment. Weaknesses in this program may allow impaired personnel to have access to vital areas in a plant where they could commit errors.

Overtime Policies and Practices – Most companies establish limits for work hours to reduce on-the-job fatigue and the potential consequences for poor task performance. Routine authorization for work hours in excess of those recommended may result in fatigued workers. Furthermore, a practice of excluding training or meetings that occur outside of an individual's normal work schedule from work-hour limitations will also contribute to fatigue. But in many cases these have weak enforcement, especially during shutdowns when the company is under stress to re-start soonest.

Shift Scheduling – Shift scheduling may also affect the likelihood that personnel will show performance decrements due to fatigue. A change in the assigned shift or a rotating shift schedule will disrupt circadian rhythms and may increase the likelihood of errors. So, a company must choose the proper shift rotation to allow adjustments to sleep patterns.

Work processes & Supervision

Written plant policies and procedures are meaningful only when they are enforced; otherwise, worker practices are the policy. Once discrepancies are tolerated, individual workers have to use their own judgment to decide what tasks are necessary and/or acceptable. Supervisors and managers should vigorously enforce plant policies/procedures and ensure that plant policies/procedures and practices are revised as necessary to be consistent. To enforce well, a supervisor must (1) lead by example, (2) watch workers often to ensure they are following company policies and procedures, and (3) determine the correct course of action (including recommending changes to policy and procedures) when workers are found deviating from procedures.

Work environment

Programmatic causes of task environment errors are typically found in the company's processes for designing human-system interfaces or in managing maintenance activities. Other programs may also be implicated. Common programmatic causes of task environment errors include:

Industrial Hygiene and Radiation Protection – These programs are responsible for ensuring that task environments have been evaluated to identify hazards and that needed controls are implemented to minimize exposures. Weaknesses in these programs may result in personnel working in task environments that are conducive to errors.

Work Planning and Control – Weaknesses in the work planning and control system may allow work to be planned without consideration of adverse environmental conditions and performed without the necessary compensatory measures. For example, communication devices may not be provided in noisy environments to support task performance. For tasks that involve unusual physical positions or cramped workspace, additional time to complete the task may not be scheduled. Rest breaks for hot and cold environments may not be planned into the work, or additional temporary lighting may not be provided if the work site is not adequately lighted.

Procedures – Weaknesses in the company's procedure development process may result in the design of procedures that are inappropriate for the conditions in which they will be used. For example, procedures that may be used at night, outside and in the rain should be laminated and the type size should be larger to ensure the procedure can be read. Procedures that will be used in vibration conditions may also require larger type size than procedures read in the stationary environment of the control room.

Human Factors Engineering – Weaknesses in the human factors engineering program may result in the installation of new equipment or systems without consideration of task environment characteristics. For example, the impact of control room lighting on the visibility of digital displays or effects of vibration on the legibility of dials or gauges at local control stations should be considered before installation.

Communication

Successful communication requires several steps. The sender first develops the intention to communicate either verbally or in writing. The sender then composes a message that presents the meaning as clearly as possible. The receiver must pay attention to the message and then interpret its meaning. If the communication is successful, the receiver interprets the message consistently with the sender's intended meaning. Table 8 summarizes the rules for successful verbal communication.

The similarity of the meanings given to the message by the sender and receiver can be verified through feedback. An example of feedback verification in verbal communication is when the receiver "repeats back" the message and the sender either agrees with the receiver's repeat back or corrects it. Verification feedback serves an important error-checking function in the communication process. It also allows supervisory oversight of communications to catch errors before they have consequences.

A sender and receiver must both be active for communication to be effective. The sender and receiver share responsibility for ensuring successful communication. However, when companies analyze the causes of events, errors in sending messages are more often identified than errors in receiving. The reasons for the difference are unclear. A company's investigation should consider sending and receiving errors and corrective actions should address both to be effective.

Shift handover is another critical aspect of communication. The status of equipment, repairs that occurred, problems faced, upcoming chores, etc., are all key information the oncoming worker will need to know. One good guide on shift turnover is U.S. DOE 1038-93. [19]

Table 8. Rules for effective verbal (face-to-face, radio, phone) Communication (© PII, 2012)

#	Communication rules
1	Senders MUST require repeat back by the receiver; ask questions to make sure of understanding.
2	Receivers of messages MUST repeat a message they receive.
3	Communicate one task at a time.
4	Sender & receiver must use the standard name and component number (and building name and number) and/or for related reference material (drawing, procedure name/number) in each message.
5	Sender must require receiver to give feedback of foreseeable conflicts. Sender (if message is to subordinate) should follow-up in the field if possible, to make sure message was understood.
6	Receiver must request clarification anytime they believe the message is confusing.
7	Do NOT give verbal instructions to workers who do not have the demonstrated skills to correctly understand and perform the task assignment (unless you plan to supervise/coach the task yourself).
8	Communicate to the most senior member on the work crew.
9	Know your audience & change message accordingly to compensate for lack of knowledge (when appropriate; see other related rules).
10	Use units (and use approved units, such as SI) for process parameters.
11	Always count from left to right when giving instructions related to multiple choices of equipment.
12	Use approved jargon only.
13	Use the words **check**, **make sure**, and **actions** consistently (see procedure writing rules).
14	When possible, write out the task (use a special procedure or work order) rather than communicating verbally. Use formalized templates for consistency and make sure the receiver is provided all information (such as work orders and drawings) necessary for the task.
15	Perform pre-job meeting with work crew (including with contractors or construction company).
16	Perform walkthrough at the location where the work will be completed.
17	Have a backup communication method. Don't rely on one mode of communication. In emergency situations have hand signals as backup to loss of verbal communication.
18	Talk to the operator or technician who is doing the work (no delegation of work); but see Rule 8.
19	Receiver (worker or group) must report back when work is complete.
20	If confusion exists in implementing a task, the shift supervisor (and higher, if supervisor is unavailable) must be contacted to make sure he/she understands the problem.
21	Do NOT use a PA (public address system) for process instructions, since this method of communication does not allow repeat-back.
22	Use Open questions and non-confrontational questioning methods when requesting clarification.
23	When communicating remotely (by phone or radio), if the message is not understood on the second attempt at clarification the Sender must find the Receiver and communicate face-to-face.
24	For large or complicated jobs, in addition to repeat back, the sender must ask the receiver for an assessment of the pre-job briefing to ensure the workers are not confused.
25	Workers and supervisors must keep a shift log to aid in turnover between shifts. Workers and supervisors and support staff (if necessary) must have a minimum or 15 minutes overlap with a relieving shift. *See shift turnover standard for more details.*

Conclusion

Closing the gaps in these 4 areas has a greater than 100:1 payback!

Acronyms and Abbreviations

ACC - American Chemistry Council
AIChE – American Institute of Chemical Engineers
API – American Petroleum Institute
CCPS – Center for Chemical Process Safety (of AIChE)
CMA - Chemical Manufacturers Association
CSB – Chemical Safety Board
DM – Damage Mechanisms
FMEA – Failure Mode and Effects Analysis
GCPS -- Global Congress of Process safety
HAZOP – Hazard and Operability; as in HAZOP Analysis or HAZOP Study
HRA – Human Reliability Analysis
HSI – Human-System Interface
KSA – Knowledge, skills and abilities
LOC – Loss of containment
LOPA – Layer of Protection Analysis
MOC – Management of Change
OSHA - US Occupational and Health Administration
PHA – Process Hazard Analysis
PSM – Process Safety Management
RBPS – Risk Based Process Safety
SOP – Standard Operating Procedure

References

[1] *Guidelines for Technical Management of Chemical Process Safety,* CCPS/AIChE, 1985.

[2] "The Cost and Benefits of Process Safety Management: Industry Survey Results," W. Bridges, *Process Safety Progress*, AIChE, January 1994.

[3] "Necessity of Performing Hazard Evaluations (PHAs) of Non-normal Modes of Operation (Startup, Shutdown, & Online Maintenance)", W. Bridges and Mike Marshall (US OSHA), *18th Annual International Symposium, Mary Kay-O-Connor Process Safety Center, College Station, TX, October 2015.*

[4] *Guidelines for Risk Based Process Safety*, CCPS/AIChE, 2007.

[5] "Human factors and their optimization" W. Bridges and G. Collazo-Ramos, *8th Global Congress on Process Safety*, Houston, AIChE, April 2012.

[6] PII Course 5, *Writing effective operating and maintenance procedures*, © 2003-current

[7] "Best practices for writing operating procedures and trouble-shooting guides", W. Bridges and R. Tew, *13th Global Congress on Process Safety*, San Antonio, AIChE, March 2017.

[8] "Effective Procedure Writing – A Two-Part Series (Part 1: Reducing Risk of Worker Errors; Part 2: Addressing Process Safety Management Requirements for Operating Procedures)," *Chemical Engineering Progress*, 1997, W. Bridges with T. R. Williams (both of JBFA).

[9] "Best Practices for Writing Operating Procedures and Trouble-Shooting Guides", L. Madden and W. Bridges, *12th Global Congress on Process Safety*, Houston, TX, AIChE, March 2016.

[10] *Guidelines for Hazard Evaluation Procedures*, 3rd Edition, 2008, CCPS/AIChE.

[11] "API RP 571 - Damage Mechanisms Affecting Fixed Equipment in the Refining Industry", API, 2nd edition, 2011.

[12] Bridges, W.G., et. al., "Addressing Human Error During Process Hazard Analyses," *Chemical Engineering Progress*, May 1994.

[13] "Necessity of Performing Hazard Evaluations (PHAs) of Non-normal Modes of Operation (Startup, Shutdown, & Online Maintenance)", W. Bridges and Mike Marshall (US OSHA), *18th Annual International Symposium, Mary Kay-O-Connor Process Safety Center, College Station, TX, October 2015 and 12th Global Congress on Process Safety, April, 2016.*

[14] "Recipe for a Complete Process Hazard Analysis – Especially Addressing the Key Demands from US CSB", R. Tew and W. Bridges, *13th Global Congress on Process Safety*, San Antonio, AIChE, March 2017.

[15] PII Course 8, *PHA/HAZOP Leadership training*, © 2003-current.

[16] "Gains from Getting Near Misses Reported", W. Bridges, *8th Global Congress on Process Safety*, Houston, AIChE, April 2012.

[17] US Chemical Safety and Hazard Investigation Board (CSB), *Anatomy of a Disaster: Explosion at BP Texas City Refinery,* published 2006.

[18] US Chemical Safety and Hazard Investigation Board (CSB), *"Pesticide Chemical Runaway Reaction Pressure Vessel Explosion – Bayer CropScience, LP",* published 2008.

[19] "DOE-STD-1038: Guide to Good Practices for Operations Turnover", US DOE, 1998.

546668 Recycle Line from NGL Spheres to Residual Gas Header

Abstract:

HNGL plant is received 1.6 BSCFD of Haradh gas flows into the DGA Gas treating units where CO2 is removed and the produced sweet gas is directed to the NGL recovery units .Hawiyah sales gas is pretreated at HGP for CO2 removal ; hence the produced 2.4 BSCFD flows directly to the NGL units where it is mixed with the Haradh gas. The mixed gas then is processed in the NGL units producing C2+ NGL and sales gas stream. The NGL is pumped to storage spheres and subsequently shipped to NGL piping network through Shedgum Gas Plant. The Residue Gas flows to the sales gas compression area which increase the pressure allowing routing back the gas to the sales gas grid.

The purpose of this project is to install vent line from the two NGL Spheres to Sales Gas (Residue Gas) common header at Hawiyah NGL Plant. The aim of this project is to minimize the continues flaring from the NGL spheres , safe plant resources and protect the environment

Sphere PVs were venting the NGL vapor to the cryogenic burn pit when the pressure of the sphere exceeds 350 psig. In order to lower sphere pressure without losing NGL production, newly installed PVs are set at lower pressure set point of 305 PSIG to recycle the spheres vapors to NGL trains residue gas line, downstream of the brake-compressors..

During summer months July and August (if associated with maximum plant production), spheres are venting about 4.0 MMSCFD of Methane. Based on Methane price of 4.4 $/MMBTU then total saving is 1.05 MM $/year.

547776 Practical Notes on Ragagep Compliance

Process Safety Management (PSM) and Risk Management Plan (RMP) regulations have long required facilities to ensure that they comply with Recognized and Generally Accepted Good Engineering Practice (RAGAGEP). In recent years, authorities having jurisdiction have placed increased emphasis on RAGAGEP compliance. Specifically, emphasis has been placed on documentation and compliance with the industry codes and standards to which a facility is designed and maintained. This poses a unique challenge to older facilities, which have been designed to codes and standards long out of date and especially to owners and operators that do not have access to engineering documents used in the original design of their facility. For facilities using outdated codes, OSHA requires that the operating company document that "equipment is designed, maintained, inspected, tested, and operating in a safe manner."

Over time, the gradual evolution of industry codes and standards may cause facilities to fall out of compliance with the RAGAGEP requirements of the regulations. This paper provides example cases based on authors' experience with various owners and operators of older facilities. Practical methods are discussed for achieving and maintaining RAGAGEP compliance. It includes examples of comparing systems designed to outdated codes and standards against the latest codes, to ensure that they are able to maintain compliance with RAGAGEP.

548482 Propane Safety: Investigation Findings and Lessons Learned in the 2014 Philidelphia FOOD Truck Explosion

On July 1, 2014, a propane cylinder catastrophically failed and exploded on the back of a food truck in Philadelphia, PA. The explosion led to a sudden release of pressurized propane vapor and superheated propane liquid, which rapidly evaporated resulting in a large white cloud that entered and engulfed the rear of the food truck. This cloud ultimately found an ignition source within the food truck and resulted in a significant fireball burning both inside and outside of the food truck. The explosion and ensuing fire fatally injured the food truck owner and her daughter, and also caused injuries to 11 other individuals. This paper will first present the results of our investigation into the cause and origin of the catastrophic failure and explosion event, which included: (1) analyzing video footage of the explosion and ensuing fire; (2) applying propane phase diagrams to determine the expansion and resulting pressures within the cylinder prior to failure; and (3) applying blast techniques to determine the post catastrophic blast and ignition during the event. These techniques included advanced computational fluid dynamic modeling.

The paper will also discuss safety critical design features and processes in the propane industry that must be in place to ensure the safe use and filling of propane cylinders, which include: cylinder requalification; having a fixed maximum liquid level gauge; having a pressure relief valve; and filling cylinders by weight or volume.

548391 Using the Risk Analysis Screening Tool (RAST) in Process Design to Manage Risk

The Risk Analysis Screening Tool (RAST), available through the Center for Chemical Process Safety (CCPS), courtesy of the Dow Chemical Company can enable users to identify hazards and risk scenarios to better manage process safety risks. RAST is a spreadsheet based tool that uses simplified algorithms to identify and evaluate risk scenarios of a unit operation. The risk scenario identification is based on generic HAZOP deviations applied to the unit operation. An engineer can use RAST to identify risks to proposed process designs or plant modifications in the design phase of a project, before P&IDs are developed and a full process hazard analysis study is done. This can save time in process design by weeding out higher risk process alternatives as the process flow diagram is being developed.

The utility of RAST is demonstrated by application to actual case studies. RAST is applied to the runaway reaction at T2 Laboratories in 2007 and to the chlorine release at DPC Enterprises in 2003 showing how the hazardous events and consequences could have been analyzed during the design phase of the projects.

Drug Product vs. Process Safety: Integrating Both into the Design Process from Concept Development to Start-up

Jeffrey Heil, P.E.
Lead Process Engineer
DPS Group Global
959 Concord Street, Suite 100
Framingham, MA 01701
Jeffrey.Heil@dpsgroupglobal.com

Prepared for Presentation at
American Institute of Chemical Engineers
2019 Spring Meeting and 15th Global Congress on Process Safety
New Orleans, LA
March 31 – April 3, 2019

Drug Product vs. Process Safety: Integrating Both into the Design Process from Concept Development to Start-up

Jeffrey Heil, P.E.
Lead Process Engineer
DPS Group Global
959 Concord Street, Suite 100
Framingham, MA 01701
Jeffrey.Heil@dpsgroupglobal.com

Keywords: Process Safety, Process Design, Biotechnology, Inherently Safe Design

Abstract

Process safety has been around for over 350 years but may easily be overlooked or treated as an afterthought in the design process, often leaving hazard reviews and safety valve considerations to the end of detailed design. In the chemical industry, emphasis on safety is generally in respect to process safety, but in the growing biotechnology industry the emphasis is often shifted to product and patient safety. Due to this subtle yet significant difference in focus, it is often challenging for the design team on these projects to secure the necessary project resources to ensure both robust product and process safety in the facility being designed, often on an accelerated schedule. Leveraging experience from working in the oil and petrochemical industry for over 10 years, an in-depth understanding of the project design cycle, and by asking some simple questions along the way; process safety in design can be assured for even the most challenging projects, without a significant impact on project resources.

1 Introduction

1.1 Brief History

The first safety valve, developed in 1681 by Frenchmen Denis Papin, protected the first ever pressure cooker, one of which was used to make meat jelly for England's King Charles II. [1] Papin first invented the steam pressure cooker in 1676 to make the jelly and other food products but then realized no one would be willing to operate the cooker given the risks. Having identified the hazard, he set out to invent a way to mitigate that hazard while ensuring the King got his jellied meat pies.

Following development of the first weighted lever style safety valve invented by Papin, little progress was made until the early 1800's when locomotive steam engines gained notoriety both as a superior means of transportation and a hazardous machine often operated beyond its limits in the name of more power and speed. Safety valves were revised throughout the early 1800's to make these locomotives safer and formally modernized in 1856 with the invention of the first spring actuated pressure relief device that could not be modified to a different pressure, while in service [2].

Modern process safety, as we can still see it in existence today, dates to the early 1900's with the development of modern oil refineries and all the hazards that they include. With these refineries and their complex chemical processes came a wealth of experimentation, safety system design, and industry supported safety initiatives. The goal was the same as it is today, ensuring safety to the employees, the environment, and the facility. The problem with many of these early safety initiatives is that different companies had different ideas of what safe meant and relied on historical data to predict the future. The notion that "it has been running like this for 20 years and will continue for another 20 years" lead to a raising number of deadly process safety incidents, particularly throughout the 1970's to 1980's (i.e. Bhopal, India Methyl Isocyanate release and Piper Alpha explosion in the North Sea.). Plants were operating to the status quo, not improving their practices new technology became available. In parallel, urban sprawl forced industrial plant operation closer to populated areas, increasing risk to the community. The OSHA process safety management (PSM) initiative was introduced into law in the United States to address this very problem [3]. The PSM standard was performance-based meaning that as new technology was developed and safer systems could be designed, those safer designs had to be implemented; i.e., the status quo was no longer good enough. The PSM standard was ratified in 1992: hence, even though the oil and petrochemical industries in the United States have been around for about 100 years, they have only been operating under this standard for 26 years.

Contrast this with today's biotechnology companies. Gentech, one of the earliest, received FDA approval for the insulin drug Humulun in 1982. Like any new technology, investment and growth in biotechnology was slow to start, not taking off until early 2000's. There are many metrics for tracking growth of the Biotechnology industry such as venture capitalist investment and the number of companies in operation, but the high number of mergers and acquisitions over the past 30 years makes such comparisons incomplete. The Nasdaq Biotechnology Index (NBI) was established in 1993 with a baseline index value of 200 and provides a good metric for tracking industry growth over time. After holding steady in the low 200-300s throughout the 1990's, the NBI surged to 1,300 in 2000 as the Biotechnology bubble was forming. The index has steadily risen to a peak of 4,003 on 01-July-2015 and is hovering currently around 3,500 (Q1 2019), exemplifying the continued growth of the industry over the past 20 years. [4]

Pharmaceutical companies (e.g., Aspirin, invented by Bayer in 1899) have been around almost as long the oil companies have and are sometimes confused with biotech. Biotechnology companies are adolescent at best, having typical company life of less than 20 years with little process safety oversight beyond building code requirements as they do not meet the hazard requirements of the OSHA PSM Standard. Not meeting the hazard requirements of the PSM standard does not exempt these companies from a responsibility to ensure a safe working environment. If anything, their proximity to high density population areas and minimal operating experience places them at a higher risk with respect to process safety, but in many cases the operators and designers have only worked in these types of facilities and may not fully appreciate the safety concerns present. Coming from a background where process safety is paramount and the product quality, while important, takes second place, the author has made it his personal mission to increase awareness throughout the Biotechnology industry one project at a time.

1.2 Process vs. Product Safety

In the chemical industry, emphasis on safety is always in respect to process safety; ensuring that the public, employees, environment, and facility (in that order) are sufficiently protected from inherent hazards of a process. "Safety" refers to containment of the process and protection of personnel from harm. If a hazardous material cannot escape the process; it cannot be breathed in, released to the environment, set on fire, or result in damage to equipment that may cause injury to nearby personnel. The previously described process safety management standard formalized the set of tools used by these industries to ensure current best practices are used when it comes to process safety in hazardous environments. The goal of the PSM standard and implications of it being "performance based" is summarized as follows:

> *"... Recognizing the dangers associated with highly hazardous chemicals, Congress mandated (by the Clean Air Act Amendments of 1990) that OSHA set requirements for the management of highly hazardous substances to prevent and mitigate hazards associated with catastrophic releases of flammable, explosive, reactive, and toxic chemicals that may endanger workers. As a result, OSHA developed the Process Safety Management (PSM) standard (issued in 1992), which covers the manufacturing of explosives and processes involving threshold quantities of flammable liquids and flammable gasses (10,000 lbs.), as well as 137 listed highly hazardous chemicals. The PSM standard (29 CFR 1910.119) requires employers to implement safety programs that identify, evaluate, and control these hazards. Unlike some of OSHA's standards, which prescribe precisely what employers must do to comply, PSM is a "performance-based" standard-it outlines key features of safety programs for controlling highly hazardous chemicals, and employers have the flexibility to tailor their safety programs to the unique conditions at their facilities..." [4]*

The biotechnology industry on the other hand often has a more focused emphasis on product safety. Product safety ensures that all current good manufacturing practices (cGMP) are being followed with the goal of protecting the end-user or patient from hazards of a contaminated biological drug product. Product safety was important for the pharmaceutical industry because the medicines being produced were intended for use in the human body and would chemically alter the bodies function. Product cross contamination could result in serious health side effects, especially with highly potent or specialized substances. Biotechnology processes work by growing viruses and manipulating this viral material to deliver highly specialized medicines to the patient. Poor manufacturing conditions could allow unintended bacterial growth alongside the desired product, leaving the future patient with a potentially deadly virus or bacterial infection instead of a cure. The US Food and Drug Administration (FDA) passed the first cGMP regulations in 1969 [6], making them older and with many respects better understood in the industry than the PSM standard is in the oil and petrochemical industry. The practiced engineer will note several similarities in the quoted statements describing the PSM standard and the cGMP standards.

> *"The CGMP requirements were established to be flexible in order to allow each manufacturer to decide individually how to best implement the necessary controls by using scientifically sound design, processing methods, and testing procedures. The flexibility in these regulations allows companies to use modern technologies and innovative approaches to achieve higher quality through continual improvement. Accordingly, the "C" in CGMP stands for "current," requiring companies to use technologies and systems that are up-to-date in order to comply with the regulations. Systems and equipment that may have been "top-of-the-line" to prevent contamination, mix-ups, and errors 10 or 20 years ago may be less than adequate by today's standards." [7]*

Product safety requires process containment too, but in many cases the consequences are less likely to cause bodily harm so long as product quality control procedures ensure quarantine of the affected drug product. Note that this statement applies to the biotechnology industry, the consequences of a loss of containment in the pharmaceutical industry can be as deadly as in a chemical plant, depending on the substance released. Fifty years ago, when the cGMP's were adopted, pharmaceutical operators already knew about process safety and adopted it as readily as oil refineries and industrial chemical plants did. Twenty years ago, biotechnology was in its infancy and there were few large-scale biological processes in operation. Ten years ago, most biological processes ran at atmospheric pressure using water-based chemistry. The risk associated with potential process safety events was minimal in comparison. Today, many biotech processes use advanced solvent-based chemistries needing large volumes of hazardous flammable solvents, a key element in the severity of an incident when a loss of containment occurs. Add to this unique process chemistries such as dichloroacetic acid, methyl chloride, hydrochloric acid, and sodium hydroxide; all of which can easily corrode stainless steel, a common metal alloy used in biological processing, and unique hazards become a reality even at small scale.

2 The Design Process

Designing processes to produce goods takes time, time that the future producer of those goods doesn't always like to part with because "time is money". Each month spent designing or building the facility to manufacturer a product is a month of revenue missed or often in biotech, a month of trials that could be running. A typical design process includes several independent stages or phases. An ideal project would separate each of these phases with well-defined checks creating structure, an opportunity for detailed reviews, and adding time to the project. Initial concept development, feasibility, and design of large-scale chemical or industrial facilities often involves acres of real estate, spans years, and costs millions of dollars. By contrast small-scale bioprocessing facilities, often less than 40,000 ft^2, are fully designed in terms of weeks or months and leverage parallel work paths where possible.

The impact of such a short design schedule is that time for staggered activities, separated by well-defined stage-gates, is often sacrificed in the name of the project schedule. It is possible to crash a project schedule, completing the design early, but doing so limits the amount of design work that can be completed based on previously completed design elements. If a project schedule allots for 6-8 weeks between establishing the basis of design for a facility and documents that are to be issued for vendor bid, scheduling one week for a hazard analysis and another week to implement the resulting design changes (25% of the schedule) is a challenge. By the time the hazard analysis is complete, many aspects of the design have been progressed and resulting changes will result in unbudgeted rework. The same is true for the design of key safety systems such as relief devices or process interlocks that require a complete system design to be finalized. The engineer can anticipate credible overpressure scenarios and safe operating limits for a process but cannot complete these design elements until piping to and from the relief device is complete and the hazard analysis is approved, accepting the level of protection assigned to each process step. It is for these reasons that it is not uncommon to see relief devices for a process being sized and planned for as one of the final project tasks. Explaining to an architect that a storage tank must be moved to meet a flammable liquid set-back identified during a hazard review or that another roof penetration is required for a relief device tail pipe two weeks before the completion of a design are just two examples of real-world challenges that project teams may be faced with.

Understanding the different design phases or stage-gates, the tasks that occur during them, and their impact on the ultimate facility design is important if one is to create a plan for addressing process safety on such a short schedule. Table 1 summarizes one example of a design process, key stage-gates, and design elements to be considered in each phase. Note that jumping a phase and for example combining schematic design and design development (the most common violation of the design process) into one phase can confuse the separation of tasks between these phases and eliminate checks important to the design process.

Table 1 – Key Aspects of the Design Process

Phase	Process Design	Process Safety
Feasibility Study	• Identify product to be manufactured and major unit operations • Establish block flow diagram of process • Identify facility requirements • Confirm business case for proposed facility	• Identify hazardous materials used in the process • Identify hazardous operating conditions (i.e. high pressure) • Establish acceptable risk criteria for operation of the facility

Phase	Process Design	Process Safety
Concept Design	• Breakdown individual processing steps to the equipment level • Develop process flow diagrams for each major process step • Develop preliminary heat and material balance • Develop equipment list and identify long lead equipment (to prioritize design development tasks) • Detailed layout of process equipment • Identify utility requirements and size utility systems • Size and layout major process headers	• Develop a comprehensive hazardous and flammable materials inventory / determine if additional design precautions are required • Identify potential material compatibility concerns and material specifications for each process step • Complete a code review in collaboration with the full design team to ensure all building, fire, and design codes are being followed • Ensure facility citing requirements are met in the layout
Design Development	• Develop initial P&IDs for the process • Identify and document key control loops and automation requirements • Finalize design of all process equipment (size, pressure rating, operating conditions) • Complete detailed sizing and specification of major process headers and utility lines • Finalize process heat and material balance • Prepare equipment and piping specifications and datasheets	• Confirm inherently safe design where possible (pressure profile versus equipment design) • Identify safe operating limits for each unit operation, add control points as required • Review P&IDs for relief devices and confirm their discharge locations • Complete scenario analysis and preliminary sizing for all relief devices. Based on preliminary sizing, define relief device inlet/outlet piping • Confirm piping specifications versus services • Confirm vessel sizing versus hazardous material inventory and update as required • Complete a preliminary hazard analysis on the full process at the end of the design development stage

Phase	Process Design	Process Safety
Detail Design	• Incorporate all recommendations from the design development preliminary hazard analysis • Finalize all equipment and line sizing • Fully specify all process components including equipment, lines, valves and instruments	• Ensure that all actions from the preliminary hazard analysis are incorporated into the design • Review and finalize relief device sizing • Complete sizing and design of effluent handling system for relief device discharges • Double check the set points of key process variables versus the operation's safe operating limits, previously defined. • Review final equipment specifications against material compatibility matrix and equipment specifications

3 Safety Throughout the Design Process

Process safety should be considered in the earliest stages of design. One principle of process safety design is that inherently safe design, creating a system that physically cannot fail, is the only way to prevent an incident from occurring. The easiest example of an inherently safe system is designing a pressure vessel to withstand maximum system pressure, even if all process controls were to fail. While an easy goal on paper, inherently safe design is not always possible and often never fully achievable. A vessel can be designed to the proper pressure, but an unforeseen corrosion issue could reduce the design pressure making the vessel susceptible to premature failure. In anticipation of this and to protect against it, engineers design-in multiple layers of protection to a process starting with inherently safe design at the concept level and ending with proper emergency response procedures developed during commissioning and simulated during plant operation. Continuing with the vessel example, minimizing the quantity of material stored and having an inventory management plan to support smaller on-hand inventories is another way to reduce hazards associated with the process. Additional design principles or layers of protection used are summarized in Figure 1. Different layers of protection against a process safety event are established during concept development and built-up throughout the design process.

Figure 1. Typical Layers of Protection Applied to a Design

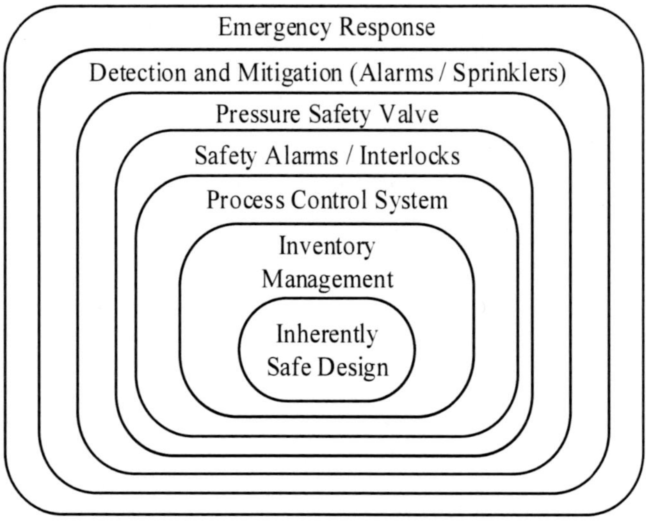

3.1 Concept Design

Concept design is one of the earliest stages of the design process. During concept design types of process equipment are first selected, their location mapped out, and consideration given to how materials will be heated up, cooled down, transferred, and stored when not in use. By asking a few questions as these fundamental design decisions are made, the design predisposes its self to being inherently safer.

- What is the worst-case operating condition, can the system be mechanically designed to handle those conditions?
- Where will hazardous materials be present and can that equipment be installed away from high concentrations of personnel?
- What hazards do the materials pose (toxic, flammable, corrosive)?
- How much of a potentially hazardous raw material must be on-hand? Weekly vs. monthly delivery may reduce stockpiles by 75%.
- Is there a safer way to produce the desired product?
- Is there a safer way to complete this unit operation?

3.2 Detail Design

Concept design develops enough detail to prove that the processing goal is obtainable and business case can be made. Concepts developed such as system location, connectivity, and required unit operations are further developed into a constructible design that can ultimately be operated to produce a finished product. In detailed design, detailed questions are needed as more layers of protection are added. These questions are to be asked informally throughout design and will be formalized in a process hazard analysis (PHA), completed during the detailed design phase.

- Where are the areas of high pressure, temperature, and hazardous material inventories; can they be reduced?
- Are there any specific code requirements or industry best practices for this system?
- What are the expected operating parameters, how can they be exceeded, identified ahead of time, and mitigated?
- Are all systems protected by a mechanical pressure relief device (if needed), do those devices discharge to safe locations?
- How and where could a loss of containment occur, how can it be quickly identified (leak or fire detection)?
- How will the consequences of an event be mitigated once started; how can the event be stopped (process shutdown or fire protection)?
- How can the public, employees, and environment be protected from the consequences of an event if it were to occur (emergency management)?

3.3 Start-Up

Following detailed design and construction, a process must be started, commissioned, qualified, and transitioned to manufacturing. Throughout this process, design assumptions must be confirmed, formal procedures verified, and changes to the design noted so that their impact on the process can be determined. Process safety considerations during start-up must include the full team from engineering to operations and maintenance.

- How does the as-built design compare to the final design, what are the impact of any changes?
- Were changes made to the operating envelope for any of the systems (pressure, temperature, level, concentration)?
- Has a pre-start-up safety review (PSSR) and as-built PHA been completed?
- Do operating procedures match the actual operation of the process?
- Have emergency drills been completed, are there lessons learned from these?

Addressing process safety throughout the design process, versus at the end of design when final design reviews are completed; minimizes late stage design changes, results in a better designed facility, and ensures safe facility operation for years to come.

4 Conclusion

Process safety is a well-established element in the design of oil and petrochemical facilities, taking a front row seat in the design process and being an area of focus once operations begin. Product safety holds a similar role and has been common place for a similar period in pharmaceutical and recently bioprocessing facilities. While the gap between these two elements of design has been closing in recent years as more pharmaceutical companies enter the bioprocessing arena and bioprocessing operations grow, a considerable gap remains. Based on the authors experience designing processes and facilities in both of areas, a few causes for the gaps are easily identifiable:

- Historically, bioprocessing facilities have not contained hazardous processes. Areas of pressure were limited to gas delivery systems (i.e. nitrogen and compressed air) and storage tanks contained aqueous based chemistries (i.e. bioreactors) minimizing the hazards with a loss of containment of these products.
- The shortened design timeline for bioprocess facilities challenges the ability to incorporate process safety into the design at distinct stage gates. The schedule does not allow for a discrete process safety design effort while other design work is on hold.
- There is a general lack of knowledge regarding safety system design in the biotechnology industry. This goes back to the opening statement that in many cases engineers responsible for these systems don't know what they don't know.

Product and patient safety are well defined requirements for biomanufacturing companies and are considered in the earliest stages of the drug discovery process. Applying similar principles to process safety throughout the design lifecycle, and not at distinct stage gates, will ensure a smooth transition from process development to design, validation, and ultimately commercial manufacturing.

5 References

[1] F.M. Editors of Encyclopedia Britannica. "Denis Papin." *Encyclopedia Britannica*, Encyclopedia Britannica, Inc., 18 Aug. 2018, www.britannica.com/biography/Denis-Papin.

[2] Ramsbottom, J. "On an Improved Safety-Valve." *Process Institute of Mechanical Engineers*, vol. 7, 1856, pp. 37–47.

[3] US Department of Labor Final Rule, "Final Rule on Process Safety Management of Highly Hazardous Chemicals; Explosives and Blasting Agents", 1910.109, *Fed Register #: 57.6356*, 24 Fen. 1992.

[4] NASDAQ Biotechnology Index, NBI, https://finance.yahoo.com/quote/%5Enbi/.

[5] *Process Safety Management Introduction*. Occupational Safety and Health Administration, www.osha.gov/dsg/psm/index.html.

[6] US Food and Drug Administration, *Good Manufacturing Practices (GMPs) for the 21st Century - Food Processing*, "GMPs Section One: Current Food Good Manufacturing Practices", 09 Aug. 2004, www.fda.gov/food/guidanceregulation/cgmp/ucm110907.htm.

[7] US Food and Drug Administration, "*Facts About the Current Good Manufacturing Practices (CGMPs)*", 25 Jun. 2018, www.fda.gov/food/guidanceregulation/cgmp/ucm110907.htm.

546706 LOPA and Sil Risk Analysis As Critical Tools for PSM Culture and Accidents Prevention

Industrial facilities that operate with dangerous products, whether chronic or acute, are increasingly common to be present and part of society at large, often allowing an aggregate of increasingly rich and economically beneficial professions and specialties to the community, be it directly or indirectly. Accidents in these types of units, especially those operating with hazardous products of acute consequence (those that have the capacity to cause fatalities when released to the environment), such as flammable, toxic, explosive products, due to the possibility of occurrence of undesirable events and the behavior of the effects resulting from releases of these types of product, could cause serious damage to the people and / or the installations that are in its vicinity. The various types of accident studies are known as EAR (Risk Analysis Studies), some of which are used more widely and others are specific to specific response needs.

According to BERNSTEIN (1997), the word "risk" derives from the ancient Italian risicare, which means "to dare". While DUARTE (2002) describes that the meaning of the word risk is an expression of thought that has a lot to do with the social, scientific and technological evolution of man. As BILLINTON and ALLAN (1992) note, society has great difficulty distinguishing between a hazard, which can be prioritized in terms of its severity, but does not take into account its probability, and risk, which considers not only the dangerous event, but also their probability of occurrence. KAPLAN (1997) explains that when a person asks "What is the risk?", 3 (three) questions are actually being questioned: What can happen? How can this happen? And if this does happen, what are the consequences? The answers define the scenario, frequency and consequence.

Within the agencies that tend to manage corporate risk, it is noticeable that the risk associated with a particular accident scenario is calculated according to its frequency of occurrence and its severity. Frequency is related to the temporal chance of occurrence of the initiating event, which leads to the release of dangerous product, such as LPP (Liquefied Petroleum Gas) emanation by ball rupture through overpressure of this equipment or, as a second example, the intrinsic rupture of a stretch of line. The consequence, also called severity, considering personal safety, is the capacity that the physical effects, due to the release of dangerous product, associated to the analyzed initiating event, can cause fatality (s), related to the potential of damage related to each effect analysis. The risk, qualitatively or quantitatively observed, and defined in a simplified way, is evaluated as the multiplication of frequency of occurrence and severity. The total social risk related to a particular facility, which can be defined as the risk that a community outside the facility is subject to by its presence in its vicinity, can be calculated as the sum of the risks of all the initiating events arising of the same. In order to do this, for each accident scenario, not only the generating causes, known as Initiating Events, but also the likelihood of ignition, the probability of affecting people, and several other parameters are considered, being one of them very important and permissible of analyzes, which are the failures

safeguards. These, if they act effectively when demanded, can prevent the accident from occurring, since the Initiating Event happens.

Hazardous product risk studies, such as qualitative (Hazard Operational Preliminary Analysis (APPP), Hazard Operability (HAZOP, among others) or those known as quantitative analyzes (Vulnerability Analysis and Quantitative Risk Analysis per se), in addition to being increasingly requested by regulatory agencies, have often been seen as an internal requirement of companies, representing one of the pillars of industrial safety management, in order to keep their risks at low or acceptable levels. However, these analyzes are usually performed in stages

until the final definition of the risk, comparison with acceptability criteria, and later mitigation, if necessary, each of these stages having a minimum time required for its execution, which may make the analysis complete rather time-consuming. Situations similar to these are detrimental to Risk Management when, in fact, what is desired is not a complete risk analysis, but only the achievement of a specific response, such as if my identified risk scenario is effective and properly protected.

The extent of possible damage related to each identified accident is delimited by the intensity of the physical effect causing the damage, and the relationship between the intensity of the physical effect and the corresponding damage is established through the vulnerability models.

On the other hand, semiquantitative risk analyzes such The extent of possible damage related to each identified accident is delimited by the intensity of the physical effect causing the damage, and the relationship between the intensity of the physical effect and the corresponding damage is established through the vulnerability models.

On the other hand, semiquantitative risk analyzes such as LOPA (Layer of Protection Analysis) and SIL (Safety Integrity Level), even though they are less demanded by chemical and petrochemical companies than qualitative or quantitative steps can have beneficial and effective results in terms of knowledge of all the characteristics of the analysis scenarios and of their protection barriers and, in a practical way, much faster to be developed. In addition, they can provide results that do not make up the analysis range of qualitative and / or quantitative studies in depth, such as knowledge and prior review of safeguards, many of which are known as Independent Protection Layers (CIP or IPL - Independent Protection Layers).

Therefore, with the anticipation of an increase in the number of future installations operating with hazardous products and the current installations that already do so, the semi-quantitative studies of LOPA represent an enormous value for the observance of the existing protections in each scenario of process accidents, also serving as the basis for the determination of new Independent Protection Layers or even the reduction of their number, tending to simplify the steps of determining the frequency of the scenario., even though they are less demanded by chemical and petrochemical companies than qualitative or quantitative steps can have beneficial and effective results in terms of knowledge of all the characteristics of the analysis scenarios and of their protection barriers and, in a practical way, much faster to be developed. In addition, they can provide results that do not make up the analysis range of qualitative and / or quantitative studies in depth, such as knowledge and prior review of safeguards, many of which are known as Independent Protection Layers (CIP or IPL - Independent Protection Layers).

Therefore, with the anticipation of an increase in the number of future installations operating with hazardous products and the current installations that already do so, the semi-quantitative studies of LOPA represent an enormous value for the observance of the existing protections in each scenario of process accidents, also serving as the basis for the determination of new Independent Protection Layers or even the reduction of their number, tending to simplify the steps of determining the frequency of the scenario.

546375 Controlling Static Discharge Using Antistatic Additives

Controlling Static Discharge using Antistatic Additives.

Abstract

The presentation will discuss controlling static discharge in non-fuel, non-conductive/low-flash point fluids. The scope of the presentation will concentrate on industrial producers, distributors, consumers of these fluids. Furthermore,furthermore, the presentation covers several industries where these fluids are most commonly found. Common Static-Accumulating Flammable Liquids That May Form Ignitable Vapor-Air Mixtures:

- VM&P naphtha
- Cyclohexane
- n-Heptane
- Benzene
- Toluene
- n-Hexane
- Xylene
- Ethyl benzene
- Styrene

Static discharge has caused catastrophic explosions and fires (U.S. Chemical Safety Board, Static Spark Ignites Explosion Inside Flammable Liquid Storage Tank No. 2007-06-I-KS).

Issues

- Nonconductive flammable liquids can accumulate static electricity during transfer and storage.
- Static sparks can readily ignite flammable vapor-air mixtures inside storage tanks.
- Material Safety Data Sheets (MSDSs) often do not adequately communicate hazard data and precautions.

Key findings:

- Safety publications omit antistat options. (see Process Safety Beacon, August, 2016)
- US affiliates of multinational corporations have different guidelines for antistatic use than their parent companies.
    - NFPA®77 and the American Coatings Association's, *Generation and Control of Static Electricity in Coatings Operation* has guidelines that are different than EU regulations ATEX- Appareils destinés à être utilisés en ATmosphères EXplosibles and ESIG's *Flamability: Safe Working With Industrial Solvents.*
    - The presentation will highlight these differences and how they contribute to the disparity in antistatic additive applications.
- Normal Bonding and Grounding May Not Be Enough! Companies that handle, transfer, and store flammable liquids should contact manufacturers to determine if these liquids can accumulate dangerous levels of static electricity, and if they can form explosive vapor-air mixtures inside storage tanks. If so, *extra precautions*—beyond normal bonding and grounding—may be necessary. Antistatic Additive is an extra precaution.

Recommendations: Knowledge members can use

- Who adds the Antistat?
- How much Antistat is added?
- How is the Antistat added?

546677 Led Your Process Safety Culture, Lead Beyond Hierarchies

The Oil & Gas industry is trying to advance at the next crossroads: "How to achieve that leaders without skills and without training in the management of culture, transform their own culture?"

Our research presents three key ideas that leaders from all over the world

can apply immediately to lead risk management with effectiveness:

frequency, consistency, meaning.

When specific leadership competencies to manage risk have not yet been developed,

Leaders can use this three concepts of easy adoption.

Increasing frequency of interactions with people will allow us to generate safer operations. In full attention economy and contexts of increasing complexity, for risk management, more communication is better communication. But it is not enough to improve the frequency of interactions. That frequency must be accompanied by consistency. The challenge is to achieve possible consistency between what we say and what We do. Leading by example is one of the central keys of risk management. The lack of consistency provides symbolic permissions to normalize deviations. We must maintain a high frequency of face-to-face interactions, but in addition, we must be consistent at every opportunity. And finally, to give meaning to our collaborators. It is necessary to ensure the understanding of risks. Each collaborator need to build his or her own sense of vulnerability and Operational Discipline.

We need workers who operate according to procedures and not adjusted to what the bosses say. This is what meaning allows.

548192 The Importance of Using Color Scale in Process Safety

Everyone knows the importance of color. From physics to fashion going through the automotive industry she will always be present. There is no world without color. Often, they come up on the scale form or not, but the variation in the scenarios are always important. It is so on the traffic sign and the ph tape. From the Diamond of Hommel to across the street. Colors, colors and colors. In addition to not necessarily needing ididomas, such as color of the sky and molten metal have a variety of shades. The main idea of the work is to choose a suitable color to be used in communication within a work area, choosing which and how many elements can evaluated suggest a color scale from the chosen one and present its contribution in the process safety.

550372 Flammable Gas Release Modeling for Real-Time Analysis Using Adversarial Variational Bayes

In this study, we have suggested a method to calculate high-accuracy damage range from flammable material in petrochemical complex. There are many flammable and toxic substances in chemical processes. Computational Fluid Dynamics(CFD) is one of the best way to calculate and simulate scenarios of release range of various fluids or gases. However, it is not used to calculate damage range, because of a lot of time to simulate one scenario. Thus, lots of researchers have enthusiastically studied the surrogate model or reduced order model to get the results of release range in real-time [1-3]. Recently, deep learning based generative model has receive attention because of its superior performance of non-linear manifold estimation [4]. Herein, we propose surrogate model based on adversarial varational Bayes (AVB) [5] for real-time analysis of probability of death. The proposed surrogate model can extract the latent space of high-dimensional flammable and toxic substances dispersion contours. Furthermore, it can generate the estimated contour image from variables of interest much faster and more precise than CFD data memorization with linear interpolation because it does not memorize the data but train the non-linear manifold regarding exact posterior distribution on given observation which basis is predefined in standard variational autoencoder (VAE) [4, 6].

The acrylonitrile was studied about risk assessment for health damage like cancer hazard.[7] But the component has flammable and toxicity, so we should calculate range of damage. The main scenario in this study is the acrylonitrile released from underground model, and the result is shown different damage following hole size, mass flow rate, wind velocity, and wind direction. The superior performance of the proposed model was exemplified by comparing with other surrogate models.

[1] Chen, T., Hadinoto, K., Yan, W., & Ma, Y. (2011). Efficient meta-modelling of complex process simulations with time–space-dependent outputs. *Computers & chemical engineering*, *35*(3), 502-509.

[2] Wang, K., Chen, T., Kwa, S. T., Ma, Y., & Lau, R. (2014). Meta-modelling for fast analysis of CFD-simulated vapour cloud dispersion processes. *Computers & Chemical Engineering*, *69*, 89-97.

[3] Long, K. J., Zajaczkowski, F. J., Haupt, S. E., & Peltier, L. J. (2009). Modeling a Hypothetical Chlorine Release on a College Campus. *JCP*, *4*(9), 881-890.

[4] Na, J., Jeon, K., & Lee, W. B. (2018). Toxic gas release modeling for real-time analysis using variational autoencoder with convolutional neural networks. *Chemical Engineering Science*, *181*, 68-78.

[5] Mescheder, L., Nowozin, S., & Geiger, A. (2017, August). Adversarial Variational Bayes: Unifying Variational Autoencoders and Generative Adversarial Networks. In *International Conference on Machine Learning (ICML)* (pp. 2391-2400). PMLR.

[6] Kingma, D. P., & Welling, M. (2013). Auto-encoding variational bayes. *arXiv preprint arXiv:1312.6114*.

[7] Haber, L. T., & Patterson, J. (2005). Report of an independent peer review of an acrylonitrile risk assessment. Hunam & Experimental Toxicology, 24, 487-527.

546615 Modeling and Analysis of Flammable Gas Dispersion and Deflagration from Offshore Platform Blowout

The operational safety of offshore platform is challenged by a series of adverse factors including harsh environmental conditions and process accidents. Blowout is a type of major accident that may result in catastrophic consequences. This paper uses a computational fluid dynamics (CFD) model to simulate the blowout accident on an offshore drilling platform. The simulation includes the dispersion of flammable gas and the resulting deflagration based on dispersion results. Through the modeling, the blowout gas dispersion processes under different wind speeds and directions are analyzed, and a deflagration scenario due to the ignition of flammable gas is predicted in which the impacts of jet fire, overpressure and heat radiation are assessed. The simulations indicate that wind speed and direction have a great impact on the dispersion behavior of blowout gas. During the deflagration, the maximum overpressure is 280 KPa, the jet fire mainly distributes in the derrick, and the maximum radiation flux of thermal radiation field is 2300 W/m^2. Essentially, the outcomes from simulations could help to conduct a prior risk assessment and make an emergency response plan. Additionally, it could give a reference for the inherent safety design of offshore platform to mitigate the unexpected consequences due to blowout accidents.

548682 Application of Inherently Safer Design Tool (*i*-SDT) for Hazards Assessment and Safer Operation

Inherently ***S***afer ***D***esign ***T***ool (*i*-SDT) is a newly developed semi-quantitative metrics for hazards identification and characterization[1]. It is built on property platform, probabilistic analysis of historical accidents information and characteristic equations of safety parameters (i.e., flammability, toxicity, explosiveness, reactivity, etc.) which are developed in-house to assess potential hazards under various operating conditions. Designers can identify dangerous components and vulnerable equipment of the investigated industrial process and can take necessary corrective actions by applying the inherent safety principles[2] (e.g., minimization, substitution, moderation and simplification). The final output of this *i*-SDT tool is a cluster safety parameter score (C_{SP}) which provides insights regarding the development of safe design and operational parameters to manage the hazards using a very limited amount of process information and safety incidents information.

In this work, the application of the proposed *i*-SDT tool is explained for assessing different alternative routes of an industrial process supply chain. The proposed *i*-SDT safety metric identifies the chemical components with highest risk and the major accident-prone operational units. The detail analysis using *i*-SDT tools shows how different operating conditions can affect the safety performance of the given process. The ultimate object is to find the safer route while considering the techno-economic and environmental performance. It provides Pareto curve to compare alternative routes and gives the freedom to the end-users choosing their safer operating route/technology and process conditions. Therefore, safety can be considered as an important selection criteria for choosing best alternative from the beginning.

Final cluster safety parameter scores (C_{SP}) obtained from the analysis enables designers to compare between available technologies and provides information on various hazardous situations. It can also provide in depth knowledge regarding the cause behind the hazardous behavior. The proposed *i*-SDT metrics allows designer to do the safety analysis simultaneously while doing the initial process synthesis which will reduce the future modification cost of any add-in safety features. This work shows how the unique features of *i*-SDT tool can be applied in a real life scenario while doing inherent safer design during early stage of design.

1. Kazi, M.–K, Eljack F. T., & AlNouss, A. Application of a property-based inherent safety quantification framework for integrating risk assessment into process safety life cycle. *AIChE Spring Meeting, Orlando, Florida, USA, 2018.* (156c), p. 2260.

2. Kletz, T. A., & Amyotte, P. (2010). Process Plants: A Handbook for Inherently Safer Design, Second Edition: Taylor & Francis.

546631 An Approach to Understanding Worker Interactions with Procedures Under Normal Working Conditions

Incidents in high-risk industrial settings are often attributed to the way in which procedures are used. Despite the identification of procedures misuse, little research has been conducted on how workers have interacted with procedures in normal conditions. To understand these interactions a behavior assessment technique has been developed which can describe how a worker interacts with a procedure to complete tasks in everyday work situations. interactions under normal conditions. Using the concept of work-as-imagined/work-as-done, the techniques describes the interactions at a macro whole-of-task level as well as an individual step level. At the macros level, assessment considers if the task was completed correctly and safely (Tier 0). At the micro step-level, assessment considers if the step was completed as it was imagined (Tier 1) or if the step was done in an alternate way (Tier 2). For the alternate approaches, assessment considers if the worker struggled or required assistance, completed the steps out of order or with iterations or if the step was not completed as intended.

Two studies, one involving industrial tasks Second Life® (Study 1), the other at Shells Robert, LA training facility (Study 2), were used to test and develop the behavior assessment technique. In Study 1, the behavior of 13 participants from offshore drilling conducting tasks in either day or night conditions was assessed. In Study 2 the behavior assessment technique was used to test if experience accounted for behavior difference in 25 participants from high-risk industries. Results demonstrate the success of applying the assessment techniques through identifying how the behavior of the participants changes according to their experience and familiarity with the tasks the were performing. Further analysis identified recommendations for design and work flow changes of both written procedures and specific equipment. In conclusion, the initial results suggest the assessment technique is an important contribution to the methods use for understanding early interventions to improve safety in the process industry.

PROCESS HAZARD ANALYSIS (PHA) PREPARATION: A ROLE BASED OPTIMIZATION

Susan S. Frizzell, PE
Adley Services, LLC
Lafayette, LA
Susan@AdleyServices.com

Biography:

Susan Frizzell, PE is the President of Adley Services. Shortly after high school, Susan began her career alongside her father on a drilling rig. Upon graduating from the University of Louisiana with a BS in Mechanical Engineering, she obtained her MBA and began working for a firm performing engineering and process safety reviews on refineries and other land facilities in the New Orleans area. After Susan relocated to Lafayette, LA, she began engineering offshore facilities, complimented with process hazard analysis facilitation. During her almost 20-year career, she has worked upstream, midstream, downstream, and all points in between. She is a licensed professional engineer and an executive member of various organizations.

Abstract

A PHA meeting consumes large resources and can place a strain on departmental manpower. Teams are often faced with the want to be more efficient in PHA conflicting with the need to be thorough and accurate. The key to solving this paradox is efficient preparation. By examining the lessons learned from past PHA's, a role by role analysis of best practices with regards to preparation will allow a PHA team to reduce the costly time spent in the meeting and yet still stay true to the methods and deliver a great, safe result.

When it comes to an optimal PHA, preparation is everything. For example, in a 6-member team PHA, 60 minutes of misused time equates to 6 hours of misused manhours. Whether it is managerial, operations, engineering, or even facilitation, there are lessons learned from previous PHA's that when shared, will not only make the PHA more efficient and the facility safer, but the industry as whole better. The goal being that those reading this be able to take back to their facility these practical preparation tips to share with their present and future PHA team members so that the resources of the facility can be better spent.

1. PHA Preparation Optimized

To fully optimize the Process Hazard Analysis (PHA) preparation, the team needs a clear vision of the end product: "To identify, evaluate, and control the hazards" [1, 2] of an operation or process to minimize the likelihood of occurrence and the severity of consequences of release of a dangerous substance [3]. With that in mind, the team can focus time and energy preparing for the PHA on factors within their control to meet that goal. An efficient, cost-effective PHA begins with including the different roles in the preparation. The PHA planning team should be comprised of management, engineering, operations, the scribe, and the facilitator.

An active commitment to reduce risk begins by assigning tasks to each role prior to the PHA. From the start, management should provide support and direction for the PHA team. They set the tone for obligation and importance. Second, engineering involvement in the preparation ensures accurate and complete process safety information (PSI) is available and accessible. Without correct information, the hazards may not be fully identified. Next, operations should walkdown the facility and identify areas of concern. They have the most intimate knowledge of the workings of the process, which is extremely valuable to the PHA when properly communicated to the team for hazard recognition. The scribe should have the technical skills to ensure equipment is functional and to properly and efficiently document the team's findings. Lastly, the facilitator needs to coach the team before and during the PHA to ensure the methodology is followed.

Successful PHA preparation can improve risk identification and limit cost through appropriate management direction, current and detailed documentation, and clear expectations of the timeline leading up to the PHA. We identify ways that companies can assist in preparation in the months leading up to a PHA and the benefits that preparation has on hazard recognition, team morale, and, most importantly, the bottom line.

2. Management – "*Set the Tone*"

It is important for management to set the tone of the PHA. More specifically, management needs to set the tone before, during, and after. In a PHA, the attitude, the posture, and the dedication of the team is usually a direct reflection of management's attitude, posture, and dedication. In our experience of both facilitating and participating in PHAs, we've seen attitudes of management in a variety of flavors. The following are observations we've made in the past regarding a successful PHA and one not as successful.

The corporate culture and support infrastructure set by management begins well before the first day of the PHA. A typical PHA is done every 5 years [1]. Given that schedule, it is imperative that management starts out on the right foot and begin planning well in advance. A management component that is very serious and very organized will typically induce an attitude of seriousness and organization to the team members. For example, a management team that waits until the last moment and rushes preparation will typically have PHA members that delay and do little preparation, if any at all, for the PHA. PHAs are very expensive and very time consuming. Lack of adequate and efficient preparation will be very damaging to an efficient use of PHA budget and resources. When management in advance sets the tone that a PHA needs to be done correctly and efficiently, the PHA team members tend to take preparation seriously. Just as it is well documented that an effective safety culture of a company starts at the top with management, effective preparation for a PHA also starts at the top with management [4]. The management teams we've seen in the past that do this most effectively set the tone of preparing and doing a PHA as part of your job, not as additional work on top of your daily duties. In other words, management sets the tone that PHAs are in your job description and that we are all PHA team members. For example, a PHA is not additional work given to engineers, it is in an engineer's job description.

The other things we have observed that effective management does by setting the tone is setting up a support team for the PHA team members. These PHA team members will have to transition out of their daily duties, be present at the PHA, and transition back into their regular duties. Often, their project managers and/or functional managers may give some resistance to that. It is imperative for upper management to set the tone with the managers of the team members that they need to have a plan for them to transition into the PHA and transition out of it with regards to their regular duties. Most of these facilities run around the clock all year long and the work never stops. Therefore, it is imperative for management to have a plan for these employees to be present at the PHA while it is going on. A PHA which is constantly stopped for employees dealing with issues outside of the PHA is an unacceptable waste of resources. For example, if a PHA consists of ten team members and one team member must stop the PHA to deal with outside problems unrelated to the PHA, the nine other team members are wasting company resources. When management sets this tone early on that we are dealing with real human lives and significant company resources, the employees and their managers tend to put greater weight on a PHA. And sometimes, the employees just need the help of upper management to persuade lower and mid-level management that their presence, uninterrupted, is necessary to conduct a PHA.

During the PHA, management's role is typically only at the very beginning. It is very effective during the opening remarks of a PHA for a member of management to also make an opening remark. During this address, management will set the tone on several things. It is not uncommon for management to begin opening remarks with a thank you for the team members for their hard work in preparing and committing of their time. Some members of management stress employees' safety; others may stress the importance of safety to the community or environment. Opening remarks is a great time for management to state known worst-case consequences initiated within the limits of the PHA scope or define the assumptions regarding severity. This guides the team towards consistent company-wide risk rankings and avoids the team paranoia over risk ranking too high or too low. It's not management's position to force upon the team a subjective belief; it is management's position to reaffirm the team uses best judgement and not be hesitant of ranking the risks as they believe they should be. Finally, management can give the PHA team members feedback from previous PHAs as to the wording, substance, and ranking of the recommendations to assist with review and completion post-PHA.

Once the PHA begins, management's role is very limited. We've seen effective managers from time to time pop-in, review recommendations, and discuss with team members their thoughts on such. Effective managers will pick a recommendation, look at the supporting data, and ask specific questions of how the team arrived at its conclusion. This serves the purpose of correcting any doubts that the PHA team may be having and serves the purpose of reaffirming to the PHA team what they are doing is correct. Management can also review recommendations with only the facilitator at specific progress intervals of the PHA. The facilitator can then bring suggestions back to the team without the direct watchful eye of management. This minimizes the intimidation factor but gives the team the opportunity to clarify wording or details.

We find in PHAs that the PHA team members want to be very cooperative with management. That is why it's imperative for management to not only set the expectations and tone, but also provide a supportive means and network for a high-quality PHA to be conducted. By management providing a support team and stressing the importance of a PHA during the preparation phase, the PHA team members are motivated to attend the PHA prepared with their full attention. It goes back to the old saying, garbage in, garbage out. Anyone who has sat in a PHA will attest to the opinion that PHAs are grueling and a marathon of being attentive. The PHA team members tend to be very technically sound and understand the PHA process and goal. The biggest hurdle is usually morale and minimizing outside distractions. A great opening statement by management and periodic check-ins to keep morale and encourage the team members goes a long way. Nothing lifts the moral of battle wary troops like a visit to the battlefield by the general. After the PHA is over, feedback and encouragement go a long way in instilling confidence for the PHA team members that will carry over to their next PHA. Although hardly any of this is required by the statutes, these are observations we have noticed with regards to effective management participation in PHAs.

3. Engineering – "The Devil is in the Details"

As an engineer, it is understood that your job is to ask the questions that have not been answered, review the material that has not been presented, and observe that which is not obvious. PHAs are no different. One of the duties of the engineer during a PHA is to gather all the process safety information (PSI). However, we've observed the more effective engineers will take this list, check how current it is, and review the design pressures and temperatures, such as pump discharge pressures. This review helps determine if the PSI is accurate. It also serves the dual purpose of observing any abnormalities that will need to be discussed during the PHA. Looking at this before a PHA is very effective when the engineer's eyes are fresh, and his mind is clear. Once the PHA is ongoing, this task becomes a little more complicated with lower morale and PHA fatigue. This holds true to the majority of the engineer's tasks.

Drawings, drawings, drawings. One of the largest time and money wasters we observe in PHAs is showing up to a PHA with either outdated drawings or needing drawings that are missing. It is imperative for the effective engineer to not only review the drawings to make sure they are up to date and accurately reflect what is present in the field, but it is equally important to review which drawings have been gathered and submitted to ensure that there are no surprises during the PHA. Anyone who has sat in a PHA and heard the words "these are not the most up to date drawings" or "where are all the other drawings at", will attest that it is going to be a long day. The PHA facilitator, the scribe, and pretty much everyone else in the room relies on the drawings being both accurate and complete in their preparation. If the drawings submitted for the PHA are either outdated, incomplete, or just wrong; it will have a ripple effect for the rest of the team both before and during the PHA. If the reader of this paper takes nothing away but one thing, please note that accurate and complete drawings are imperative for an effective PHA team.

PHA team members include those who are present in the room full time and other team members who may be called upon as needed. Effective engineers anticipate ahead of time what possible other resources will be needed during the PHA. For example, engineers for adjacent facilities need to be made aware of an upcoming PHA and given a heads-up as to be ready to answer questions regarding upset conditions and their effects. Likewise, engineers for skid packages or other equipment requiring expertise of vendor engineers need to also be made aware of an upcoming PHA and request their availability if called upon. Similarly, engineers of other disciplines may have better knowledge of failure mechanisms or a specific design review already conducted which is beneficial to the PHA team. It is not uncommon in a PHA to have a question that needs to be answered by a vendor engineer, subject matter experts, or an adjacent facility engineer. Examples of information needed include possible flow rates or pressure to/from receiving/feeding units, adequacy of safe guards in packages provided by vendors, or electrical area classifications. Oftentimes, a simple phone call to these engineers asking if they would be available by phone during the PHA is enough. Other times, effective engineers will require that other personnel be available to come in as requested. PHAs involve a lot of moving parts with each part having its own calendar or schedule. Giving the other possibly necessary personnel a heads up will ensure that should their presence become necessary, the PHA team will not have to recess until they are available.

Effective engineers will typically review MOCs and make phone calls or e-mails to update their current status. The adage that change is constant is very applicable to MOCs. MOCs may have been implemented, postponed, in progress, or completed. It is important to update the status of an MOC. One that has been canceled may require different attention than an MOC that is in progress or completed.

Effective engineers provide the technical expertise of the specific process under review [1]. Before a PHA, effective engineers will take a closer look to evaluate operating trends and upset conditions looking for a pattern that is not readably apparent. Engineers should discuss the facility and how it runs day-to-day with operations prior to the PHA. They should review procedures to ensure they are current and understood. It can be an awkward moment during a PHA when the engineer says, "we've never had a problem with that", and the operator rebuts with "we have a problem with that every day". Engineers know that engineered controls are the strongest defense against hazards. They look for ways to solve problems without relying on administrative controls.

Finally, effective engineers are present during a PHA. This does not mean they are sitting in a chair, it means they are focused, engaged, and completely dedicated to the PHA while they are in that room. Engineering is a life of dealing with problems. To be an effective PHA team member, it is important to make sure the engineer has support to deal with problems while he/she is in the PHA. Effective engineers can have a clear mind because they know that if normal, everyday problems occur, someone else is taking care of them. An engineer that is constantly worrying about what great surprises are awaiting them when they check their email at the end of the day, are not present during a PHA.

4. Operations – "The Ambassador of the field"

The operator is the ambassador of the conditions in the field. The operator, not engineering, management, or the facilitator, is the eyes and ears of the field. As an ambassador, it is important that the operator speak to his constituents (i.e. opposite crews), learn how to effectively communicate, and be present for the meeting of the PHA team. Effective operators speak to their coworkers and get a history of problems, concerns, or other issues they may be having that might not be well documented. Effective operators also walk around and observe conditions of the facility and document such. Documentation being key. Effective operators also review previous incidents not only in their facility, but other similar facilities or neighboring facilities to bring up any concerns that may not be readily apparent when reviewing documentation during a PHA. A great ambassador becomes the eyes and ears of his constituents by listening to what the other operators have to say and physically observing the condition of the facility.

Other things effective operators will do is review operating procedures and note deficiencies. Sometimes operating procedures are written in a way that no one understands them making them ineffective as a safeguard. Other times, there is no written procedure, forcing operations to react to conditions "the way we always do it." If operators are not following the procedures because they are vague, ambiguous, confusing, or out of date, it is up to the operator during the PHA to bring that to the attention of the team members. Frankly, effective operators understand that management and engineers assume that everything they write is perfect, clear, and easy to follow. It is up to the operator to explain why that may not always be the case. Some of the most effective operators we have encountered in a PHA not only describe what the problems are but are very specific about why these are problems.

The next characteristic of the operator as an ambassador for the field conditions is communication. For an operator to communicate with a PHA team, they must speak the same language. Effective operators know that to communicate with a PHA team, they must become a master at reading P&IDs. Being able to understand how the elements of a facility and its attributes are represented on a P&ID are extremely important for an operator. Effective operators can look at a P&ID with little or no guidance and interpret it to see if it is an accurate reflection of what is going on in the field. When the engineer asks a question regarding the P&ID, effective operators are on the same page. If the reader wanted to take away one important thing from this section on operators, please take away that an operator being a master of a P&ID is extremely important to be an effective operator in a PHA.

5. Scribe – "The Show Must Go On"

In American football, a great offensive lineman only receives attention when things do not go as planned. The same can be said about a scribe in a PHA. An effective scribe will give the outward appearance of very little responsibility or having very few complicated tasks. Nothing can be further from the truth. With the goal of quickly and accurately recording information [4], effective scribes understand that a lot of thought and preparation must go into making sure that nothing goes wrong, and if something does go wrong, the PHA suffers minimal interruption. Also, effective scribes are great at being very efficient in the tasks they complete.

At the end of the day, a scribe's responsibility is to ensure that the audio/video equipment works as planned. Although some facilities may be tasked with providing projectors and so forth, the scribe cannot scribe unless all this equipment works. Therefore, an effective scribe will oversee the equipment being used to make sure the parties who are responsible for providing it are communicating and that all the items are compatible. The scribe will ensure that both the scribe's computer and the display have the proper connectors and cables needed. An effective scribe not only brings the connectors and cables needed, but also bring the connectors and cables that were not anticipated as being necessary. An example is the conference room being used for the PHA has HDMI and ethernet and Wi-Fi...or at least that what was told to the PHA facilitator. In the past, we have shown up to conference rooms to find that the HDMI port is in fact a VGA port, or that the HDMI cable is three feet long and the scribe's position is ten feet away. Very effective scribes keep with them a toolbox of supplies such as VGA to HDMI converters, extra batteries for mice and Bluetooth keyboards, and extra cables. Sometimes, the right connectors and cables are present but are malfunctioning. The truly great, effective scribes seamlessly work through these problems and the show continues without interruption. Along those lines, a great practice for scribes to have is to not only back up the PHA file regularly, but also have a written procedure that has been reviewed regarding the backup. We always joke that we PHA the PHA. In other words, what are the risks for file corruption or data loss. Which risk are we willing to accept, and which risk are we going to mitigate? Knock on wood, so far, it's been very effective.

A lesser problem, but a problem none the less, is having to wait on a scribe to catch up. Although not as catastrophic as the PHA having to be stopped completely while technical issues are worked out, a scribe who is not able to keep up with the PHA team stifles the free flow of thought and can leave the door open for other interruptions such as taking phone calls and checking email. Very effective scribes will have experience and practice with the software, know the shortcut keys, and have a vocabulary list prepared for technical terms and names. A typical vocabulary list will be something as simple as a word document open where the scribe has already checked the spelling of names and technical equipment and as the scribe goes along, they can copy and paste the names and equipment labels very quickly without having to correct for spelling. A truly exceptional scribe will have already created templates with standard verbiage for safeguards, recommendations, and comments. For example, when the team decides to make a recommendation, the scribe can copy and paste it into the software and the team can edit the few words that need to be changed to modify the recommendation for the specific application. What this does is it allows the team to make a recommendation and move on to the next node without getting bogged down in the weeds of language choice. We find it is always easier to give a group of people a starting point for a recommendation as opposed to generating the language of a recommendation by consensus of the group. Along those lines, effective scribes review the drawings, MOCs, and other information to become familiar with the vocabulary in the drawings and have an inventory of what PSI they have and what node it corresponds to. Therefore, if a team member erroneously refers to a drawing that does not exist or refers to a drawing or MOC by incorrect name, the scribe will be quick to point out that the item does not exist in the inventory.

Effective scribes will also begin opening vital documents for a specific node as soon as the review of that node begins. Therefore, when the team would like to reference a specific document such as a drawing, MOC, or incident report, the document will be readily available. A common, effective means of doing this is having a second monitor only seen by the scribe, in addition to the screen shared with the projector. The scribe can work on his/her monitor opening documents or correcting language. As the team is ready to view specific information, he/she can drag the window over to the extended screen that is being displayed overhead. This allows the scribe to work without being micromanaged by the team and minimizes the distraction of the team from the scribe flipping through windows or searching for files. Again, the overall theme for an effective scribe is to keep the free flow of thoughts and the team moving. Anytime the team must pause or take a break because of the scribe performing its duties, the team's momentum and productivity can be lost.

Finally, truly effective scribes are great at covering the intangibles. This can include filling all the supplies or doing things to the environment in the room to lift morale. For example, effective scribes will typically have a box of random office supplies, whether it be pens, hi-liters, or post-it notes. At the beginning of each day and after each break, he/she should refill the table or the workspace of the PHA team with extras. So, if a team member would require a specific color pen, they would simply grab it and not have to pause the PHA. This gets back to the overall theme that an effective scribe keeps the evaluation moving forward without interruption. Another intangible that we often see effective scribes will do is they will place motivational posters around the room, such as large check-off list of nodes to visually see progress, to hang inspirational quotes, and so forth. PHAs are really an endurance test of morale and motivation and these little intangibles that the scribes do are of very limited cost, but of great benefit. In all honesty, we often do not realize how great and effective our scribe is until we do a PHA with an ineffective scribe. When you do a PHA with an ineffective scribe, you feel like you're stuck in the mud and going nowhere fast. Although realistically, the delay may add just a few minutes a day, with regards to morale and motivation, it feels like an eternity.

6. Facilitator – "What's Your Plan, Coach?"

Like a good coach, an effective facilitator enables great people to do great things. Effective facilitators setup the team members to be successful and, during the PHA, guide them to produce a high-quality product. Although a facilitator's duty is to ensure the methods are adhered to and the regulations concerning a PHA are complied with [1], a great PHA facilitator is so much more.

A great PHA facilitator, like a coach, ensures that the team is prepared come game day. Effective PHA facilitators will do such things as set deadlines for collection and development of information and suggest further actions to better prepare for game day. Not only do they set the deadlines, an effective facilitator will check in with the team members prior to those deadlines to ensure that progress is proceeding as planned and that the deadline will likely be met. Effective facilitators are great at helping team members overcome obstacles and hurdles in collecting this information. A team being well prepared on game day when they take their chairs and listen to management's opening remarks is no accident. Also, during the preparation, an effective PHA facilitator will ask the hard questions of the team members, such as: Do you have a transition plan for someone to fulfill your role during the PHA? Or, are there any allergies, physical, or dietary restrictions a team member may have so that these can be addressed well ahead of the PHA? To be an effective PHA facilitator, it really comes down to having the team prepare for the work they anticipate doing and addressing as many possible obstacles as one can prior to the PHA commencing. The litmus test for a PHA facilitator is simple - once the PHA commences, no surprises.

In addition to ensuring the rest of the team is preparing for the PHA, the effective facilitator puts in just as much effort as the rest of the team into preparing the PHA file, determining the node segregation, and sharpening his skills as a technical guide. On the day of the PHA, the PHA file should be prepopulated with node descriptions, drawings, common safeguards, etc. The facilitator should have reviewed the node list and noded drawings with other member(s) of the team to ensure the segregations make sense and all parts of the process are included. The effective facilitator should also review recent changes to regulations or process conditions that might affect the hazard identification. He should gain a basic understanding of the process under review along with the potential hazards specific to the facility. Effective facilitators have a list of questions to pose to the team during the appropriate node to ensure all hazards have been addressed.

Again, following the theme of the PHA facilitator being a coach, once the PHA commences, it is game day. It is the PHA facilitator's job to adhere to the methods and regulations regarding PHA facilitation. However, great, effective PHA facilitators understand that the true challenge in a PHA is morale, motivation, and focus. Effective PHA facilitators will typically start off the PHA with a "rules of the road" presentation. They do not assume that everyone in a room regularly participates in PHAs and is familiar with the process. They review the PHA process, the end goal of risk mitigation, the timeline, the daily schedule, acceptable and unacceptable behaviors, and expectations of participation. An effective PHA facilitator will go over these rules of the road and get each team member to commit to their role, to fully participate, and to strive towards the goal. It creates a sense of responsibility and obligation among the members. Once the PHA commences, the effective PHA facilitator will have a sense of when progress is slowing down and when to step in to move the team along. The effective PHA facilitator knows that the team must be given opportunity to discuss causes and consequences, but also has goals and objectives that need to be met within the time allotted.

As the team leader, the facilitator should plan on how to react to differences among team members. The great PHA facilitators have a way of getting conflicts resolved without upsetting or demeaning anyone in the room. Typically, in an argument, somebody must be wrong and somebody must be right. However, the great PHA facilitators have a way of allowing both parties to express their beliefs, guiding them to understand the beliefs of others, and concluding with a common solution. A great way to kill the free flow of ideas in a PHA and stifle productivity is to tell someone, "you're wrong, please be quiet". Great PHA facilitators know how to do that in a way that allows the person to maintain dignity and continue to perform in a PHA.

Another attribute of great PHA facilitators is that they are systematic and thorough. They have a plan of attack for each node. This usually includes the node description and limits review, a review of the appropriate drawings, written list of deviations to encourage the team, and review of MOCs, previous recommendations, and incidents related to the node. The effective facilitator also has a plan in place to limit idle time when critical information necessary to properly review a node is found to be missing. The action is documented as a follow-up for a specific team member to complete prior to the next day allowing the PHA to move forward. Any outstanding to-dos are documented and presented to the team members at the end of each day prior to dismissal. The PHA facilitator is responsible for meticulously documenting the team's findings, which cannot be done without adequate preparation of a game plan.

Finally, great PHA facilitators understand they have an obligation to future PHA facilitator to send their team off with a sense of accomplishment and pride in PHA participation. If the team members of the PHA leave with a bad taste in their mouth or feel discouraged, their participation in the next PHA may not be to the level it could be. Great PHA facilitators compliment their team members and make sure they leave the PHA on a positive note. Therefore, when the team members approach their next PHA, the new facilitator will not be dealing with baggage leftover from the previous PHA. It is a courtesy that great facilitators give to and expect from other facilitators. PHA facilitators are often asked by team members during a PHA, "how can you do this every day?" The response we often use is, "what can we do better to make you volunteer for the next PHA?" Soliciting feedback from team members besides a multiple-choice survey is imperative to be an effective facilitator. The feedback needs to be of substance that can be acted upon and improved upon. We truly love doing PHAs and at the end of every PHA, we express that to the team and solicit their feedback to what we can do better to make them truly love doing it too.

7. Summary

In conclusion, when it comes to an optimal PHA, preparation is everything. Each team member plays an important and specific role. Management sets the tone with their level of commitment to preparation, resources, and action. Engineering ensures the quality of information available to the team and contributes a deep understanding of the process design. Operations represents the field conditions, the procedures, and the most intimate knowledge of the process. The scribe provides accurate documentation through technical skill, proficiency with hardware and software, and team morale support. The facilitator leads the team to identify, evaluate, and control the hazards utilizing a systematic approach to preparation. It has been proven that a small amount of time spent preparing for each role will result in improved hazard identification, higher team morale, and reduced PHA cost.

8. References

1. Occupational Safety and Health Standards, Toxic and Hazardous Substances, Hazard Communication, Available at https://www.osha.gov/laws-regs/regulations/standardnumber/1910/1910.1200 Accessed on December 12, 2018.
2. Bureau of Safety and Environmental Enforcement, Department of the Interior, Offshore, Oil and Gas and Sulphur Operations in the Outer Continental Shelf, Available at https://www.ecfr.gov/cgi-bin/text-idx?tpl=/ecfrbrowse/Title30/30cfr250_main_02.tpl Accessed on December 12, 2018.
3. American Petroleum Institute Recommended Practice 750: Management of Process Hazards, First Edition, American Petroleum Institute, Washington D.C., 1995.
4. CCPS. *Guidelines for Hazard Evaluation Procedures*. Center for Chemical Process Safety, American Institute of Chemical Engineers, New York, NY, 2008.

551204 Building Process Safety Intelligence: What Should Every Graduating Chemical Engineer Know?

Process safety intelligence is derived from derived from a vast corpus of sources - including textbooks, handbooks, journals, conferences, incident reports, safety datasheets, and available materials and thermophysical property data. Dedicated decision-support tools enabling practitioners to harness this intelligence in their daily work, such as Knovel are an important component of chemical engineering practice today. However, many students don't realize the breadth and depth of technical knowledge needed to support process design and safety – they tend to overrely on non-specialist search, and miss pedigreed information available on specialized knowledge platforms, such as Knovel.

The undergraduate engineering space offers an excellent opportunity to plant the foundational information seeking behaviors in future process safety professionals and leaders. Knovel focuses on applied a preview of the types of insight they will be forced to leverage in industry careers, without a direct cost to students and faculty.

Common Process Safety Management Gaps Based on Occupational Safety and Health Administration Process Safety Management of Highly Hazardous Chemicals Violation Trends

Janet Persechino
Engineering Planning and Management, Inc.
959 Concord Street
Framingham, MA 01701
jap@epm-inc.com

Ryan Welch
Engineering Planning and Management, Inc.
959 Concord Street
Framingham, MA 01701
rdw@epm-inc.com

Prepared for Presentation at
American Institute of Chemical Engineers
2019 Spring Meeting and 15th Global Congress on Process Safety
New Orleans, LA
March 31-April 3, 2019

Janet Persechino
Engineering Planning and Management, Inc.
959 Concord Street
Framingham, MA 01701
jap@epm-inc.com
Keywords: OSHA, violations, regulations, process safety

Abstract
On July 17, 1990 the U.S. Occupational Safety and Health Administration (OSHA) published in the Federal Register the proposed standard *Process Safety Management (PSM) of Highly Hazardous Chemicals (HHCs)*. Approximately four months later it was enacted into law via Clean Air Act Amendments (CAAA). These amendments require that the U.S. Secretary of Labor, in coordination with the U.S. Administrator of the Environmental Protection Agency, promulgate, pursuant to the Occupational Safety and Health Act of 1970, a chemical process safety standard to prevent accidental releases of chemicals that could pose a threat to employees [1]. The full text of the final standard was issued in the Federal Register on February 24, 1992.

The OSHA standard is listed under Title 29 (Department of Labor) of the Code of Federal Regulations (CFR) as 1910.119. This standard leans towards a performance-based guidance document that outlines methods and programs to decrease the likelihood of flammable, toxic, reactive and explosive chemical releases. The intent is to reduce the risk to employees, visitors and contractors that may enter or work at chemical processing facilities.

Facilities are subject to the PSM regulation if they hold a threshold quantity of any HHC that is listed in Appendix A of the standard, with some exceptions. The violations that OSHA issues for the 1910.119 standard will cite the specific element and its relevant sub-section. Published data allows for review of the violations, and based on that information an indication of common PSM program gaps can be assessed.

A ten-year span of OSHA PSM violation data was examined (2007-2017). The data shows the top four OSHA cited PSM sections (often referred to as elements) are (highest to lowest):

1) Process Safety Information
2) Mechanical Integrity
3) Operating Procedures
4) Process Hazard Analysis

Each of these elements have known solutions, and specific corrective actions can be easily implemented given the lessons learned, the current level of expertise that exists in the process safety industry and the cost-effective software tools that are now available.

Regulatory compliance is not the only goal of adequate PSM programs, nor is it the goal of this review. The purpose of reviewing the OSHA violation information is not meant to help facilities avoid OSHA citations, but to uncover gaps found by OSHA and bring awareness and attention to some common issues, hopefully leading to focused solutions for safer, more reliable facilities.

1.0 Introduction

There are approximately 24,000 sites that are subject to the OSHA PSM standard nationwide [2]. Facilities are subject to the standard if they process any of the 130+ chemicals listed in Appendix A (of the standard) in excess of the associated threshold quantities, also listed in Appendix A, on-site in one location. They are also subject to the standard if they process flammable liquids and/or gases in quantities equal to or greater than 10,000 pounds. OSHA defines "on site in one location" as areas under the control of an employer, or affiliated employers, that have control over any group of vessels that are interconnected, or in separate vessels that are located in close enough proximity that a HHC could be involved in a catastrophic release.

Exceptions to the applicability include [3]:

- ✓ Retail facilities
- ✓ Oil/gas well drilling or servicing operations
- ✓ Normally unoccupied remote facilities; the standard defines "remote" as a facility that is operated, maintained or serviced by employees who visit the facility periodically to check its operation, no permanent employees are stationed at the facility, and is geographically remote from all other buildings and persons.
- ✓ Hydrocarbon fuel used solely for workplace consumption as a fuel if such fuels are not part of a process containing another HHC covered by the standard
- ✓ Flammable liquids with a flashpoint below 100°F stored in atmospheric storage tanks or transferred, which are kept below their normal boiling point without benefit of chilling or refrigeration

The above information outlines specific threshold quantities that trigger, and exclude, the PSM regulatory requirement. It is strongly recommended that some form of PSM be practiced at plants that may not be subject to PSM, especially those that fall into the following categories:

- Plants that store quantities of HHCs close to the threshold quantities listed in Appendix A of the standard
- Plants that utilize hazardous chemicals not listed in Appendix A of the standard
- Processes exempt from the standard, for example, those listed as exceptions in the standard, plants that have runaway reaction potential, etc.

An overview of "PSM Lite", and related fire code requirements, can be found in these conference proceedings in the paper titled *Fire Code Chemical Process Safety Requirements*. Implementation of an abbreviated form of PSM should be evaluated on a case-by-case basis which will take into account various factors that would clarify the actual risk and the residual risk that would remain after programs are incorporated into the operations.

2.0 Violation and Corrective Action Evaluation

The PSM elements are listed in the 1910.119 standard in the following order/sections:

(c) Employee Participation
(d) Process Safety Information
(e) Process Hazard Analysis
(f) Operating Procedures
(g) Training
(h) Contractors
(i) Pre-startup Safety Review
(j) Mechanical Integrity
(k) Hot Work Permit
(l) Management of Change
(m) Incident Investigation
(n) Emergency Planning and Response
(o) Compliance Audits
(p) Trade Secrets

OSHA has been administering violations for the 1910.119 standard since 1992. For purposes of relevancy, the last 10 years of violation data was examined (2007-2017). These violations were issued when OSHA completed an inspection and/or accident investigation visit to a particular facility.

Each one of the elements listed above has subsections pertaining to the requirements for that element. For example, when OSHA administers a PSM violation during a visit it is listed as: 1910119 E03 I. The nomenclature correlates to a violation for section (e) Process Hazard Analysis (PHA), part 3(i), or (3) the process hazard analysis shall address (i) the hazards of the process. This particular example indicates that the PHA did not properly address any or all of the hazards of the process. The OSHA inspection information also includes other data such as penalty dollar amounts, dates and type of inspection. The information provided in this summary reviews a few segments of this data from years 2007-2017 [4].

3.0 Violation Data Interpretation

All OSHA PSM violations were tallied over the 10-year period (2007-2017), this included:

- ✓ All citation classifications types (serious, repeat, willful etc.)
- ✓ All types of inspections/visits (accident, planned etc.)
- ✓ All types of facilities

This review does not distinguish whether violations were given during the same visit or during separate visits.

The data shows that there are 4 elements, or 4 deficiencies, that clearly stand out as the most prevalent (see Figure 1):

Most Commonly Cited PSM Elements (highest to lowest):

1) Process Safety Information (PSI)
2) Mechanical Integrity (MI)
3) Operating Procedures (OP)
4) Process Hazard Analysis (PHA)

Total Monetary Amount (highest to lowest):

1) Mechanical Integrity
2) Process Safety Information
3) Operating Procedures
4) Process Hazard Analysis

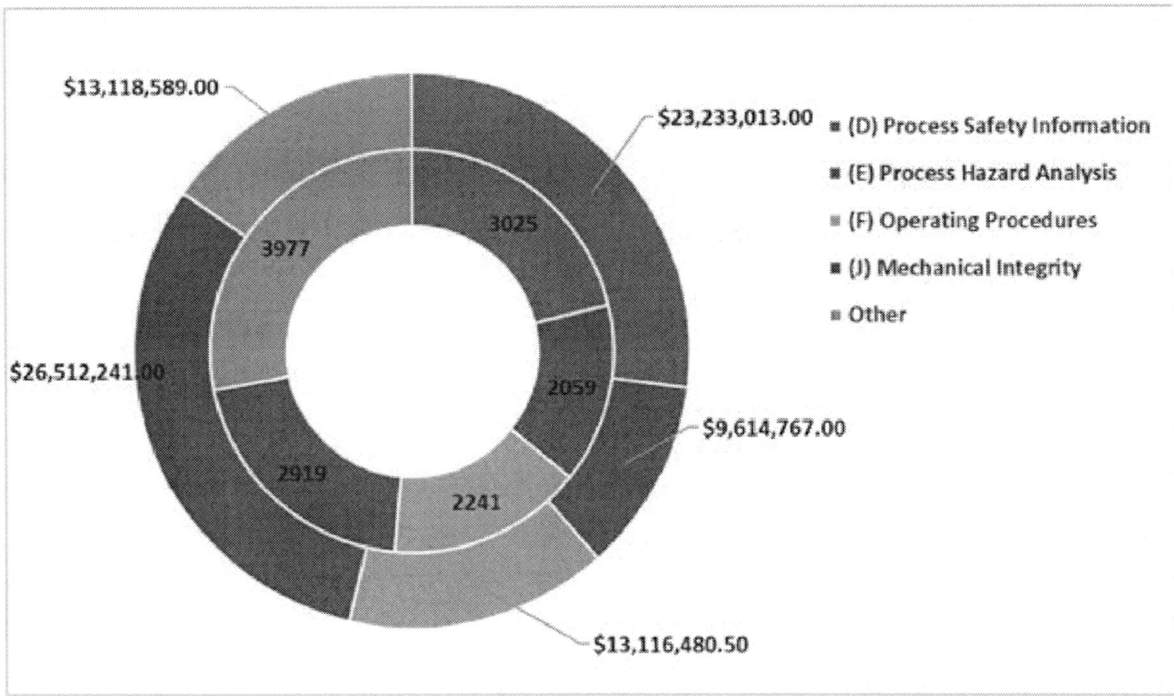

Figure 1: Total Number of OSHA PSM Violations and Total Citation Dollar Amounts by Element: 2007-2017

The top two elements (PSI and MI) have close to the same number of violations and the same goes for the third and fourth most common violations, giving the impression that the struggle to execute PSI and MI is about equal and OP and PHA both come in as close second.

It is alarming that PSI is the most cited element. Having correct and adequate PSI is the foundation for implementing, not only all PSM elements correctly, but any and all chemical safety precautions. PSI is crucial in completing an adequate PHA, the analysis that examines the hazards of the process and ensures that the proper safeguards are in place. The lack of PSI could lead to process hazards being missed, or incorrectly evaluated, during a PHA. Incomplete PSI can also lead to having inappropriate MI programs since some of these programs can be based on the hazards of the process, for example, more frequent inspections than the manufacturer's recommended practices based on the characteristics of the material being processed.

MI is the second most cited element but has the largest monetary amount, this could point to more instances of MI being noted within a single violation. OSHA will increase a fine for a single violation based on the number of "instances", or how many times they see a violation for the same element during the same visit. This review did not take into account the number of instances per violation.

MI can be difficult to implement, but new software solutions and training programs exist to help with proper execution. There are MI software products on the market that include capabilities to capture PSI, PHA and OPs, addressing all of the top-cited elements within one program.

It is helpful to utilize PSM experts for implementation and execution of PSM programs, as they are aware of the depth of information that is needed to meet the intent of the standard and for overall safety. They can often recommend products depending on your particular operation. Individuals dedicated to PSM will have access to resources that might not be available to others who are not embedded in the PSM industry. For instance, access to fee-based databases with previous incident data, one item that must be documented as part of a PHA. The expenditure for bringing in this type of expertise can provide a beneficial return on the investment when examining the average cost per violation and average number of violations per visit (Figures 2 and 3).

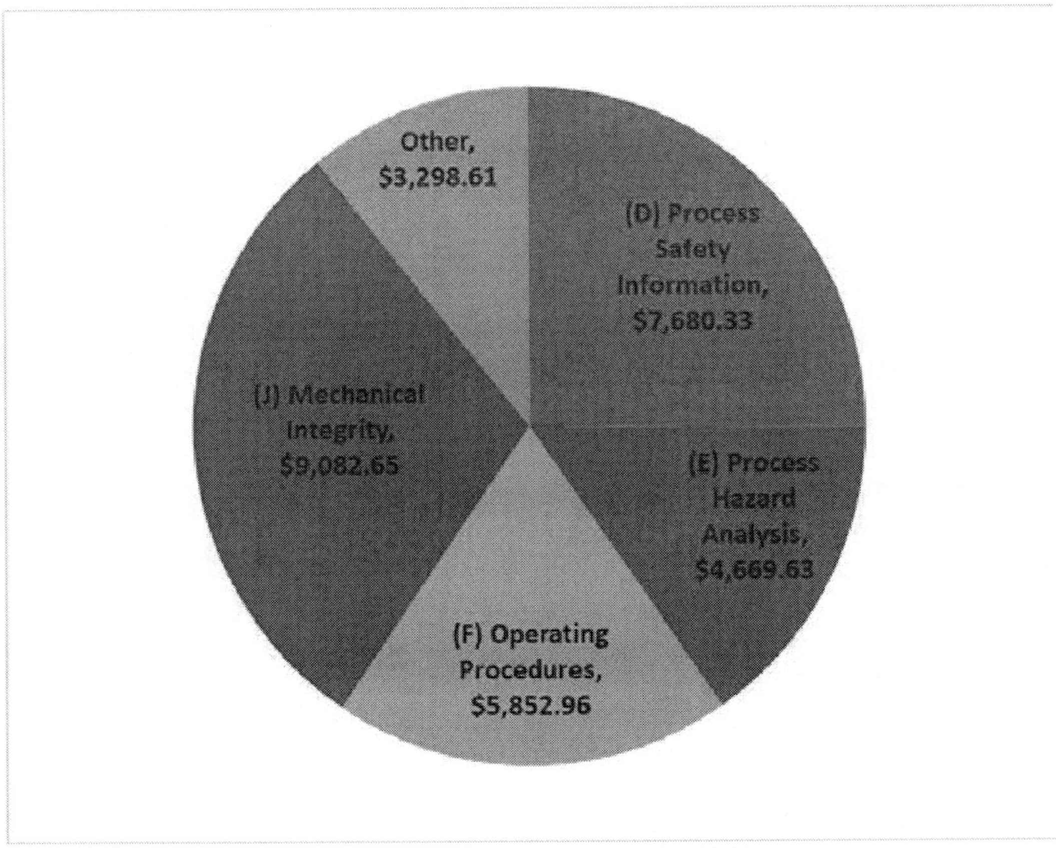

Figure 2: Average Cost per Violation by Element 2007-2017

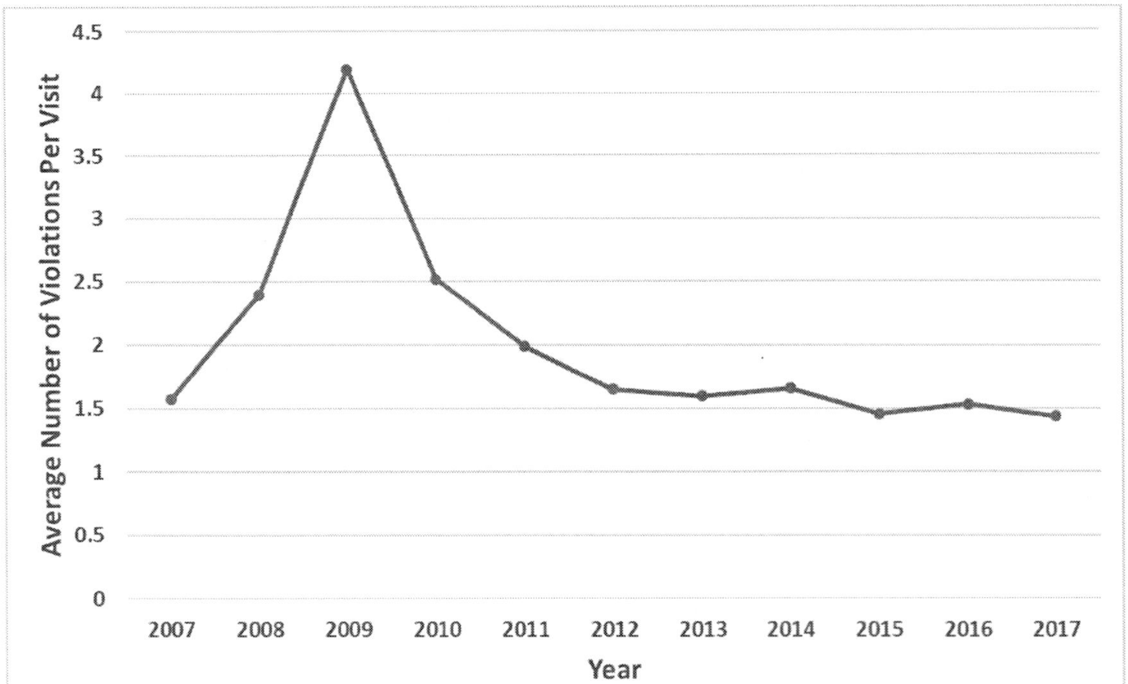

Figure 3: Average Number of Violations per Visit 2007-2017

(To note about Figure 3: the spike in the number of violations in 2009 could be due to the National Emphasis Program (NEP) in which OSHA extended to include a pilot on PSM-covered facilities from the 2007 NEP which only included refineries. The NEP went on from 2011 to only include chemical facilities and not petroleum refineries)

The citation costs are only one aspect to consider when reviewing the return on the investment. Other benefits that can be quickly realized once the programs have been established include but may not be limited to:

- ✓ Improvement in operating efficiencies
- ✓ Increased employee morale
- ✓ Lower turnover rates
- ✓ Safer working conditions
- ✓ Uptime gains
- ✓ Insurance benefits
- ✓ More consistent product quality
- ✓ Decreased likelihood of an incident

A deeper dive into the subsections of the most cited elements yielded the results in Figure 4.

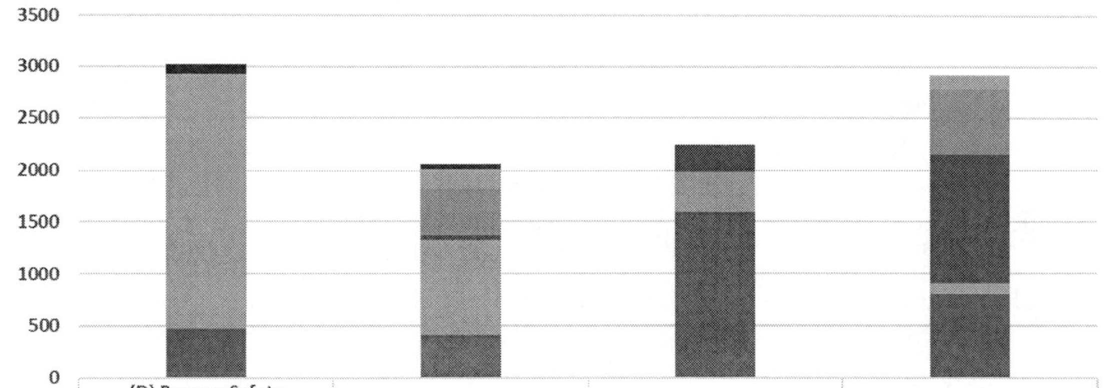

	(D) Process Safety Information	(E) Process Hazard Analysis	(F) Operating Procedures	(J) Mechanical Integrity
Blank	101			
07		48		
06		186		135
05		463		640
04		37	255	1235
03	2452	918	391	105
02	418	26	23	798
01	54	381	1572	6

Figure 4: Sub-section Cited in the Top 4 OSHA Most Cited PSM Elements 2007-2017

Figure 4 shows that the biggest gaps over the last 10 years include the following sub-sections of these elements:

1. Process Safety Information (d)
   - Equipment Information (section 3)
2. Mechanical Integrity (j)
   - Written Maintenance Procedures (section 4)
   - Inspection and Testing (section 2)
   - Correcting Equipment Deficiencies (section 5)
3. Process Hazard Analysis (e)
   - Inadequate coverage of the hazards of the process (section 3)
4. Operating Procedures (f)
   - Develop written operating procedures (section 1)

There is a common theme among these four (4) subsection violations. While PSI was the most cited element, the most cited PSI sub-section is Equipment Information, which ties into the MI portion of PSM. Based on this, deficiencies in equipment related information and MI ranks very high on the list. MI related programs help to ensure that chemicals remain inside the vessels, pipes, etc. A review of the OSHA data shows that the industry may be struggling with proper execution of this critical piece of process safety.

As noted above, MI programs are complex and can be difficult to establish, but there are current solutions on the market to help correct these deficiencies, such as economical software solutions that improve compliance by organizing maintenance tasks. The right solution can alert staff to overdue maintenance, and store the maintenance history for each particular piece of equipment. Some tools can also track inventory of spare parts and link Management of Change documents related to the process and/or the equipment. Consultants versed in the regulations and MI techniques can help establish appropriate programs and find solutions useful to your specific operation.

Development of operating procedures for both process and maintenance activities has a high number of citations in the MI and OP elements. It is essential that these procedures are comprehensive, accessible, updated and easy to interpret. Mistakes in operating steps, or failure to recognize abnormal conditions, can quickly lead to catastrophic releases.

There are various writing and auditing techniques that can be used to ensure that these procedures include the correct information and address all stages of the process (start-up, normal conditions, abnormal conditions and shut down). Software tools currently on the market can also assist in organizing this critical information. This approach, while time consuming up-front, is time well spent in providing a safer and more efficient workplace.

Once authored, it is easier to audit the documents for accuracy and audit the operators using a behavior-based system to determine that procedures are being followed correctly.

Vast benefits can be gained using this approach, and many PSM consultants offer services to help author these documents and provide suggestions for document control and review processes.

4.0 Resources

Despite the intent for OSHA PSM to decrease the likelihood and severity of accidents, the OSHA data is showing that PSM gaps still exist in the most fundamental PSM elements. These gaps present an unacceptable risk. This is evidenced by the accidents that continue to occur. The various incident investigations performed by the U.S. Chemical Safety and Hazard Investigation Board often attribute inadequate PSM programs as the root cause.

There are many resources available for assistance with PSM implementation and improvement for plants of all types and sizes. There has been an increase in consulting organizations providing guidance and software solutions for PSM. The list of resources readily available to even the smallest of plants are much more prevalent than they were 10 years ago.

OSHA has published documents with guidelines for PSM for storage facilities and for small businesses. They also have a PSM resource page: https://www.osha.gov/SLTC/processsafetymanagement/. The Center for Chemical Process Safety has many publications, classes and tools to help with all aspects of PSM. Some of these resources are free, or provided at reasonable cost, and CCPS/AIChE (American Institute of Chemical Engineers) membership increases those benefits.

Small businesses often find it difficult to execute PSM. In 2016-2017, in almost the same number of inspections, OSHA administered 57 PSM related violations to companies with under 100 employees and only 18 violations to companies with over 100 employees. OSHA also points to a study that estimates "*employers with 1-25 employees are 47 times more likely to have a release and 17 times more likely to suffer an injury, per employee, than facilities with 1500 or more employees.*" [5]

There are a few PSM consulting firms that target small to medium companies, the companies that cannot dedicate resources to PSM, nor have the expertise in-house to meet the intent of the standard. Many of these consulting firms not only focus on OSHA compliance but becoming a partner to help embed PSM concepts into the plant's overall safety culture. Bringing in these third parties is worth the cost because they are embedded in the PSM realm and can bring along expertise from other industries, the latest information and years of experience.

There are cost effective software solutions, some aimed at small businesses from a price point and feature standpoint. These solutions can improve PSM while improving the bottom line through improved efficiency, production uptime gains, insurance benefits and violation and incident avoidance. The smaller chemical processing companies are yet to realize that these solutions exist at very reasonable prices, and the small upfront investments have very beneficial returns that easily justify the expenditure.

It is ideal to use the same consulting firm for various locations to standardize PSM solutions and audits across multiple sites that may have different risk. This type of PSM implementation helps to normalize the oversight and tracking to give a better picture of risk at the corporate level.

5.0 Conclusion

The research objective of this paper was to spot the potential for common gaps and issues in implementing PSM across the United States at all types of chemical processing industries using OSHA violation data, and to use that information to identify opportunities for solutions and preventative/corrective actions. The OSHA violation picture is only one piece of the gap analysis puzzle, but it is one place that collects enough data to increase awareness of PSM weaknesses nationwide regardless of the type of operation. This information can aid in creating education/awareness programs and campaigns aimed at improving PSM execution.

The top four cited violations are PSI, MI, PHA and OPs. Solutions to the deficiencies revealed by the OSHA data are well-known and PSM experts are aware of the many practical and feasible methods that exist to close these deficiency gaps. The return on investment on items outside of the safety aspect are usually realized soon after the solutions are put in place, including: more efficient and consistent work procedures (which can lead to improved product quality), increased up-time and insurance benefits to name only a few.

A review of the OSHA data can help quantify the costs that can be expected from OSHA PSM violations and citations. However, other impacts of violations and/or PSM incidents that are not so easily quantified include the business, legal, liability, company reputation and the impact that the injuries have on families. Sometimes plants are not able to rebuild in the same location or regain the customer share they may have lost due to an incident. Additionally, the OSHA violations are public record and can be accessed by lawyers, local authorities having jurisdiction, insurance companies, activists, etc. Companies may also find it difficult to obtain insurance due to OSHA violation or incident history. While these impacts are difficult to quantify, they must be considered in the overall cost to the business.

PSM implementation is a continuous improvement journey with new lessons learned, new technologies and new resources being created every year. Using the plethora of tools available can help close the deficiency gap.

References

[1] U.S Department of Labor OSHA 3132 Publication *Process Safety Management* 2000
[2] Washington State Department of Labor & Industries July 2010
[3] U.S Department of Labor OSHA 3133 *Publication Process Safety Management Guidelines for Compliance*, 1994
[4] OSHA Violation Data 2007-2017
[5] OSHA *Process Safety Management for Small Businesses* 2017

546725 Safety Performance Evaluation: A Benchmark of Practices across Various High-Risk Systems

The concept of performance has invaded all management spheres and imposed itself as a strong driver of decision making. Following the funding work of Kaplan and Norton (1996), performance appears as a multidimensional concept which evaluation and use require adapted tools and well calibrated decision processes.

Safety performance does not escape these requirements as it englobes various dimensions (workplace safety, process safety, environmental safety), can be approached through various evaluation paradigms (auto evaluation, external assessment through audits and KPIs) and may be used by several stakeholders (risk owner, regional and national authorities, public perceptions). Indeed, different combinations of these three dimensions allow for a large variety of performance governance paradigms carrying various understandings and practices with regard to the way performance should be assessed and used.

By comparing safety performance management in three different risk systems (process, railway and health) in Europe, this works aims at providing safety management actors with a description of the richness and variety of safety performance understandings and practices. Insights gained from this work should enhance safety managers' abilities to grasp the strengths and weaknesses of their approaches and explore the possibilities of a beneficial benchmarking.

19 AIChE Spring Meeting
15TH GLOBAL CONGRESS ON PROCESS SAFETY
NEW ORLEANS • 2019

Lessons learned from analyzing an explosion accident at Shanghai Secco chemical plant

Hanhai Dong
China University of Petroleum (East China)
No. 66, West Changjiang Road, Huangdao District, Qingdao, China
donghanhaiupc@126.com

Yi Liu
China University of Petroleum (East China)
No. 66, West Changjiang Road, Huangdao District, Qingdao, China
liuyi@upc.edu.cn

Dongfeng Zhao
China University of Petroleum (East China)
No. 66, West Changjiang Road, Huangdao District, Qingdao, China
zhaodf@vip.sina.com

Jian Chen
China University of Petroleum (East China)
No. 66, West Changjiang Road, Huangdao District, Qingdao, China
836771528@qq.com

Yanchun Wang
China University of Petroleum (East China)
No. 66, West Changjiang Road, Huangdao District, Qingdao, China
Wychun25@163.com

Wei Wang
China University of Petroleum (East China)
No. 66, West Changjiang Road, Huangdao District, Qingdao, China
15621097585@163.com

Prepared for Presentation at
American Institute of Chemical Engineers
2019 Spring Meeting and 15th Global Congress on Process Safety
New Orleans, LA
March 31 – April 3, 2019

Lessons learned from analyzing an explosion accident at Shanghai Secco chemical plant

Hanhai Dong
China University of Petroleum (East China)
No. 66, West Changjiang Road, Huangdao District, Qingdao, China
donghanhaiupc@126.com

Yi Liu
China University of Petroleum (East China)

Dongfeng Zhao
China University of Petroleum (East China)

Jian Chen
China University of Petroleum (East China)

Yanchun Wang
China University of Petroleum (East China)

Wei Wang
China University of Petroleum (East China)

Keywords: event sequence diagram; Multilayer Bow-tie Analysis; FLACS

Abstract

On May 12, 2018, six workers from Shanghai Aijinke Engineering Construction Services Co., Ltd. died on the spot of an explosion when they were maintaining a benzene tank in the tank farm at Shanghai Secco Petrochemical Co., Ltd. The accident was characterized as a disastrous accident caused by hot work in a confined space, which resulted in massive casualties and a direct economic loss of 11.66 million yuan. This paper aims to analyze the accident occurrence process in a chronological order firstly by using the event sequence diagram (ESD). Multilayer Bow-tie Analysis was adopted to explore and classify the causes of the explosion of the benzene tank, along with the corresponding consequences, and the preventive measures. The analyses show that the primary inducements were the unsafe behavior of on-site workers, hazardous conditions of in-tank substances, risky factors in the construction environment, and deficiencies in management. Casualties, fire and explosion, equipment damage, and property losses were all included as consequences. The FLACS software was used to simulate the variation of explosion pressure and explosion temperature varying with time when benzene vapor was ignited at the tank bottom during the accident, and a flow field diagram for the explosion was achieved correspondingly. The results obtained are of direct significance for investigation of similar accidents in the future and are instructive to equipment safety evaluation as well. Based on the qualitative and quantitative analyses, we emphasize three considerable reasons leading to the accident, which includes inapposite flammable gases detection points before the hot work, failing to test the flammable gases during the hot work, and deficiency of the Management of Change (MOC). Finally, the paper concludes with a summary of lessons learned from the accident as well as suggestions for accident prevention and safety improvement.

1 Introduction

The term confined space refers to closed or semi-closed facilities and places with limited entrance and exit, and poor ventilation, which may lack oxygen or contain flammable, explosive, toxic, or harmful substances. The main causes of accidents in a confined space include fire, explosion, poisoning, asphyxiation, and physical injury [1]. According to the statistics of the Ministry of Emergency Management of the People's Republic of China, from January 2012 to April 2018, a total of 64 accidents involving three or more deaths in a confined space occurred in China, resulting in 225 deaths and 82 injuries [2]. Among them, on April 8, 2014, three people were killed and two injured by an explosion during the repair and welding of the desulfurization tank at a factory in Wuhai, China. On March 18, 2015, the operators of a chemical company in Binzhou, China, suffered an explosion accident during an operation in a confined space, resulting in four deaths and two injuries. Hot work in a confined space is the common cause of the two accidents above.

Due to the poor ventilation, a large amount of flammable gas may be accumulated in a confined space. When construction workers perform hot work inside, they will face potential safety risks and problems, such as difficulty in escaping, delayed personnel rescue, and inconvenient communication. Hot work in a confined space can easily lead to fire, explosion, and personal injury accidents, which threaten not only the personal health of operators but also the property of enterprises.

Researchers have done a lot of risk analysis and exploration for the particular operating environment in a confined space. Selman et al. [3] analyzed the mortality rate of confined space operations in similarly industrialized countries from 0.05 to 0.08 per 100,000 workers and pointed out that the main risks are toxic atmospheric hazards and physical hazards. Selman et al. [4] divided the emergency rescue of confined space accidents into three levels: individual self-rescue, non-entry rescue, and progressive rescue. Botti et al. [5] applied the Internet of Things (IoT) technology and proposed a holistic approach to analyzing the procedure of confined space operations. Meanwhile, many studies adopted experimental and numerical simulation methods to observe the evolution of characteristic parameters during the diffusion and detonation of flammable gas in a confined space. The results obtained are helpful to the quantitative analysis of confined space accidents. Bang et al. [6] established a theoretical model to predict the pressure rise caused by the deflagration in a confined space. He et al. [7] presented an LES turbulence model to simulate the diffusion process of LPG in the underground confined space.

This paper mainly analyzes a disastrous accident caused by hot work in a confined space at Shanghai Secco Petrochemical Co., Ltd (Secco). On May 12, 2018, the workers from Shanghai Aijinke Engineering Construction Services Co., Ltd. (Aijinke) triggered the explosion of flammable gas when they were performing inspection and maintenance inside a benzene tank. The serious accident resulted in the immediate death of six on-site workers and direct economic loss of 11.66 million yuan. Through a series of qualitative and quantitative analysis methods, we reveal the root causes of the accident, simulate the development process of the explosion, and propose corresponding rectification suggestions. Here we want to emphasize that the reason why we write this paper is not to pursue responsibility and punishment but to reflect on how to avoid the recurrence of similar accidents.

2 The occurrence of the explosion accident

On the morning of May 12, 2018, Aijinke workers arrived at the utility tank farm of Secco to prepare for maintaining TK-75-0201 benzene tank (0201 benzene tank). Before the start of the operation, a Secco field operator used a hand-held gas detector to measure oxygen and flammable gas at the manhole outside the tank. The on-site supervisor of Aijinke, the on-site supervisor of Secco, and the on-duty monitor of Secco utility tank farm signed the work ticket successively without verifying the measurement data and checking the personal protective equipment (PPE) and explosion-proof tools of the on-site workers according to the requirements specified in the operating permit.

At 1:15 p.m., six workers of Aijinke continued to dismantle the floating pontoon in the tank and two workers worked outside for monitoring. At the scene, another Secco field operator was in charge of regular oxygen and flammable gas measurement. At 3:25 p.m., a sudden explosion occurred in the 0201 benzene tank.

After the explosion, Secco immediately launched their emergency response plan and reported the accident information to the response center of Shanghai Chemical Industry Zone Management Committee (SCIZMC), the fire brigade, the medical center, and the production operations monitoring center of Sinopec. At 3:50 p.m., the fire brigade put out the fire by spraying water. At 5:50 p.m., the on-site rescue work ended and a total of six Aijinke workers in the tank died on the spot. Figure 1a shows the scene of the 0201 benzene tank after the explosion.

Figure 1. The 0201 benzene tank and floating pontoons

2.1 Description of the timeline

The occurrence of accidents is a complex system problem, and it gives useful information on how accidents arise in practice [8]. In March 2018, Secco found that the breather valve of the 0201 benzene tank leaked slightly, which caused the VOC concentration to exceed the standard. After inspection, it was judged that the floating pontoon leaked. In order to find the root causes, we describe in detail the fundamental behaviors of the primary responsible party and the contractor prior to the explosion.

- On April 16, Secco established a maintenance order and a work list in the SAP system.
- On April 19, Secco began to empty the 0201 benzene tank, performing four days of steam dilution and eight days of nitrogen replacement.
- On April 24, Secco informed Aijinke to formulate maintenance plan.
- On April 26, Secco carried out the Job Hazard Analysis (JHA) around the maintenance work.
- From May 1 to 7, Secco opened the manhole of the tank for natural ventilation and inspection, and found that more than half of the floating pontoons had effusion.
- On May 8, the relevant departments of Secco held a special meeting and decided to change the original plan to replace all the floating pontoons.
- On May 9, Secco asked Aijinke workers to enter the tank and destructively dismantle all floating pontoons at the regular OSBL equipment maintenance meeting.
- From May 9 to 11, Aijinke workers entered the tank and dismantled the floating pontoons. A total of 38 floating pontoons were dismantled, as shown in Figure 1b.

2.2 Analysis by event sequence diagram

Event Sequence Diagram (ESD) is an accident scenario-oriented simulation method, which treats events as the basic research object, and takes into account the domino effect and risk transfer effect [9]. ESD is also an intuitive reflection of the accident scenario. It provides a more literal representation of the system state, indicating the behavior of crucial process variables, and even operator and hardware state changes [10].

In order to further illustrate and point out the critical links of the accident, this paper makes an intuitive description of events by using ESD. As shown in Figure 2, the basic events of accidents are divided into initial events, judge events, action state events, and termination events. At the same time, the path indicated by the dashed line represents the potential events that may occur in this accident, and helps to make a comprehensive and logical analysis of the whole process. The graphic symbols in the ESD are explained in the Appendix.

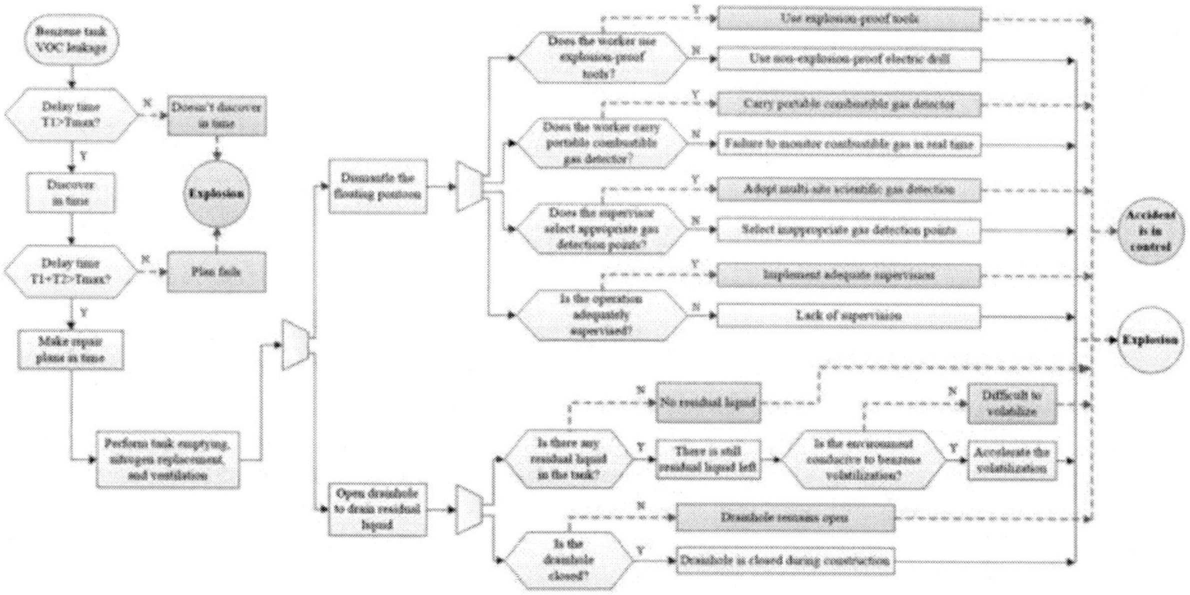

Figure 2. The ESD of the accident

As can be seen from the ESD, there are nine pivotal judgment nodes on the occurrence of the event chain, which can determine the direction of the incident. Among them, whether to discover leakage in time and whether to make a repair plan in time are the nodes at the time aspect. When the detection instrument and worker cannot find leakage and formulate the response plan in the maximum allowable delay time, they will directly lead to the explosion. Enterprises need to enhance equipment reliability and inspection frequency, so as to reduce emergency response time. Whether there are residues in the tank and whether environmental conditions are conducive to benzene volatilization are two nodes at the environmental aspect, which are also the leading causes of accidents. They require the workers to make a detailed hazard identification of the construction environment, and improve scientific detection. The critical nodes regarding human factors include whether the drainhole is closed, whether the on-site workers use explosion-proof tools, whether they carry portable combustible gas detector, whether they select appropriate detection points, and whether the supervisors take adequate monitoring [11]. Strengthening safety training before the operation, improving individual risk identification ability, perfecting risk control measures, and enhancing safety supervision level are essential means to avoid accidents caused by human factors.

All of these factors, organizational and operational, contributed to the accident, with a hierarchy of casualties among them. Through the comprehensive description, ESD can not only find out the nodes where the problem lies, but also guide the prevention of similar accidents. Therefore, enterprises should pay attention to the basic events expressed by nine judgment nodes and take appropriate interventions to block the development path of dangerous accidents.

3 Multilayer bow-tie analysis

3.1 The process of MBTA

Among many techniques available, bow-tie analysis is prevalent because it organically combines the two sophisticated risk analysis methods of Event Tree Analysis (ETA) and Fault Tree Analysis (FTA) [12], systematically summarizes the causes and consequences of accidents, and further integrates the prevention and mitigation of accidents or risks [13]. Bow-tie analysis has been proven to be a reliable and efficient tool, mainly used for process safety analysis, risk management, and accident risk analysis. It helps to understand which significant event combinations will lead to the top events, and which security barrier failures would escalate the top event to a certain consequence.

Traditional bow-tie analysis mainly includes top event, causes, consequences, preventive barriers, and mitigation barriers, which shows a lot of information. However, the single-level structure lacks logic and is not conducive to people's deep understanding. Thus, this paper establishes a Multilayer Bow-tie Analysis (MBTA), which resolves the top-event-related ETA and FTA individually into hierarchical structures, to deeply analyze the root causes, specify the severe consequences, and propose the corresponding safety barriers and scientific disposal methods. MBTA uses graphics to visually display the relationship between different levels and avoid repeated analysis effectively. Figure 3 shows the MBTA model, in which the code explanations are listed in Table 1.

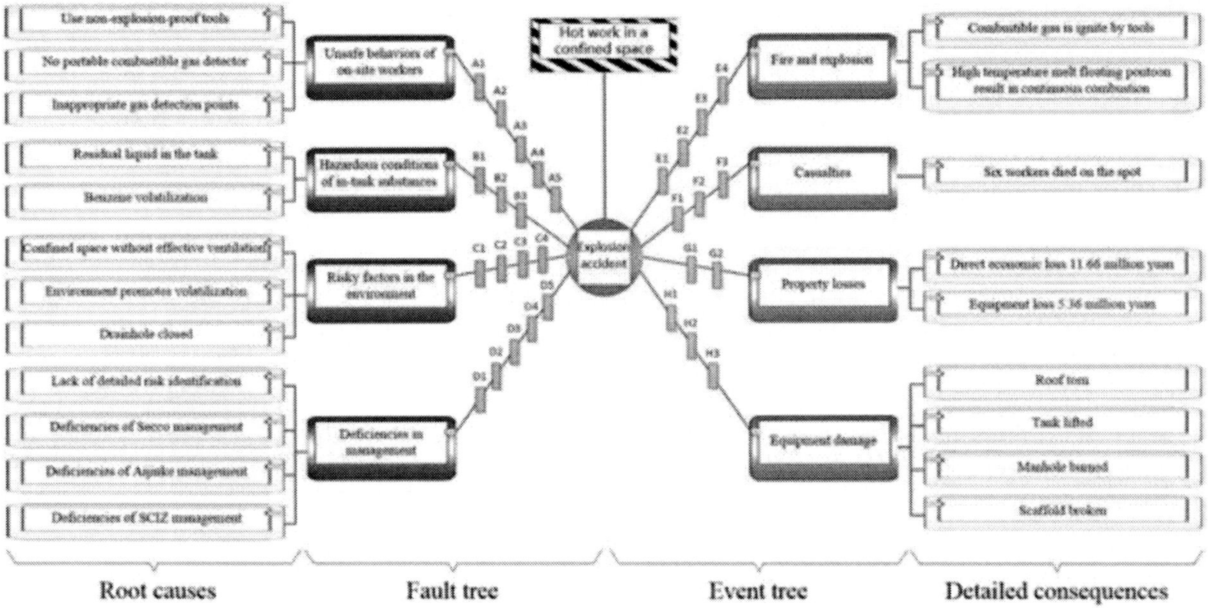

Figure 3. The MBTA model

Table 1. The explanations of MBTA

Code	Safety barrier	Code	Safety barrier
A1	Use explosion-proof tools	D4	Implement the main responsibility
A2	Carry portable combustible gas detector	D5	Improve contractor management
A3	Strengthen operation training	E1	Call the fire brigade in time

A4	Enhance supervision and assessment	E2	Spray water to extinguish fire
A5	Set up warning signs	E3	Fire isolation
B1	Remove the residual liquid	E4	Launch the emergency plan
B2	Hazardous material isolation	F1	Call for medical treatment
B3	Combustible gas detection in the tank	F2	Search and rescue the wounded
C1	Forced ventilation	F3	Launch the emergency plan
C2	Open drainhole	G1	Insurance claim
C3	Continuous detection during operation	G2	Launch the emergency plan
C4	Take environmental risk assessment	H1	Dangerous equipment isolation
D1	Compliance with rules and regulations	H2	Equipment repair
D2	Increase the intensity of punishment	H3	Equipment replacement
D3	Strengthen process safety management		

In the above MBTA model, we adopt the analytic hierarchy process (AHP) to extract the common characteristics of 12 causes of accidents and classify the causes into four categories: unsafe behaviors of on-site workers, hazardous conditions of in-tank substances, risk factors in the construction environment, and deficiencies in management. Moreover, we could conclude the direct and indirect reasons for the accident.

3.1.1 Direct reasons

Benzene is a colorless and transparent liquid with a strong aromatic odor. It has a boiling point of 80.1 ℃, a saturated vapor pressure of 10 kPa at 20 ℃, a flash point of -11 ℃, an explosion limit range of 1.2%-8.0%, and minimum ignition energy of only 0.20 mJ. Benzene is a highly flammable substance. The explosive mixture of benzene vapor and air can burn and explode when exposed to ignition source or high temperature.

At the construction site, the temperature is 24.7 ℃, and the pressure is 100.68 kPa. Under these conditions, benzene is easy to volatilize and form a gas explosion environment in a confined space. During the process of dismantling, residual benzene in the floating pontoons leaked onto the bottom of the tank and was not cleaned up in time. Because there was no effective ventilation in the tank, benzene vapor mixed with air to form an explosive environment and benzene concentration reached the explosion limit. The non-explosion-proof tools used by the operators produced ignition energy, which brought about the deflagration of the flammable gas. The high temperatures generated by the combustion melt the other aluminum alloy pontoons, resulting in sustained combustion for a short period.

3.1.2 Indirect reasons

Aijinke failed to abide by the relevant safety production rules and regulations. Managers did not fulfill their duties of safety management and supervision. Aijinke failed to give safety technical information to workers following the operating rules before the construction. Without implementing and supervising PPE, carrying portable gas detector, and using explosion-proof tools, operators were arranged to work in a confined space. When significant changes occurred in the job conditions, Aijinke did not promptly request workers to stop working and implement the Management of Change (MOC).

Secco had some omissions in safety risk management, professional management, and special operation management. In addition, Secco did not follow the safety production rules and regulations. On-site inspectors failed to carry out gas detection in accordance with specifications. After the change of operation, the manager did not ask the contractor to stop the process in time and revise the construction plan according to the regulations. Secco failed to report the maintenance plan to SCIZMC as required.

SCIZMC failed to implement the safety management system fully and did not discover that Secco performed maintenance operations without permission.

3.2 Lessons learned from MBTA

This paper puts forward a series of preventive measures, such as strengthening operation training, removing residual benzene in the tank, improving risk assessment of the environment, and enhancing process safety management. It is noticed that the environment in a confined space is harsh and there are many potential unsafe factors. Furthermore, the contractor's maintenance work cycle spans the statutory weekend. Based on the two reasons above, the enterprise should upgrade the level of hot work and adopt a broader system of operation approval and management. On-site workers need to stop working immediately when safety problems are found in the construction process and report to their superiors by relevant regulations.

Through refining the consequences of accidents, we summarized that detailed safety emergency plans and energy isolation methods are universally applicable in various measures to control and mitigate the impacts of accidents. They require Secco and Aijinke to formulate a comprehensive emergency response plan before, including fire extinguishing, personnel search and rescue, equipment maintenance, property damage remediation and so on. When an accident occurs, workers have to stay away from dangerous substances and places immediately.

As with many disasters, what ended up in a sequence of equipment damage and human errors started mostly as a management problem. Many unsafe practices were tolerated and even encouraged. The risk management of confined space operations should be divided into ten steps according to relevant regulations, namely training, pre-operation planning, job approval, hazardous energy isolation, ventilation, gas detection, personal protection and tool safety, warning signs, monitoring, and emergency response, so as to strengthen process safety management comprehensively. It is necessary to enhance the issuance of management of work tickets for hazardous operations such as hot work and confined space operations. Enterprises must strictly inspect illegal operations and supervise safety precaution measures to ensure that the operation process is safe and controllable. For changes occurring in the construction process, enterprises ought to strictly implement the MOC, analyze the potential risk and possible accidents, formulate corresponding safety measures, and carry out safety education for all operators.

MBTA provided a detailed analysis of the accident and established the corresponding safety barriers. However, there are still some certain subjective problems, and its content should also be flexible. Therefore, we encourage employees, accident investigators, case learners, and other relevant personnel to improve and develop the content of MBTA continually, thereby effectively expanding the application scope and effectiveness of MBTA.

4 Quantitative simulation of the explosion

The analysis of the accident should include not only qualitative analysis but also quantitative analysis [14]. We use FLACS software to simulate and analyze the occurrence of the benzene explosion. According to the relevant information, the 0201 benzene tank is an internal floating roof tank with a volume of 10,000 m^3. Its diameter is 30 m and height is 19 m. The design temperature range and pressure range are -10 – 65 ℃ and -0.49 –1.96 kPa, respectively. The tank adopts box-type aluminum alloy internal floating pontoons, and a manhole with a diameter of 0.6 m is set at the position of 0.75 m above the ground. Figure 4 shows the 0201 benzene tank model.

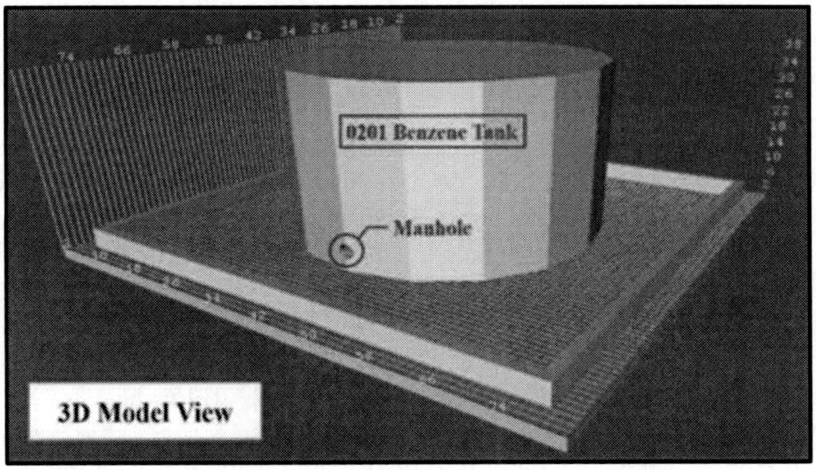

Figure 4. The 0201 benzene tank model

The simulated environmental conditions are fixed at the temperature of 24.7 ℃ and the pressure of 100.68 kPa, which are the same as the construction environment. We make reasonable assumptions: (a) The residual benzene in the tank evaporates continuously into the confined space and forms a uniform mixture of flammable gas. When the volatilization concentration of benzene reaches the lower explosion limit (LEL), the ignition source that persists during the construction operation detonates the flammable gas and results in an explosion. (b) Six on-site workers dismantled the float pontoons at the bottom of the tank, so the ignition source is set at the height of 0.75 meters from the geometric plane center, as shown in Figure 5 point I. There are several detection points of pressure and other characteristic parameters in the model of benzene tank, as shown in Figure 5, points M1-M5. (c) The benzene tank is windless, and the flammable gas accords with the real gas state equation. (d) The gas explosion process satisfies a series of fundamental equations, such as the continuity equation, momentum equation, and energy equation.

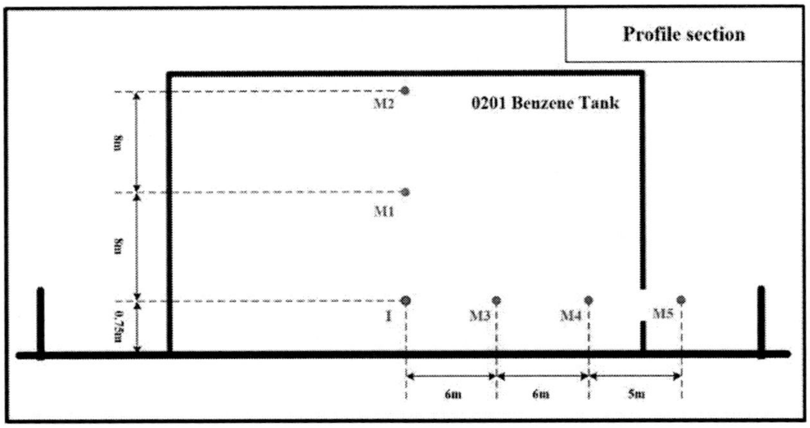

Figure 5. The profile section of the model

We selected the maximum profile section of the tank as the research object and intercepted the temperature field and pressure field at four different moments in the explosion process, as shown in Figure 6 and 7. The simulation results intuitively describe the development process of the explosion. When the flammable gas in the tank was ignited, the flame diffused around in the form of a spherical wave and reached the bottom of the tank at 0.35 s, as shown in Figure 6a and 7a. Since the radial distance of the tank is less than the vertical axial distance, the flame propagated horizontally to the wall, as shown in Figure 6b and 7b. At 1.52 s, the flame reached the manhole and sprayed out to the outside of the tank in the form of jet flow. During this period, the pressure and temperature increased continuously, which are the leading causes of the manhole being burned, as shown in Figure 8a. When the flame propagated to the top of the tank at 2.11 s, the uncombusted gas was recirculated by the viscous effect and blocking effect of the wall, which further caused the recirculation vortex. The recirculating vortex enlarged the contact area of the flame front, resulting in flame acceleration and pressure increase, as shown in Figure 6c and 7c. Figure 6d and 7d illustrate the flammable gas burned out at 3.24 s. Simultaneously, the explosion temperature and pressure reached the peak value contributing to the tearing of the roof and the lifting of the tank, as shown in Figure 8b and 7c.

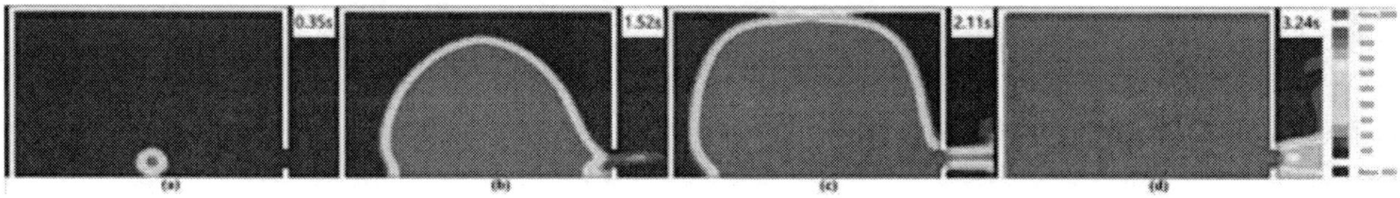

Figure 6. Temperature simulation of the explosion

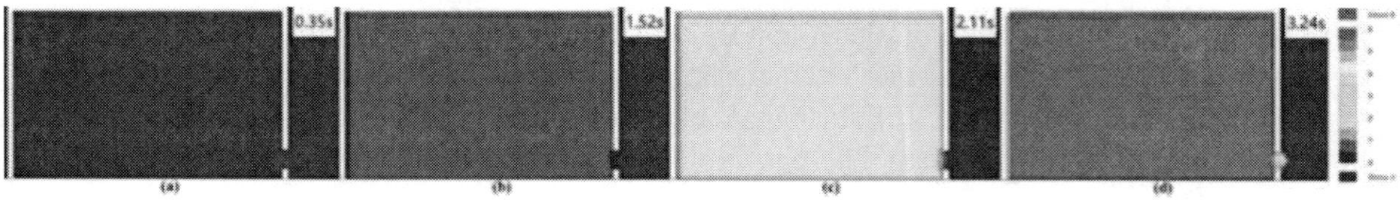

Figure 7. Pressure simulation of the explosion

Figure 8. The results of the accident

Figure 9 reveals the change of temperature and pressure recorded by the detection points M1-M5 during the explosion. Table 2 describes the data of the detection points, including maximum and minimum explosion pressures, maximum and minimum explosion temperatures, peak data arrival times and so on. Combining the data of detection points and the pressure field diagram in Figure 7, it is shown that the pressure in a confined space will be uniform immediately for a certain time and further proves the accuracy of the analysis of the flame propagation process. In particular, the external detection point M5 was affected by the propagation of explosion flame in the tank at 1.6 s, resulting in a peak explosion overpressure of 0.031 bar and a peak temperature of 1895 ℃. Subsequently, due to the pressure rise in the tank, the pressure at the detection point M5 dropped to negative pressure until the flammable gas burned out.

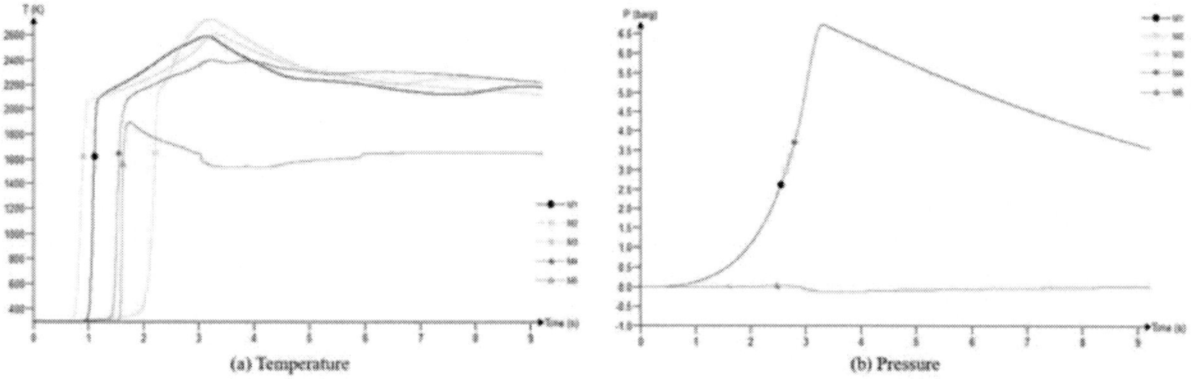

Figure 9. Parameters change of the detection points

Table 2. Data from the detection points

No.	Temperature				Pressure			
	Tmax/K	Time/s	Tmin/K	Time/s	Pmax/bar	Time/s	Pmin/bar	Time/s
M1	2589.4	3.11	298.1	0.04	6.75	3.24	0.00	0.04
M2	2732.2	3.24	298.1	0.04	6.76	3.24	0.00	0,04
M3	2609.3	3.28	298.1	0.00	6.76	3.24	0.00	0.00
M4	2306.2	3.25	298.1	0.00	6.75	3.24	0.00	0.00

M5	1895.5	1.68	298.0	0.69	0.05	1.65	-0.15	3.29

Although there is only a small amount of residual benzene, the benzene volatilization forms a flammable gas with the air inside the tank, which generates a substantial destructive power when ignited. The simulation results show that the flame propagation velocity of the flammable gas increases constantly, with the peak value reaching up to 41 m/s. Since the flame is propagated in the form of subsonic velocity, it can be judged that there are mainly the phenomena of deflagration occurred in the process of explosion [15]. Soloviova-Sokolova et al. [16] calculated the flame burning rate of the benzene vapor was 0.6 m/s at 298 K, 0.1 MPa and 0.6 eq., which clearly illustrated that the peak propagation velocity of the benzene vapor in a confined space would be several times higher. Meanwhile, the explosion pressure in a confined space is usually greater than that in an open area due to lack of heat loss and volume expansion.

When workers are engaged in hot work in a confined space, they should carry out comprehensive risk identification, select several representative flammable gas detection points, and detect the flammable gas in real time. Workers ought to stop work immediately and withdraw in an orderly manner once the flammable gas exceeds the standard. Through the analysis of the explosion process, the flame propagates to the wall in a very short time. Thermal radiation and explosive shock wave are the primary causes of casualties in the tank. Owing to the special working environment in a confined space, the phenomenon of mass casualties and injuries often occurs under the effect of high temperature and high pressure, which needs people's sufficient attention.

5 The subsequent event

The explosion accident of Secco has aroused widespread concern, and 22 responsible persons have also been punished. But unfortunately, lessons from the accident did not prevent a subsequent similar event. On November 26, 2018, three contractor workers of the Ningbo Engineering Company (NEC) and one worker of Secco inspected the discharge tank of cracking gas compressor in the olefin separation unit. Without ascertaining the concentration of oxygen in the tank, a contractor worker of NEC entered the tank for construction and fainted unexpectedly. A Secco worker immediately went into the tank to rescue him, but he also fell into the tank. The accident resulted in the death of two workers who had been rescued after asphyxia.

It is the second confined space operation accident in the same company within six months. The main causes of the accident are still the deficiency of contractor management and risk identification in a confined space. While expressing sadness, the public should profoundly reflect on why accidents continue to occur, what is the role of punishment and accountability, and how relevant regulations and reform measures can be truly implemented. Drawing lessons from the previous accidents is necessary. Enterprises ought to raise the safety awareness of their employees and use strict supervision to ensure safe production of chemical plants.

6 Conclusions

This paper provides an introduction of an accident caused by hot work in a confined space, together with the detailed ESD-based description of nine pivotal judgment nodes leading to this disastrous accident. It concludes that enterprises should take suitable intervention measures to block the pathway of the accident. The MBTA, covering the causes and consequences of the accident, offers proactive and reactive measures based on an intuitive analysis. Meanwhile, it emphasizes the importance in selection of appropriate detection points, real-time detection of in-tank operation, and implementation of MOC. Also, we apply the FLACS software into quantitative simulation of the explosion process in the tank. The results contribute to the guidance of accident investigation and safety measures selection. Finally, it suggests that enterprises adopt technical remedies and better safety management practices, since they are the essential basics for effective implementation of risk control measures.

Appendix

Symbol	Annotation
	Initial event: Beginning event of the ESD.
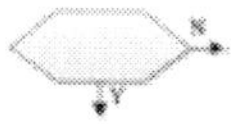	Judge event: The event has two possible outcomes that determine whether subsequent events occur.
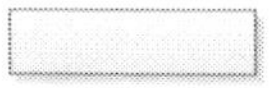	Action state event: Describe the action state of the event.
	Termination event: Ending event of the ESD.
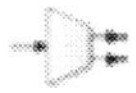	Output AND gate: Input event lead to multiple output events occurring.

7 References

[1] D.Burlet-Vienney, Y.Chinniah, A.Bahloul and B.Roberge, Design and application of a 5 step risk assessment tool for confined space entries, Safety science 80 (2015), 144-155.

[2] The Office of the State Council Security Committee Notified 8 Cases of Blindly Rescuing in a Comfined Space Leading to Casualties Enlargement, Available at http://www.chinasafety.gov.cn/default.htm Accessed on June 8, 2018.

[3] J.Selman, J.Spickett, J.Jansz and B.Mullins, Safety science 109 (2018), 333-343.

[4] J.Selman, J.Spickett, J.Jansz and B.Mullins, Safety science 113 (2019), 78-90.

[5] L.Botti, V.Duraccio,M.G.Gnoni and C.Mora, A framework for preventing and managing risks in confined spaces through IOT technologies, ESREL (2015), 3209-3217.

[6] B.Bang, C.Ahn, J.Lee, Y.Kim, M.Lee, B.Horn, D.Malik, K.Thomas, S.James, A.Yarin and S.Yoon, Theoretical, numerical, and experimental investigation of pressure rise due to deflagration in confined spaces, International Journal of Thermal Sciences 120 (2017), 469-480.

[7] B.He, X.Jiang, G.Yang and J.Xu, A numerical simulation study on the formation and dispersion of flammable vapor cloud in underground confined space. Process Safety and Environmental Protection 107 (2017), 1-11.

[8] I.M.Shaluf, F.Ahmadun and A.M.Said, Fire incident at a refinery in West Malaysia: the causes and lessons learned, Journal of Loss prevention in the Process Industries 16.4 (2003), 297-303.

[9] S.Swaminathan and C.Smidts, The event sequence diagram framework for dynamic probabilistic risk assessment, Reliability Engineering & System Safety 63.1 (1999), 73-90.

[10] J.Zhou, G.Reniers and N.Khakzad, Application of event sequence diagram to evaluate emergency response actions during fire-induced domino effects, Reliability Engineering & System Safety 150 (2016), 202-209.

[11] E.Zarei, M.Yazdi, R.Abbassi and F.Khan, A hybrid model for human factor analysis in process accidents: FBN-HFACS, Journal of Loss Prevention in the Process Industries 57 (2019), 142-155.

[12] N.Khakzad, F.Khan, P.Amyotte, Dynamic risk analysis using bow-tie approach, Reliability Engineering & System Safety 104 (2012), 36-44.

[13] M.B.Mulcahy, C.Boylan, S.Sigmann and R.Stuart, Using bowtie methodology to support laboratory hazard identification, risk management, and incident analysis, Journal of Chemical Health and Safety 24.3 (2017), 14-20.

[14] C.Zhu, J.Zhu, L.Wang and M.S.Mannan, Lessons learned from analyzing a VCE accident at a chemical plant, Journal of Loss Prevention in the Process Industries 50 (2017), 397-402.

[15] I.C.Tolias and A.G.Venetsanos, An improved CFD model for vented deflagration simulations–Analysis of a medium-scale hydrogen experiment, International Journal of Hydrogen Energy 43.52 (2018), 23568-23584.

[16] J.V.Soloviov-Sokolova, V.A.Alekseev, S.S.Matveev, I.V.Chechet, S.G.Matveev and A.A.Konnov, Fuel 175 (2016), 302-309.

Strengthening Process Safety Culture throughout Learning from Each Other´s Vulnerabilities

Luciana Poli SIlva
PETROBRAS
Av. Republica do Chile, 65, 4th floor, aisle 402.
lucianapoli@petrobras.com.br

Prepared for Presentation at
American Institute of Chemical Engineers
2019 Spring Meeting and 15th Global Congress on Process Safety
New Orleans, LA
March 31 – April 3, 2019

Strengthening Process Safety Culture throughout Learning from Each Other´s Vulnerabilities

Luciana Poli Silva
PETROBRAS
Av. República do Chile, 65, 4th floor, aisle 402.
lucianapoli@petrobras.com.br

Keywords: andragogy, dialogue, leadership, safety leadership culture, trust, vulnerabilities.

Abstract

A fundamental approach to any program that targets the implementation of a process safety culture is to work within reality, going beyond procedures and compliance. To facilitate the emergence of the unspoken is essential to stimulate the dialogue among leaders that brings about their vulnerabilities and opens new perspectives to address them effectively. This paper presents a pilot program designed to increase the Process Safety Leadership Culture in a Brazilian Energy Company. Based on a benchmark made within the industry and in andragogic principles, the program created a trust ambiance among peers and inspired leaders to recognize and share their vulnerabilities, as well as to take actions towards the strengthen the role of being a safety culture leader.

1 Introduction

A lot has been said and written about Leadership Process Safety Culture, but there is a gap between what the compliance procedures dictate and what managers really do. When it comes to the Brazilian Energy Industry, the pressure leaders are under in order to achieve the results they are expected to can be considered significant. How to balance the production challenges with a safety culture that must be among the core values of the operation? Are these factors compatible?

The pilot program entitled '*Strengthening Process Safety Culture*' dealt with reality, using andragogic approaches to bring about the complex factors forty managers face in their routine, such as the existence of several levels separating top managers from the operational team, privileging safety over production goals, and being an effective representative of the Company´s policies. All these factors can jeopardize the safety culture from becoming stronger every day.

This paper describes the implementation of a methodology designed to create a safe environment, in order to build trust among participants and at the same time, make them reflect and dialogue about how they can evolve from compliance to commitment.

In the coming sections, we will present the central concepts used to develop the program – Culture, Safety Culture and Leadership - exploring its implication on an actual work environment.

In the sequence, we will talk about what it takes to make the unspoken, unwritten rules emerge. From there, how we stimulated the increase in the self-conscious level as well as in the collective one, building a matrix entitled 'Personnel Influence on Safety' to diagnose how one really sees himself concerning his leadership in safety culture.

We will then present the method used to address the issues managers pointed out, in order to transform the discussions into actions to strengthen the safety process leadership culture.

In the conclusion, we will present our findings and discuss possible future approaches to amplify the safety leadership culture further.

2 Safety Culture and Leadership: definitions and implications

There are many possible definitions for culture. Taylor (2015) says culture has to do with the behaviour patterns that are encouraged, discouraged or tolerated over time in an organization. Schein (2015) says that culture is "the total sum of everything an organization has learned in its history in dealing with the external problems – which would be goals, strategy, how we do things – and how it organizes itself internally."

Since we humans are gregarious social beings, we are naturally wired to fit in with our community. We use our perceptions to read what the tribe accepts and we adapt by behaving accordingly, because we want to fit in and be rewarded. Therefore, we reinforce the existing norms. This process is perpetuated over time; established behavior influences the behavior of new employees and so on.

Most cultural sources are non-verbal. Taylor (2015) says it is a myth to believe that culture reflects the values statements that are declared in the company's annual reports. Most commonly, the statement will be the organization´s intents, not the culture as it actually is.

If we want to change culture, it is necessary to have a clear direction and a focused, effort-driven leadership to lead people towards the goal culture. Culture has to do with what is encouraged and allowed. For Taylor (2015), three major sources send cultural messages: behaviours - of other people, especially important, recognized ones; symbols - artefacts, events and decisions that are meaningful to the community; and systems - mechanisms that manage people and tasks.

In this sense, culture has to do with the manifestation of what is valued in the organization.

According to Lafraia (2014), culture shapes reality. Cultural lenses structure the perception of the people involved in the organization.

When it comes to safety culture, its meaning may vary from one organization to another, or even in different departments of the same company. It is perfectly possible to have diverse lenses that translate the meaning and the importance of safety practices in diverse ways.

OGP, in its 452 Report (2013), says that a positive Safety Culture has to do with a culture in which safety is a core value for everyone who works in the organization, and it is pursued in an everyday basis. In order to build a Safety Culture, there are factors that must be considered, such as "leadership commitment to safety; employee involvement and motivation; responsibilities and accountability; production and bottom line pressures versus quality issues; and actions, or lack of action, to correct unsafe behaviours and unsafe conditions"; among others.

In order to build a consistent safety culture, leaders must educate their teams by demonstrating, by their practices, that some policies are fundamental. Safety culture should use this lens, rather than a command and control opaque one. Leaders must educate people to understand why safety is a core value that concerns the most precious gift: one´s life.

Safety culture is not about only following the rules, but making the team understand that the rules serve a greater principle. This approach can influence values and beliefs and change the individual and team´s behavior. The challenge leaders face is to conciliate the educational role with the policeman role, guaranteeing that compliance is required and everyone in the organization, starting by himself, values that as an unnegotiable principle.

In order to develop a strong safety culture, some elements must be considered (OGP, 2013):

Elements of a Strong Safety Culture

- an informed culture: the company has quality data concerning safety and uses it well;
- a reporting culture: people feel they can report safety concerns without worrying about being punished;
- a learning culture: the organisation actually learns from its unsafe conditions and changes them, banning any equipment, practice or whatever takes place that hazards safety;
- a flexible culture: the company changes its command chain if necessary to adapt;
- a just culture: people understand what is acceptable and unacceptable. What is not tolerated is dealt fairly.

For Peter Drucker (Cohen, 2018), leadership can be defined as the lifting of a man's vision to higher sights. Throughout leadership, people perform in higher standards, they are inspired and evolve in multiple levels.

When it comes to safety leadership, we can quote Ram Charan (2007) when he says that "a leader who does not produce other leaders is not a great leader." A safety leader should act in a way to enhance in everybody´s minds and actions a safety leadership commitment.

In order to make it possible, it is crucial that the leader himself is aware of his beliefs and constantly monitoring his behaviours. It takes time and a lot of effort to build a strong safety culture. Despite that, a safety culture is fragile and easily threatened by wrong messages. (OGP, 2013)

In the next section we explore the method used to address self-consciousness about safety.

3 Creating a trust environment that favors Safety Process Leadership Culture

According to Malcon Knowles, who develop the paradigm of andragogy, adults have a learning style characterized, among other factors, by changing one´s concepts according to new experiences one lives; the pursuit for autonomy; increasing one´s own motivational level to match the social role assumed; and acquiring knowledge to immediately solve practical issues (Muresan, 2013).

Considering this practical approach, we began the program to strengthen safety leadership culture by making each participant reflect on its daily practices concerning management, as well as the safety implications involved in routine decisions.

To start this reflection, we did a benchmark and decided to invite a senior top manager from another company in the Energy Industry to give a testimony. He made an honest statement about his team perception of his management practices. Concerning safety process, he mentioned that he realized that, in order to influence people, he must first change himself. He then shared with an amazed group of participants some practical examples of safety interventions he changed in his auditing procedures, for example, in order to be coherent with the message he wanted to pass forward.

The testimony had an impressive impact on the managers. A remarkable phrase that was repeated by managers was: "*Simplify. Be the Powepoint. Less Presentations, More You.*"

After the testimony, there was a presentation on the invited company's safety philosophy, about its journey into safety leadership. The practical steps mentioned were followed by videos with managers from different parts of the world talking about the way they see safety culture benefits.

With that background environment set, we forwarded to a self-perception diagnosis. The goal was to stimulate the increase in the individual conscious level as well as in the collective one. We used a matrix entitled ‘Personnel Influence on Safety’ that was obtained in the benchmark process with the above mentioned company.

It aimed to diagnose how one really sees himself concerning its leadership in safety culture. It involved two dimensions: the personal influence towards oneself; and the personal influence concerning other people´s safety.

The figure below illustrates the matrix:

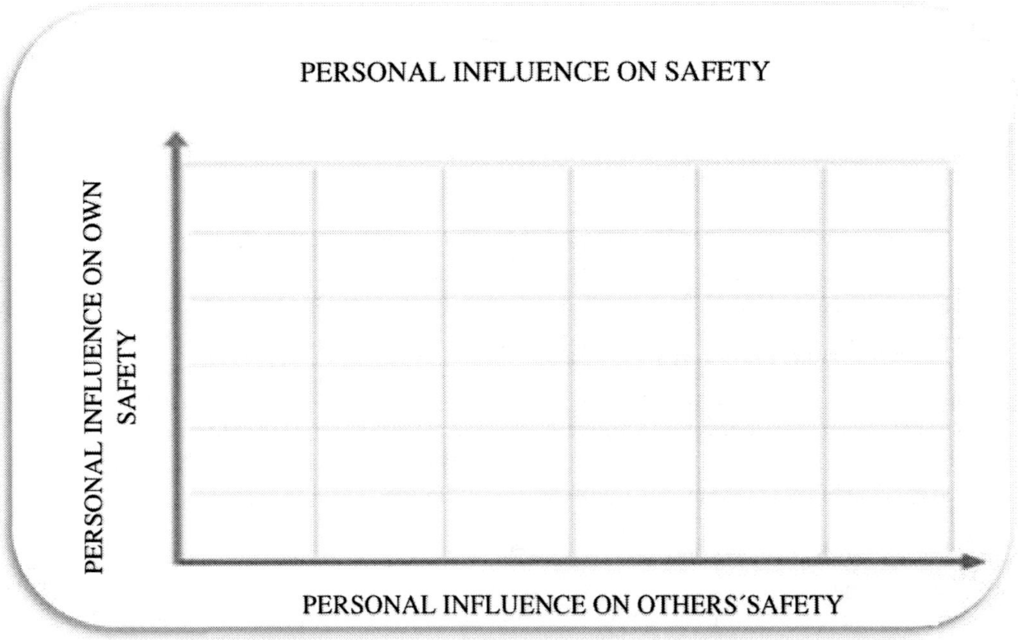

Figure 1. Personal Influence on Safety Matrix

We used the Self-perception as a tool to promote reflection and consciousness. We asked the participants to fill in the matrix in two different papers: one in his scrapbook (explained below) and another one on a plain sheet to hand over to the program facilitator. All the matrices were compiled and the result is shown in figure 2. As indicated, most participants initially evaluated themselves as influencing more themselves than others.

The coordination staff evaluated the matrix result as requiring attention from the participants. The staff observed the fact that some managers attending the program were obese, others worked extra hours, and most of them had reported having trouble balancing personal and professional life. Therefore, to promote deeper self-reflection, the consultant who was facilitating the program posed some questions to the participants regarding personal care. The remarks made were an invitation to honestly look at oneself and feel if the level of personal influence corresponds to what one would recommend to a loved person. Quoting Brené Brown (2015), the consultant said "Treat yourself like someone you love. Do you do it on a regular basis? Do you put yourself first?" Those remarks made some participants change their self-evaluation.

The intention was to call leaders ´attention to the fact that genuine influence on safety starts with their own safety, and goes beyond process or operational safety. To be legit and truly be a safety leader, one should take good care of oneself.

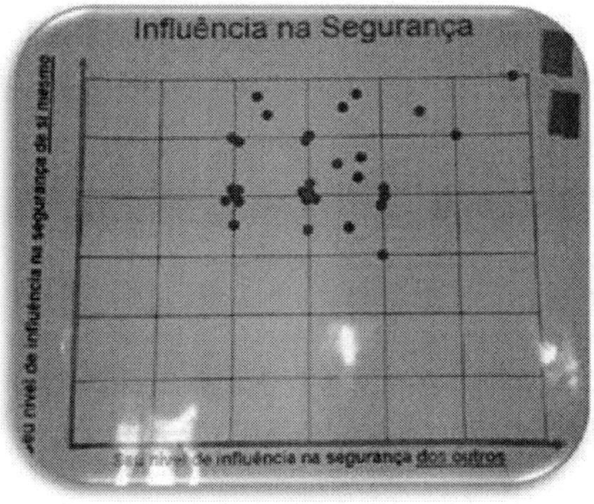

Figure 2. Compiled Personal Influence on Safety Matrix with the results from all the participants

We also developed a scrapbook that was given to each participant as a job-aid material to be used both during the program and after. The content of the scrapbook followed the program plan and was an invitation to self-analysis. The participants had special pages to do some journaling about what they had just listened to or dialogued about. There was also a special page dedicated to register insights. Figure 3 illustrates the scrapbook.

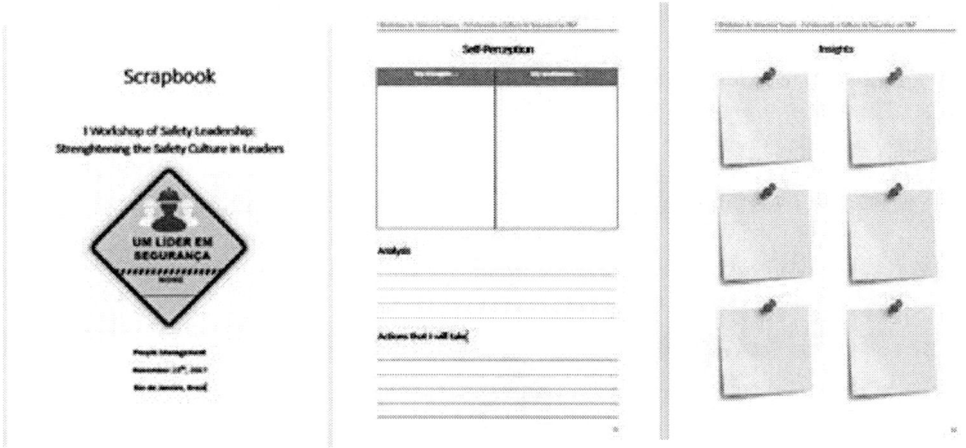

Figure 3. Pages from the Scrapbook used to stimulate self-consciousness and to plan actions

4 Exposing and Addressing Vulnerabilities

4.1 Control Room

We will now present the method used to address the issues managers pointed out, in order to transform the discussions into actions to strengthen the process safety leadership culture.

We prepared a special control room, where we designed a pathway decorated with images of accidents from the company and from the industry, and the number of fatalities. We also put an alarm sound playing during the route and videos showing accidents. These resources were intended to create an emotional ambiance to take the analysis to another level.

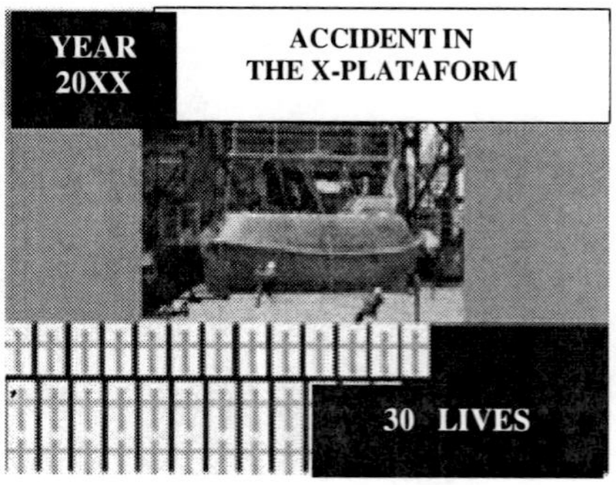

Figure 4. Picture of the Control Room and example of poster from accident used to decorate it.

Besides the planned journey passing several industry accidents, the Control Room was decorated with red material, to create an atmosphere of urgency. After going through the posters and watching the videos, the participants were invited to give their testimony of accidents they´ve been through.

The group tone got emotional as some of them shared unpleasant stories they had lived. The goal of inducing the reflection to an in-depth level was successfully achieved.

After that point, the participants went back to the regular room and dialogued about some questions that were in their scrapbook. The questions were related to the *Elements of a Strong Safety Culture* preconized by OGP (2013), mentioned in section 2 of this paper.

The methodology privileged social interaction among participants and the active listening of each other's considerations. Below is the list of the questions addressed to the participants.

Questions to Reflect and Dialogue:

1. Am I able to influence my operational team?
2. How many hierarchical levels are there between the operational team and me? To what extent does my speech actually makes sense to them?
3. How much of what is said by the operational team do I believe genuinely applies to me? Do I believe I receive good quality feedback?
4. Of the security actions implemented in my management, do I receive direct feedback on the effectiveness or otherwise of its implementation?
5. What degree of tolerance do I accept as a leader, regarding violations of safety standards?
6. Do I, as a leader, make it mandatory that the urgent inspection recommendations are executed immediately?
7. Do I, as a leader, take every opportunity to motivate, mentor and develop my team to enhance individual safety awareness and safety leadership?
8. Do I praise when I see a safe behavior?
9. Do I perform Behaviour Audits in a protocolled way, to meet the Scorecard, or do I do it because I believe in the effectiveness of what can arise from it?
10. Is Safety a value or is it a Priority for me as a leader?
11. Do I have the courage to step in and stop an activity when I see someone at risk? Do I always intervene?
12. Do my behaviours as a leader have safety as a foundation?
13. When I find that the environment is unsafe, do I privilege safety over other goals?
14. Safety comes first in my management?
15. How do I, as a leader, ensure that safety in my management is dynamic enough to ensure continuous improvement in operational procedures?
16. Thinking Safety as a chain, how can I, as a leader, be an effective link between operations and the company policy?

4.2 Results

In order to capture the precious discussions and share insights among the participants, in the final part of the workshop there was a collective presentation based on what each small group discussed about the questions above. Each group had to have at least one delegate, and others could participate if they wanted to.

Some of the findings reported were, for example: "*I have to go to the platform more often*"; "*Behaviour audits can be very useful, I should profit more from them*"; "*There are too many levels between my operational team and I. I have to get together with them on a regular basis.*"

After sharing insights, all managers filled in the Personal Influence on Safety Matrix again. The group reported that, this time, the self-perception was lower, since many managers realized they need to invest more efforts and more quality time in Safety, in both dimensions – with oneself and towards others.

After that, each participant revised and complemented his individual action plan (figure 3). They also wrote a personal commitment as a Safety Process Culture Leader.

At the very end of the program, we asked everyone to do a check-out by sharing with the entire group one action of the action plan. This intended to create a deeper commitment, as well as possible insights to others. Also, it was a practice that could be used with the managers team.

According to the group evaluation, presented in the word cloud below, the program succeeded in strengthening the safety culture in leaders, who mentioned as the top benefits from the program: (*Safety as a*) Value; Communication; Life; Commitment; Example; Integration; Consciousness; and Synergy.

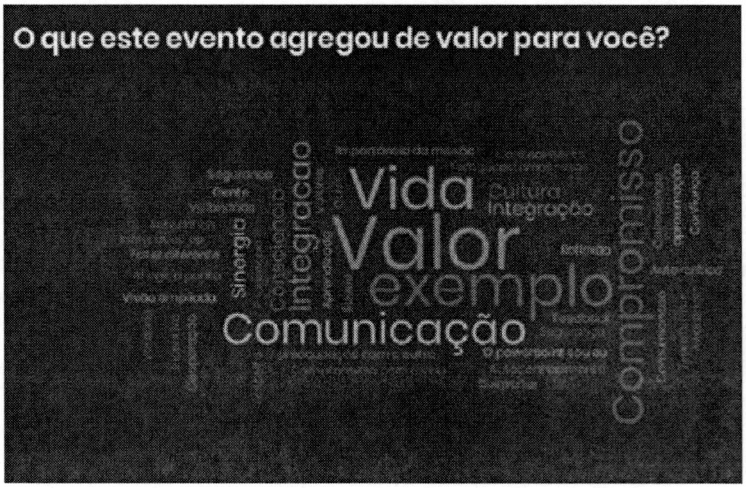

Figure 5. Evaluation: Word cloud containing the managers´ perception about the contribution brought by the program to participants.

5 Conclusions

The feedback given by the program´s participants was encouraging. It indicated that testimonies from partners, combined with relevant questions based on the daily problems managers face concerning safety, created a trust ambiance among managers.

The activities planned based on andragogic approaches favored dialogue and in-depth self-analysis. The use of the same instrument (Personal Influence on Safety Matrix) twice, as well as the individual scrapbook, facilitated the self-consciousness process and lead managers to review their initial assessment and to elaborate an action plan.

Leaders play a decisive role in strengthening the safety culture in the organizational world. Bringing about their vulnerabilities and having honest conversations with peers definitely helps managers to find ways to evolve from compliance to commitment.

Strengthening the safety leadership culture is a process that does not come to an end. It is an on-going process that must be continuously taken care of by leaders who value life and put safety first. Future projects that aim to implement effective actions in Safety Culture should take into consideration the company´s culture, as well as daily operational problems and safety as a core value to preserve life and business.

6 References

B. Brown, "*Daring Greatly: How the Courage to Be Vulnerable Transforms the Way We Live, Love, Parent, and Lead*", Penguin Random House, NY, 2015.

C. Migueles, J.R.B. Lafraia, G. C. Souza, "Criando o Hábito da Excelência: Compreendendo a Força da Cultura na Formação da Excelência em SMS", Quality Mark, Rio de Janeiro, 2014.

C. Taylor "*Walking the Talk: Building a Culture for Success*" Random House Business Books, London, 2015.

M. Muresan, "*Using Cybergogy and Andragogy Paradigms in Lifelong Learning*", 5th World Conference on Educational Sciences – WCES. Romania, 2013

OGP, *Shaping Safety Culture through Safety Leadership.* Report N° 452. International Association of Oil & gas Producers. October, 2013.

R. Charan, *Ram Charan on Leadership*,

Available at https://hbr.org/ideacast/2007/01/harvard-business-ideacast-27-r.html. Jan, 2007 (accessed Dec 4, 2018)

T. Kuppler, *Edgar Schein on Culture*,
Available at https://www.leadershipandchangemagazine.com/edgar-schein-on-culture/ Accessed on January 2, 2019.

W. Cohen, *Drucker's Top Tips to Successful Leadership*
Available at https://www.managementmattersnetwork.com/strategic-leadership/columns/druckers-top-tips-to-successful-leadership. Jun, 2018 (accessed Dec. 4, 2018)

7 Acknowledgments

I would like to thank Petrobras who fully supported this project, as well as Shell Brasil, for a very profitable partnership. A special thanks to my Human Resources, HSE and E&P work colleagues, who had a decisive role in making this project a reality.

548424 Development of Human Reliability Analysis Method for Oil and Gas Operations

The oil industry has grown in recent decades in terms of quantity of facilities and process complexity. However, human and material losses still occur due to major accidents at the facility. The analysis of these accidents reveals that many involve human failures that, if prevented, could avoid such accidents. These failures, in turn, can be identified, modeled and quantified through Human Reliability Analysis (HRA), which forms a basis for prioritization and development of safeguards for preventing or reducing the frequency of accidents. The most advanced and reliable HRA methods have been developed and applied in nuclear power plant operations, while the petroleum industry has usually applied Quantitative Risk Analysis (QRA) focusing on process safety in terms of technical aspects of the operation and equipment. A joint project from the B. John Garrick Institute for the Risk Sciences and a big company from the oil and gas industry is focusing on developing a HRA method focused on this industry operations, to bridge the existing gap between HRA and the Petroleum industry. The Phoenix HRA methodology was used as a basis, which has three layers i) a crew response tree (CRT), which models the interaction between the crew and the plant; ii) a human response model, modeled through fault trees, that identifies the possible crew failures modes (CFMs); and (iii) "contextual factors" known as performance influencing factors (PIFs), modeled through Bayesian networks. In addition to building on such a structure, the method development will rely on interviews with key personnel, visitations to installations, and analysis of past incidents data.

547327 Studying the Dependence of Operator Performance on Step Level Complexity

In the chemical, oil, and power industry, procedures are used by operators to perform their work correctly without compromising safety. Unfortunately, many procedures fail in adequately conveying information to their users. Heuristics tell us that technical jargon, vague language, and operator training all play a role in determining the effectiveness of procedures. To what extent these variables play a role is still not fully understood. Knowing this information will allow plants to more effectively write procedures. To counter this, researchers at Texas A&M's School of Public Health and Mary Kay O'Connor Process Safety Center are investigating the underlying factors that cause operators to deviate from procedures.

One possible factor may be that complex steps hinder an operator's ability to perform a task. This work aims to understand why certain steps are more complex than others. To do this, a set of procedures on how to operate a gas chromatography mass spectrometry (GCMS) machine is written. A study is performed in which participants use a set of procedures to analyze samples through gas chromatography. Using a scoring rubric, the complexity of each step is determined. The procedures have been designed to obtain an even distribution of complexity. Participants for this study are chemical engineering students with little to no prior experience in handling a GCMS machine. Thus, the experience, education, expertise, and training of the participants is controlled. The performance of each participant is reviewed and is grouped into one of four quadrants. For this study, a decrease in operator performance is expected as the complexity of each step increases.

548400 Prevent Them All or Stop Them Small: Tripod Best Practices in Reducing Chemical Accidents and Enhancing Response Preparedness

Prevent Them All or Stop Them Small: Tripod Best Practices in Reducing Chemical Accidents and Enhancing Response Preparedness

The connection between industry, government, and public safety, which we call the Tripod, is vital when implementing the 3-P's - Prevention, Protection, and Preparation. As identified by AICHE's Vision 2020 program, there are significant opportunities for organizations that address industrial process safety to work together. Responsible Collaboration between industry, regulatory and investigatory authorities, labor organizations, communities, and technical organizations promotes mutual understanding of risks, effects process safety and can reduce chemical accident risk in both small and large communities. Community based safety partnerships and leadership are fundamental to the success of such efforts.

For the past five years, a series of geographically focused Chemical and Ammonia Safety Days have cultivated and showcased chemical emergency prevention and planning efforts between Industry, NGOs such as the Ammonia Safety Training Institute (ASTI), and a myriad of public agencies including fire departments, health departments, LEPCs, SERCs, local regulators, and state and federal agencies such as EPA, CSB, DHS and OSHA. Within the context of Chemical Accidents and Releases, the focus of these efforts has been to Prevent Them All or Stop Them Small.

These Safety Day programs are designed for and with local community stakeholders from industry and first response agencies with a goal of seeding a permanent, engaged geographic collaboration for accident prevention and response readiness ownership and partnership. Agenda and curriculum elements include technical training focused on local risk, first responder field practicums and case studies, accident prevention program prioritization, practice and accountability, bowtie analyses of event propagation and management, and emergency response planning centered on the first 30 minutes and one-plan approach. The programs often include field demonstrations and exercises for fire departments and facility responders.

As a best practices demonstration, the presenters for this session will share the developed programs, recruitment methodologies and the demonstrated results of these efforts. In addition to Responsible Collaboration, other Vision 2020 principles embraced and promoted by the Safety Day Approach include Disciplined Adherence to Standards, Intentional Competency Development, Enhanced Application & Sharing of Lessons Learned, and Enhanced Stakeholder Knowledge. As a bonus, the societal theme of Harmonization of Standards is enhanced by the community of practice element of engaging federal, state and local regulators as well as industry stakeholders and process safety professionals.

548394 Risk Analysis Screening Tool – CCPS Project 280

A spreadsheet based Risk Analysis Screening Tool (RAST) is available through the Center for Chemical Process Safety (CCPS), courtesy of the Dow Chemical Company. The tool allows users to identify hazards and understand risks including causes, consequences and probabilities. RAST can be used for existing or proposed installations to facilitate MOC reviews, assess compliance with risk criteria, or evaluate proposed designs.

The users can input the type of equipment being used along with the equipment characteristics, chemical data, process conditions, and the plant layout. Using these inputs, the tool is used to help identify hazard scenarios and evaluate the consequences of representative loss of containment scenarios. RAST also allows the users to specify their own hazard scenarios based on their own knowledge of the process. Users can also input their own risk ranking matrix. With the hazard scenarios and the risk matrix, the user can select those scenarios of most concern and use the tool to create a Layer of Protection Analysis (LOPA) spreadsheet for the process.

As an example a comparison of RAST calculated scenario identification and consequence results to actual case studies from the CSB will be presented. Courses on how to use RAST, and how to adjust the pre-loaded default parameters are available.

550863 Impacts of Climate Change on Process Safety Management in Oil and Gas Operations

Impacts of Climate Change on Process Safety Management in Oil and Gas Operations

Chizaram Nwankwo

Coventry University

Project objectives and goals

The aim of this study is to analyse the impacts of climate change on process safety management in oil and gas operations. The objectives for this study include the following: -

- To evaluate past, present and future trends of climate change around various global regions
- To establish the relationship between climate change and process safety management in oil and gas operations
- To proffer adaptation and mitigation measures for climate change impacts on process safety management in oil and gas operations

In the last two decades, climate change has emerged as one of the most controversial topics that has been argued on several fronts (Hulme 2009). The idea of global warming and the negative impact of human-driven activity on the global weather has been fiercely debated (Cook *et al.* 2016). However, Field *et al.* (2012) argue that the facts and figures tend to reiterate the continuous occurrence and proliferation of climate change, as well as its increasing impacts in different industrial sectors. Indeed, the warming of the climate is unequivocal (Panteli and Mancarella 2015). According to IIGCC *et al.* (2017), 70% of global greenhouse gas (GHG) emissions are attributed to energy-related activities. The oil and gas industry contributes to 60% of these energy-related emissions due to production of fossil fuels as their main operational activity (Puig *et al.* 2015). Evidence from NASA (2018) suggests that there has been 1^0C increase in temperature which is close to the set mark in accordance to the Paris Agreement.

There are several case studies such as the flood in San Jacinto River, Houston Texas in 1984 and Coffeyville Refinery, Kansas in 2007. In the case of the San Jacinto flooding event in 1984, 8 pipelines were ruptured and 29 were undermined, 35,000 barrels of oil were spilled and ignited, while 547 people were injured and $23 million was incurred in damage costs (Matek 2017).

Hence, adaptive measures need to be applied to manage the severe consequences of climate change on oil and gas operations (Smith 2013). Such adaptive measures include more robust design specifications for oil and gas assets exposed to harsh climatic conditions (Cruz 2010). Ramasamy and Yusof (2015) suggest that this could be vital in offsetting the impacts of climate change on assets such as offshore rig platforms and equipment, transport pipelines, storage facilities etc. However, there is still little attention accorded to design and adaptation of oil and gas assets to offset the effects of climate change on them (Skjaerseth and Skodvin 2010). Therefore, this study seeks to analyse the impacts of climate change on deterioration of oil and gas assets.

Description of methods and results

Firstly, this study adopts a pragmatic paradigm since it explores the consequences of actions, is problem-centred, pluralistic, real-world practice oriented and allows for use of mixed models (Creswell and Creswell 2018). The area of investigation for this study is focused on impacts of climate change on oil and gas operations. The predominant approach taken for this study is a correlational research approach which tries to establish a relationship between climate change and oil and gas operations. The selected approach is due to the continuous occurrence of accidents in the oil and gas industry due to asset deterioration.

After thorough literature review of climate change trends and asset integrity issues in the oil and gas industry, this study identified a gap in knowledge that the oil and gas industry has not duly considered climate change during design of assets; hence posing a risk to oil and gas operations conducted under harsh climatic conditions. Therefore, the data collection for this study employed "concurrent triangulation" method of quantitative and qualitative data to ensure robust analysis of key variables such as climate change and oil and gas operations. Consequently, data collection tools for this study include questionnaires, interviews, archive data and documentary analysis. The data collection for this study was conducted over a period of 3 months.

Firstly, it involved piloting the questionnaire among academic and industry professionals to identify any errors, as well as testing and refining the data collection instrument. Next, the study was approved as a medium-risk project by Coventry University Ethics Committee, as primary and secondary data were collected from archive data, documents, questionnaires and interviews. The population sample for the study includes oil and gas industry professionals such as Mechanical Engineers, Facility Managers, Production and Well Engineers, Process Engineers, Health and Safety Officers, as well as Safety and Reliability Engineers. The questionnaires used for this study were designed using Bristol Online Survey (BOS) tool and covered questions around climate change and oil and gas operations. These data were predominantly collated and analysed using various software such as Zotero, Microsoft Excel and SPSS software. Statistical analysis conducted in this study include trend analysis, descriptive statistics, Pearson's correlation, Cronbach's alpha reliability analysis and Relative Importance Index (RII). Lastly, the findings of this study were benchmarked and justified with peer-reviewed academic literature.

Potential for application of results

This research could aid in curbing several accidents caused by asset deterioration due to increasing number of climate change events. It will help the oil and gas industry to avoid the enormous financial, legal and moral costs associated with accidents. Also, this study will provide a deeper insight on current climate change statistics and proffer adaptation and mitigation measures on how to deal with these problems. Similarly, it could broaden the scope of researchers to start incorporating impacts of climate change on various industrial sectors and subsectors asides the oil and gas industry.

References

Cook, J., Oreskes, N., Doran, P.T., Anderegg, W.R., Verheggen, B., Maibach, E.W., Carlton, J.S., Lewandowsky, S., Skuce, A.G., and Green, S.A. (2016) 'Consensus on Consensus: A Synthesis of Consensus Estimates on Human-Caused Global Warming'. *Environmental Research Letters* 11 (4), 048002

Creswell, J.W. and Creswell, D.J. (2018) *Research Design: Qualitative, Quantitative, and Mixed Methods Approaches.* 5th Edition (International Student Edition). Sage Publications

Cruz, A.M. (2010) *Vulnerability of Oil and Gas Infrastructure to Climate Change and Extreme Weather Events* [online] available from <http://indico.ictp.it/event/a09141/session/35/contribution/25/material/0/0.pdf> [5 March 2018]

Field, C.B., Barros, V., Stocker, T.F., and Dahe, Q. (2012) *Managing the Risks of Extreme Events and Disasters to Advance Climate Change Adaptation: Special Report of the Intergovernmental Panel on Climate Change*. Cambridge University Press

Hulme, M. (2009) *Why We Disagree about Climate Change: Understanding Controversy, Inaction and Opportunity*. Cambridge University Press

IIGCC, CERES, and IGCC (2017) *Global Climate Disclosure Framework for Oil & Gas Companies* [online] available from <http://www.iigcc.org/files/publication-files/Global_Climate_OG_DisclosureFramework.pdf> [14 November 2017]

Matek, B. (2017) *CHARACTERISTICS OF OIL AND GAS PIPELINE ACCIDENTS CAUSED BY CLIMATE CHANGE INTENSIFIED HURRICANES* [online] Baltimore, Maryland: Johns Hopkins University. available from <https://jscholarship.library.jhu.edu/bitstream/handle/1774.2/46088/Benjamin%20Matek.pdf?sequence=1&isAllowed=y> [26 June 2018]

NASA (2018) *Global Climate Change* [online] available from <https://climate.nasa.gov/> [9 May 2018]

Panteli, M. and Mancarella, P. (2015) 'Influence of Extreme Weather and Climate Change on the Resilience of Power Systems: Impacts and Possible Mitigation Strategies'. *Electric Power Systems Research* 127, 259–270

Puig, D., Haselip, J., and Naswa, P. (2015) *Adaptation to Climate Change in Colombia's Oil and Gas Industry: Recommendations to Promote Risk Management* [online] UDP-CRD-OLADE-01(EN). Denmark: UNEP DTU Partnership. available from <www.unepdtu.org> [19 September 2018]

Ramasamy, J. and Yusof, S.M. (2015) 'A Literature Review of Subsea Asset Integrity Framework for Project Execution Phase'. *Procedia Manufacturing* [online] 4, 79–88. available from <http://linkinghub.elsevier.com/retrieve/pii/S2351978915011336> [14 May 2017]

Skjaerseth, J. and Skodvin, T. (2010) *Climate Change and the Oil Industry: Common Problem, Different Strategies*. Manchester University Press

Smith, M.H. (2013) *Assessing Climate Change Risks and Opportunities for Investors: Oil and Gas Sector* [online] Analysis of Climate Change Risks and Opportunities. Technical Report. ANU and Investor Group on Climate Change. available here. [5 April 2018]

Further Information

Chizaram Nwankwo (E: nwankwo4@uni.coventry.ac.uk)

546246 Study of the Effects of Wind Speed and Surface Roughness Under Stable Atmospheric Condition Using Aloha Model and CFD Model

Study of the effects of wind speed and surface roughness under stable atmospheric condition using ALOHA model and CFD model

Many substances released in chemical processes are toxic, flammable, or even both. When these substances accidentally release to the atmosphere, human health and the environment could be threatened. As results, a reliable quantitative risk analysis tool is crucial to provide decision makers useful information on downwind concentration and hazards. ALOHA consequence model and Computational Fluid Dynamics (CFD) numerical method are widely used in chemical industries. Validation of these models under different scenarios is a necessary preliminary step before applying them to risk analysis. CFD models generally have acceptable concentration predictions on dense gas dispersion. ALOHA model is verified with DEGADIS model, but there is lack of study on effects of wind speed and surface roughness on dense gas dispersion. This paper compared field data from the Kit Fox experiments, which were CO_2 gas releases under stable atmospheric conditions with increased roughness, with CFD and ALOHA models' predictions. Also, statistical performance indicators were utilized to evaluate the accuracy of ALOHA model. Statistical performance indicators showed that ALOHA had systematic errors for dense gas dispersion with increased roughness under stable atmospheric conditions. Comparing effects of wind velocity profile and surface roughness values in CFD model and ALOHA model, this paper provided a better understanding of effects of key parameters in both ALOHA and CFD models. Furthermore, engineers would have better quantitative risk assessments under stable atmospheric conditions when they are using ALOHA consequence model and CFD model.

548222 Consequence Assessments and Collaboration to Streamline Project Siting

Collaboration is often sought after across disciplines, but all too often disregarded in favour of linear project development. For siting and consequence, as with other areas where design interacts with environment, health and safety constraints, this often leads to more costly design due to increased project iterations.

Stantec recently completed a consequence assessment for a compressor station to assess flammable extents during an emergency blowdown. At the same time, the client was experiencing siting issues due to sound emissions from the same process. The facility was grappling with competing factors: blowdown the facility in time to avoid catastrophic failure of their infrastructure, avoid excessive flammability extents from the blowdown stack location, and avoid extraneous noise exposure to nearby receptors. Using transient consequence modelling tools to model the blowdown sequence, Stantec was able to simultaneously and efficiently assess the flammable extents and sound emissions from the blowdown operations.

Consequence modelling for flammability was completed using a transient pipeline blowdown modelling tool to simulate the mass, energy, and momentum throughout the pipeline system. The model first simulated normal operating conditions, and then simulated the blowdown to estimate mass flow rate, liquid mass fraction, and release temperature from the blowdown stack. Several sizes of reducers upstream of the blowdown stack were introduced to estimate the pressure drops in the piping that would be required to appropriately reduce noise levels but also achieve full depressurization of the facility in time to meet safety criteria.

Consequence modelling of flammability was completed using SLAB, while the predicted change in sound pressure level from the blowdown stack was completed using Cadna acoustic modelling software. By combining these two analyses, the assessment costs streamlined for both required studies. In addition, the acoustics study benefited from additional insight provided by the consequence modelling tool.

548348 CFD Simulation of the Effect of Wind Speed and Release Elevation on Dispersion of Flammable Fluids at LNG Export Facilities

Facility siting studies for LNG export terminals are required by federal regulations to assess the potential hazard from accidental releases of flammable fluids. Federal regulations require under 49 CFR 193.2059 the evaluation of flammable vapor-gas dispersion distances from releases from flammable fluids in the facility for a range of representative wind conditions based on historical weather data expected to produce the maximum dispersion distances. A parametric CFD study will be conducted to evaluate not only the effect of wind speed but also the effect of the release elevation on the dispersion distance from flashing and jetting fluids. Hazard distances will be compared between the 2D software PHAST and the 3D software FLACS for a set of exemplar fluids found at LNG export facilities. The ability of the FLACS software to handle detailed 3D geometry provides for the evaluation of the effect of obstructions, congestion, confinement and mitigation measures on the dispersion distances. Release scenarios typical of LNG export facilities will be presented, with a focus on jetting and flashing releases of LNG, ethylene, propane, mixed refrigerants and heavy hydrocarbons.

546616 A Comparative Study of the Optimization Models for the Gas Detector Placement in Process Facilities

Gas detector network is an important layer of protection in process facilities for prevention gas release accidents of turning into fire and explosion disasters. But traditional standards just provide basic principles of the installation of gas detectors. Recently mathematical programming gradually became an important method to solve the problem. When using the mathematical programming method to optimize the detector layout, different optimization strategies will result in different optimization target models and detector layouts. In this study, three models have been developed and contrasted, namely minimal detection time P-Median model (MITP), minimal leakage concentration P-Median model (MICP), and minimal individual risk P-median model (MIRP), for the optimization of the gas detector placement based on stochastic programming. And as a basis, possible release scenarios were identified. In order to achieve the goal of using finite discrete scenarios to represent all scenarios effectively and quantitatively, clustering analysis was used to filter similar scenarios and select representative leak scenarios. The input data of these models was obtained from computational fluid dynamics. Furthermore, an adaptive particle swarm optimization (APSO) algorithm was applied to solve these models. The gas detector network optimization in a diesel hydrogenation unit was carried out as a case study. Results demonstrate that the MICP model is superior to the MITP model and MIRP model in describing the integrity of hazardous gas leakage and diffusion distribution. But the cumulative detection time of its optimized layout is slightly longer than that of MITP. Moreover, the control effect of MIRP model on personal risk caused by hazardous gas leakage was increased by 24.2%, although the minimum number of detectors was slightly higher than that of MICP model.

549411 Reducing Industrial Risk in a Converging It/OT World: 20 Experts Share Their Strategies

Rapid adoption of Industrial IoT requires tight integration between IT and OT. CISOs and corporate OT leaders are now scrambling to develop strategies, secure budget, and implement solutions to secure their mission-critical industrial control systems against burgeoning cyber threats.

Effective integration of IT and OT cybersecurity strategies is essential to address risks successfully. To secure industrial facilities and ensure safe, reliable production, IT and OT security teams – traditionally two separate disciplines with different priorities – must collaborate and share cybersecurity and risk management best practices.

In a series of interviews in late 2017 and early 2018, more than 20 different experts on the front lines of OT cybersecurity risk mitigation shared practical advice for making control systems more secure. These experts represented a diverse range of industries including oil and gas, chemicals and refining, and power generation.

In their interviews, they shared their insights into how to leverage IT/OT convergence and develop an effective OT cybersecurity program that meets challenges in real-world ICS environments. Technology plays an important role, but as these experts also point out, more technology isn't always the solution. Just as important is the need to establish effective communication and trust between cybersecurity professionals and OT engineers.

Security project teams must make a concerted effort to create alignment between business management, the cybersecurity community, and OT professionals. Without alignment, security initiatives will fail to attain the support required to ensure success.

In this presentation, attendees will learn:

- Practical, real-world recommendations from OT cybersecurity experts
- Best practices for breaking down barriers and building trust between cybersecurity professionals and OT engineers
- How to align business management and plant operations on cybersecurity strategies

549417 The Intersection of OT Security and Process Safety

The rise of cyberattacks on critical infrastructure creates troubling concerns for security, economic stability, and personnel and public safety. However, many companies are still relatively unprepared to successfully defend their facilities from cyberattacks launched by increasingly sophisticated threat actors. Safe production is at risk.

Safety focuses on incident prevention and measures risks to personnel. Cybersecurity focuses on protecting systems from threat actors and measures risk based on the threat landscape. The increased network connectivity that exists within industrial facilities today means that cybersecurity is now closely linked to the safe operation of industrial facilities and their processes. Successful defense of cyber-physical systems from nation-state attacks requires close coordination and collaboration between cybersecurity and process safety. Every safety initiative must now consider cyber risk. Every cybersecurity initiative must now consider safety.

Considerable overlap exists between risk analysis in safety and cybersecurity. Safety and cybersecurity disciplines can learn much from each other. However, safety and cybersecurity standards and corporate organizational structures remain siloed in far too many companies. This session focuses on the importance of executing a cohesive strategy for both safety and cybersecurity to ensure safe and reliable operations.

In this session, we will:

- Examine the existing cyber threat landscape
- Discuss how established process safety best practices can be applied to OT cybersecurity
- Provide practical guidance for improving cyber defenses and reducing safety and environmental cybersecurity risks

551385 Process Safety at Offshore Aerodromes (Helidecks)New Application for Process Safety

Offshore air transport is an importance activate of the oil & gas industry. In this way, the management of air operations is essential for maintaining the safety and reliability. The present work has the objectives of divulging the methods used by
Petrobras to audit maritime units with helideck and to demonstrate the contribution of such methods to the safety of process in the air operations. To achieve these objectives, the paper was based on the Center for Chemical Process Safety (CCPS) publication "Guidelines for the Safety of Risk-Based Processes", and aims to share the Bow tie methodology for the qualitative analyzes of risks inherent in Air Operations, focusing on the barriers typically faced by oil and gas industry, in order to contribute to the process by providing a graphical and structured approach to the required barriers, both to avoid catastrophic consequences and to prevent events recognized as precursors.

Thermodynamics and Computational Methods for Relief Valves

By Umesh Mathur, P.E.
umesh.mathur@aggreko.com , mathuru@gmail.com
Aggreko Process Services
3732 Magnolia Street, Pearland, Texas 77584, USA

15th Global Congress on Process Safety (AIChE, March, 2019, New Orleans)

1. INTRODUCTION

The computation of flow through a relief valve is one of the most important aspects of safety system design. This paper discusses the fundamental question of choosing the proper thermodynamic basis for such calculations. We also discuss the discrepancies we have observed in the literature on this crucial question and attempt to dispel some of the misconceptions we have found.

Our concern with this issue arose from a study of the relevant literature on this subject, which shows that the flow of a compressed gas (or gas-liquid mixture) through a rigidly held valve or hole in a pipeline must be treated as an irreversible, isenthalpic process. This reasoning flows simply from the First Law of Thermodynamics as applied to a flow process.

For example, as early as 1947, Hougen and Watson[1] discussed "Free Expansion" of gases (p. 551) and noted as follows:

"The unrestrained expansion of a gas is known as free expansion. Under conditions of no restraint no work is done, and under adiabatic conditions no heat is added. Free expansion under flow conditions is commonly known as throttling or as the Joule-Thompson effect referred to in Chapter XII.

The Joule-Thompson effect is measured experimentally by expanding the gas slowly and in steady flow through a well insulated porous plug; in this way, potential work is lost, and no heat is allowed to enter or leave the system through the walls. The fluid flows reversibly into and out of the process, but the expansion step is completely irreversible. Since no heat is added or lost, the process takes place under conditions of constant enthalpy, according to Equation (VII-11), page 207. However a change in internal energy results if any change in flow energy pV occurs."

Similarly, in a discussion of valves and throttles, Elliott and Lira[2] observe as follows:

"A throttling device is used to reduce the pressure of a flowing fluid without extracting any shaft work and with negligible fluid acceleration. Throttling is also known as ***Joule-Thompson expansion*** *in honor of the scientists who originally studied the thermodynamics. An example of a throttle is the kitchen faucet. Industrial valves are modeled as throttles....Changes in kinetic and potential energy are small relative to changes in enthalpy as we just discussed. When in doubt, the impacts of changes in velocity can be evaluated as described in Example 2.9. The amount of heat transfer is negligible in a throttle. The boundaries are not expanding, and there is also no mechanical device for transfer of work, so the work terms vanish. Therefore, a throttle is isenthalpic."*

Finally, in a discussion of nozzles, Klein and Nellis[3] observe that:

"...The potential energy terms of the inlet and outlet flows are not usually significant. Nozzles neither require nor produce power ($W_{in} = W_{out} = 0$) and are very nearly adiabatic. Note that the kinetic energy terms associated with the flows, particularly the outlet flow, cannot be neglected as the purpose of the nozzle is to increase the kinetic energy of the fluid. With these simplifications, an energy balance on the nozzle becomes:

$$H_{in} + \tfrac{1}{2}V_{in}^2 = H_{out} + \tfrac{1}{2}V_{out}^2 \qquad (4\text{-}54)\text{''}$$

In the light of these assertions by these noted authorities, and others[6][10], it is hard to understand how the flow through a relief valve, or a hole in a vessel that is open to the atmosphere, can be treated as isentropic (***reversible*** adiabatic) in nature, as opposed to the isenthalpic (***irreversible*** adiabatic) basis that is consistent with free expansion. This is because an isentropic expansion process would necessarily show an outlet enthalpy for the gas from the nozzle or valve (or an orifice in a punctured vessel) to have a lower enthalpy than the inlet, the difference in such enthalpies being equal to the work done, for example on a rocket or airplane. However, a relief valve or vessel remains stationary and absolutely no work is done during a free expansion by the gas flowing through.

Nevertheless, several standard references have used isentropic formulas for free expansion[4],[5] in stationary valves, or nozzles, from the inlet to the throat. We believe that this methodology is flawed as it violates the steady flow energy equation (i.e., the first law of thermodynamics written for an open system). This procedure would result in an erroneous prediction of outlet gas temperature when entropy is held constant. This, in turn, would predict an erroneous outlet gas density and thus, for a desired gas relief rate, the required cross-sectional area for the valve throat and also any conduit or pipe downstream of the valve.

The practical consequence of this anomaly in predicting outlet gas properties (especially the gas density at the throat) would be that the relief valve sizing could be incorrect. Therefore, it is important that the proper thermodynamic basis (isenthalpic expansion) be used in all such computations.

Conversely, for a given relief valve, errors in throat density based on an isentropic path would also affect the relief rate. Both types of errors could lead to potentially serious process safety concerns.

For these reasons, when calculating the release rate for gas at a high pressure through a relief valve, it is important to be able to use the proper basis for gas enthalpy (and therefore accurate computations of fluid thermodynamic properties). The same considerations apply to the gas release rate calculations resulting from a rupture in a pipeline or vessel containing a gaseous mixture under pressure.

To help clarify the proper methodology for a relief valve - or a hole in a pressurized vessel - the computations that should be performed are described below. We also show a few simple examples. Note that the possibility of sonic velocity at the throat, or choking, of the valve (or at the orifice in a ruptured vessel) is explicitly considered. In both such cases, the procedure described below is subject to the following assumptions:

- The opening can be represented by a convergent nozzle of exit area equal to the area of the throat of the relief valve, or hole in the case of pipeline rupture

- The compressed gas pipeline, or leaking container, is a large reservoir in which upstream pressure and temperature are maintained constant
- The compressed gas undergoes an irreversible, adiabatic process of expansion through the nozzle.

The flow through such a device from either (a) the inlet to the throat, or (b) from the throat to the outlet follows the steady flow energy balance equation (Equation 9.21, Himmelblau & Riggs[6]):

$$\Delta E = Q + W - \Delta(H + KE + PE) \qquad (1)$$

Where:
ΔE = energy accumulation (zero at steady state)
Q = net heat transfer into the system (very close to zero for a relief valve)
W = net work done by the system (zero for a relief valve)
ΔH = change in enthalpy of the system
ΔKE = change in kinetic energy of the system
ΔPE = change in potential energy of the system (zero for a relief valve)

Clearly, for steady flow through an orifice, valve (or relief valve), ΔE, Q, W, and ΔPE are all zero. Therefore, we are left with:

$$\Delta(H + KE) = 0\text{, or}$$

$$\Delta H = -\Delta KE\text{, or}$$

$$H_{out} = H_{in} - \Delta KE \qquad (2)$$

In other words, the outlet enthalpy is equal to the inlet enthalpy minus the change in kinetic energy in going through the throat or orifice. (It should be noted that the same logic would apply for discharge through an ordinary valve, a relief valve, or an orifice plate).

If there is no change in velocity, the outlet enthalpy will be the same as that of the inlet gas. This phenomenon is generally referred to as a Joule-Thompson (J-T) expansion. For an ideal gas, enthalpy is a function of temperature only; therefore, a J-T expansion would result in no change in temperature. However, for most real gases, a J-T expansion results in a decrease in temperature. There are a few notable exceptions, however, such as hydrogen or helium gas for which the J-T coefficient has the opposite sign[10]. On the other hand, if the gas speeds up, its kinetic energy will increase and, therefore, its enthalpy will decrease in accordance with equation (2) above. This kinetic energy increase would result in a decrease in outlet gas enthalpy, and therefore temperature, even for an ideal gas.

For ideal gases, the computation of enthalpy for ideal gases is straightforward as it is a function of temperature only. For real gases, however, enthalpy is also a function of pressure, temperature, and composition and use of an equation of state (EOS) becomes necessary[2] to calculate the enthalpy departure from the ideal, as discussed below.

COMPRESSIBILITY FACTOR AND ENTHALPY FOR REAL GASES

For real gases, enthalpy is found using a suitable equation of state (EOS), such as the Soave-Redlich-Kwong (SRK), Peng-Robinson (PR), or Lee-Kesler (LK). This procedure may be summarized as follows:

At the given upstream pressure (P), temperature (T), and composition, solve the EOS for the compressibility factor (z). This requires solving a cubic polynomial in z when using any cubic EOS, such as the SRK or PR, using either an explicit analytical or numerical trial and error method. For the LK, however, a trial and error procedure is always required as the LK EOS requires solving the 8-constant Benedict-Webb-Rubin[2] (BWR) equation, which is not a simple cubic in z, once for the simple fluid and again for the reference fluid.

Using the appropriate thermodynamic expressions for the EOS, use this value of z to calculate ΔH_d, the enthalpy departure from ideal.

Calculate the ideal gas enthalpy, H_0 (function of temperature T_0 only).

Find the real gas enthalpy using:

$$H_r = H_i - \Delta H_d \quad (3)$$

Where:

H_r = real gas enthalpy
H_i = ideal gas enthalpy
ΔH_d = enthalpy departure from ideal

This procedure must be followed for calculating enthalpy at both upstream and downstream conditions.

The gas velocity at the throat may be sonic or sub-sonic. Also, as the gas speeds up in passing through the valve throat or hole (since the cross-sectional area is lower than that of the pipe), its kinetic energy increases and this results in a reduction of the enthalpy (Elliott & Lira[2], Eq. 2.53):

$$H_t = H_0 - \Delta KE \quad (4)$$

Where:
H_0 = upstream gas enthalpy[J/kg]
H_t = exit gas enthalpy at the throat or hole[J/kg], and
ΔKE = change in kinetic energy[J/kg]

ΔKE is found as follows (Elliott & Lira's Eq. 2.53):

$$\Delta KE = (u_t^2 - u_0^2)/2 \quad (5)$$

Where:

ut = throat (or hole) velocity[m/s], and

u_0 = upstream velocity[m/s].

Clearly, the outlet and inlet enthalpies are equal only if the change in velocity is negligible.

RELEASE RATE CALCULATION

The computation of the release rate from a pipe rupture should be done subject to the following equations:

Bernoulli's Equation

This equation relates upstream and downstream conditions for an arbitrary flow geometry, as follows:

$P_0 + \rho_0.u_0^2/2 + \rho_0.g.z_0 = P_e + \rho_e.u_t^2/2 + \rho_t.g.z_t + \Delta P_{friction}$ (6)

Where the subscripts o and e refer to the inlet and exit planes, respectively, and

P	= pressure[Pa]
ρ	= density[kg/m^3]
u	= velocity[m/s]
g	= gravitational constant[m/s^2]
z	= elevation above datum[m]
$\Delta P_{friction}$	= friction losses[Pa]

The friction losses have been ignored for simplicity in this presentation.

Obviously, for a hole, or throttle (or relief) valve, the elevations z_0 and z_t are the same.

Sonic Velocity

The sonic velocity in any medium[2] may be expressed as:

$u_s = (\gamma . z . R . T / M)^{0.5}$ (7)

Where:

u_s	= sonic velocity[m/s]
γ	= gas C_p/C_v ratio[]
z	= gas compressibility factor[]
R	= gas constant = 8314.46[J/(kgmol.K)]
T	= temperature[K]
M	= gas molecular weight[kg/kgmol]

In Equation (7), properties at the throat must be used for choked flow calculations. For ideal gases, the compressibility factor z is equal to 1. For real gases, however, the compressibility factor z is generally not equal to 1 and can be found only using an appropriate EOS.

C_p, C_v, and $\gamma=C_p/C_v$ ratio

Similarly, for ideal gases, the value of γ is found from the value of C_p, which is a function of temperature only, and the value of C_v is found from the following:

$C_v = C_p - R$ (8)

For real gases, however, the value of γ again requires use of an equation of state to calculate the ΔC_{pd} and ΔC_{vd} departures from ideality. The real gas C_p and C_v values are found using the ideal values and these calculated departures from ideality, by subtraction, in a manner similar to that described above for enthalpy:

$C_{p1} = C_{p0} - \Delta C_{pd}$ (9)

Where
C_{p1} = real gas C_p[J/(kg.K)]
C_{p0} = ideal gas C_p[J/(kg.K)]
ΔC_{pd} = departure in C_p from ideal gas value[J/(kg.K)]

Similarly, $C_{v1} = C_{v0} - \Delta C_{vd}$ (10)

In Equation (10),
C_{v1} = real gas C_v[J/(kg.K)]
C_{v0} = ideal gas C_v[J/(kg.K)]
ΔC_{vd} = departure in C_v from ideal gas value[J/(kg.K)]

In a conduit, orifice, or valve, the actual gas velocity, u_e, can never exceed the sonic limit, u_s. Note that u_s, in general, is different at the upstream conditions compared to the value at throat conditions. The upstream and throat temperatures and pressures of the gas may be different owing to the kinetic energy effect, Equation (4).

At a sufficiently low upstream pipeline or vessel pressure, the hole velocity would be lower than the sonic velocity limit. Above some critical limit for the upstream pressure, however, the maximum discharge flow is limited by the sonic velocity criterion at the hole.

Density

Gas density is computed using:

$$\rho = P \cdot M / (z \cdot R \cdot T) \quad (11)$$

Where:
ρ = density[kg/m³]
P = pressure[Pa]
M = gas molecular weight[kg/kgmol]
z = gas compressibility factor[]
R = gas constant = 8314.46[J/(kgmol.K]
T = temperature[K]

The gas compressibility factor is 1 for ideal gases; it is calculated from the EOS for real gases.

1.2 Velocity

The velocity in a pipe or hole (or valve throat) is found using:

$$u = W / (\rho \cdot A) \quad (12)$$

Where:
u = velocity[m/s]
W = mass flow rate[kg/s]
ρ = gas density[kg/m³]
A = cross sectional area[m²]

For a circular pipe, the cross-sectional area is defined as:

$A = \pi . D^2/4,$ (13)

Where:
D = inner diameter[m]

For non-circular cross sections, the equivalent diameter D_e is defined using the concept of the hydraulic radius which is defined as:

D_e = 4 (Cross-sectional-area/Wetted-perimeter) (14)

It can readily be seen that this definition holds for a circular pipe also: $4 . \pi . D^2 / 4 / (\pi . D) = D$

As noted earlier, for ideal gases, the compressibility factor z is equal to 1. For real gases, however, finding the gas density ρ requires first calculating z from the EOS. This is done using the procedure as described above.

It should be noted that conditions at the throat of a valve (or at the hole in a pipe) are not known. Therefore, the computation of the gas flow rate must be done using an iterative procedure. When using Microsoft Excel, this is best performed using the Solver Add-in. Solver is a non-linear optimization technique that can simultaneously adjust the unknown quantities and solve the problem by enforcing the constraints imposed by the Bernoulli Equation (6) for throat velocity, and the Steady-Flow Energy Equation (1) for hole enthalpy. Also, the calculated throat velocity from the Bernoulli equation can never exceed the sonic velocity limit, Equation (7), and this constraint must be added explicitly when the upstream pressure exceeds a threshold value.

When solving for an unknown gas flow rate, if the upstream pressure is sufficiently low, hole velocity could remain sub-sonic. However, if the upstream pressure is sufficiently high, the pressure at the hole and also the gas mass flow rate would vary in a manner that obeys the sonic velocity limit at the hole.

Example

CASE (A): Methane is flowing through a 300[mm] diameter, above-ground pipeline at an absolute pressure of 46[bar] and a temperature of 300[K]. The pipeline develops a hole with an equivalent diameter of 50[mm]. Calculate the mass rate of leakage to atmosphere and also the velocity of the gas jet emerging from the hole. The pressure inside the pipeline may be assumed to be constant owing to continuous supply from the source.

CASE (B): Rework Case (A) assuming that the pressure in the pipeline is 1.5[bar] instead of 46[bar].

Case (A) P_0 = 46[bar]

PROBLEM: Solve for release rate from a hole in a pipeline - sonic velocity at throat.

WE SHOW THE FOLLOWING DETAILED SOLUTION FOR THE IDEAL GAS CASE ONLY. Note that the solution of the real gas case requires use of complex computer programs to solve the chosen equation of state (e.g., Lee-Kesler) iteratively for the required thermodynamic quantities. Since these routines are imbedded in an outer loop for the unknowns (T and P at the throat, and the mass flow rate), the computational burden is too great for hand calculations.

The POLYMATH software[7] was used for solving this problem. All variables are defined in Table 1. The problem statement for Case (A) for the ideal gas case is as follows:

 # Initial guesses

$T_t(0) = 300$ # Initial guess for T_t[K]
$W(0) = 20$ # Initial guess for W[Kg/s]
$P_t(0) = 46$ # Initial guess for P_t[bar]

 # Aly-Lee ideal gas enthalpy constants:

A = 33298 # Ideal gas enthalpy constant A
B = 79933 # Ideal gas enthalpy constant B
C = 2086.9 # Ideal gas enthalpy constant C
D = 41602 # Ideal gas enthalpy constant D
E = 991.96 # Ideal gas enthalpy constant E
F = -120280000 # Ideal gas enthalpy constant F

 # The following three functions must be made equal to zero by adjusting the three unknown variables W[kg/s], T_t[K], and P_t[bar]:

$f(T_t) = h_t - h_{t2}$
$f(W) = u_t - u_{t2}$
$f(P_t) = u_t - u_s$

 # Initial guess ranges for non-linear equations

W(min) = 1
W(max) = 50
$T_t(min) = 200$
$T_t(max) = 400$
$P_t(min) = 10$
$P_t(max) = 46$

This problem requires solving the three simultaneous, non-linear algebraic equations (defined above) by varying T_t (the hole temperature), W (the mass flow rate), and P_t (the hole pressure):

$f(T_t) = 0$
$f(W) = 0$
$f(P_t) = 0$

In this case, the upstream pressure is high enough to cause the velocity at the hole to reach the sonic limit. This problem requires solving three simultaneous non-linear algebraic equations. The POLYMATH solution shown below for Example A was obtained using a software package called POLYMATH[7]. Note that any other similar package can also be used. We verified this numerical solution using the MATHCAD[8] and EXCEL[9] software packages also.

NOTE: The numerical solutions for Cases (A) and (B) are shown in Table 1, following the problem statement for Case (B).

Case (B) $P_0 = 1.5$[bar]

PROBLEM: Solve for release rate from a hole in a pipeline. Assume that the throat velocity is sub-sonic. This means that the outlet pressure is fixed at 1[atm] = 1.01325[bar].

When the outlet pressure is fixed, we are left with two non-linear equations. Also, for consistency with the methodology in (A) above, we use a very large upstream diameter to make the inlet velocity very small.

NOTE: If the calculated outlet velocity exceeds the sonic limit, we would revert to the problem formulation under (A) above. (This would add an equation requiring outlet velocity to equal sonic velocity, and thus again require solving three simultaneous non-linear equations.)

As noted earlier, the solution of the real gas case requires use of complex computer programs to solve the chosen equation of state (Lee-Kesler, in these examples) for the required thermodynamic quantities.

In this case, the upstream pressure is low enough so that the velocity at the orifice does not reach the sonic limit. Accordingly, this problem requires solving only two simultaneous non-linear algebraic equations. The solution shown below was obtained using a software package called POLYMATH. Note that any other similar package can also be used. As with Case (A), we verified this numerical solution using the MATHCAD and EXCEL software packages also.

The POLYMATH problem statement for Case (B) is as follows:

$f(T_t) = h_t - h_{t2}$

$f(W) = u_t - u_{t2}$

 # Initial guesses

$T_t(0) = 300$ # Initial guess for T_t[K]

$W(0) = 20$ # Initial guess for W[Kg/s]

Using POLYMATH, the solution for Case (A) shows that the gas release rate, W, for an ideal gas is 17.16[kg/s]. As noted above, identical results were obtained using the Mathcad and Excel software packages.

TABLE 1: SONIC OR SUB-SONIC FLOW THROUGH A VALVE OR HOLE

VARIABLES AND EQUATIONS

Variable	Description & Units	Explicit equations
A	Ideal gas enthalpy constant A	A = 33298.
B	Ideal gas enthalpy constant B	B = 79933.
C	Ideal gas enthalpy constant C	C = 2086.9
D	Ideal gas enthalpy constant D	D = 41602.
E	Ideal gas enthalpy constant E	E = 991.96
F	Ideal gas enthalpy constant F	F = -120280000.
P_t	Throat pressure [bar]	$P_t = 1.01325$
kcal_j	Conversion: 1[kcal] = 4184[J]	kcal_j = 4184
M	Gas molecular weight[kg/kgmol]	M = 16.0
T_0	Inlet gas temperature[K]	$T_0 = 300$
P_0	Inlet gas pressure[bar]	$P_0 = 1.5$
z_0	Inlet gas compressibility factor[]	$z_0 = f(T_0,P_0,x)$
$(c_p/c_v)_0$	Inlet gas C_p/C_v ratio[]	$(c_p/c_v)_0 = 1.32$
R	Gas constant[J/(kgmol.K)]	R = 8314.46
D_0	Diameter of pipe[m]	$D_0 = 0.3$
D_t	Diameter of Throat[m]	$D_t = 0.05$
h_0	Inlet enthalpy[kcal/kg]	$h_0 = (A*T_0 + B*C/\tanh(C/T_0) - D*E/\tanh(E/T_0) + F)/M/kcal\_j$
rho_0	Inlet density[kg/m³]	$rho_0 = P_0*M*1.e5/R/T_0$
A_0	Pipeline cross-sectional area[m²]	$A_0 = 3.14159/4*D_0^2$
A_t	Throat flow area[m²]	$A_t = 3.14159/4*D_t^2$
h_t	Throat enthalpy[kcal/kg]	$h_t = (A*T_t + B*C/\tanh(C/T_t) - D*E/\tanh(E/T_t) + F)/M/kcal\_j$
u_0	Upstream velocity[m/s]	$u_0 = W/rho_0/A_0$
rho_t	Throat density[kg/m³]	$rho_t = P_t*M*1.e5/R/T_t$
u_t	Throat velocity[m/s]	$u_t = W/rho_t/A_t$
u_s	Sonic velocity at throat[m/s]	$u_s = (z_0*(c_p/c_v)_0*R*T_t/M)^{0.5}$
u_{t2}	Throat velocity from Bernoulli Eq.[m/s]	$u_{t2} = (2/rho_t*(P_0*1.e5 + rho_0*u_0^2/2 - P_t*1.e5))^{0.5}$
dh_{KE}	ΔKinetic Energy = $\Delta(u^2)/2$[kcal/kg]	$dh_{KE} = (u_{t2}^2 - u_0^2)/2/kcal\_j$
h_{t2}	Calc. enthalpy at throat = inlet enthalpy - $\Delta(u^2)/2$[kcal/kg]	$h_{t2} = h_0 - dh_{KE}$
T_t	Temperature at throat[K]	$f(T_t) = 0$
W	Mass flow rate[kg/s]	$f(W) = 0$
P_t	Pressure at throat[bar]	$f(P_t) = 0$

TABLE 1: SONIC OR SUB-SONIC FLOW THROUGH A VALVE OR HOLE (CONTINUED)

CASE (A) AND CASE (B) SOLUTIONS (USING POLYMATH)

Variable	CASE (A) Value (SONIC)	CASE (B) Value (SUB-SONIC)
A	33300.0	33300.0
B	79930.0	79930.0
C	2086.9	2086.9
D	41600.0	41600.0
E	991.96	991.96
F	-120300000.0	-120300000.0
P_t	27.71686	1.01325
kcal_j	4184.0	4184.0
M	16.0	16.0
T_0	300.0	300.0
P_0	46.00	1.50
z_0	1.00	1.00
$(c_p/c_v)_0$	1.32	1.32
R	8314.46	8314.46
D_0	0.30	0.30
D_t	0.05	0.05
h_0	226.2098	226.2098
rho_0	29.50683	0.96218
A_0	0.07069	0.07069
A_t	0.00196	0.00196
h_t	205.2641	210.3263
u_0	8.22527	7.71172
rho_t	20.86568	0.73254
u_t	418.7383	364.6543
u_s	418.7383	427.2981
u_{t2}	418.7383	364.6543
dh_{KE}	20.94576	15.88352
h_{t2}	205.2641	210.3263
T_t	255.6215	266.1833
W	17.15555	0.52433
P_t	27.71686	1.01325

CASE (A) RESULTS

This problem was also solved using the Lee-Kesler equation of state to calculate non-ideal gas properties, using a suitable computer program. This solution is provided below, with the ideal gas results quoted alongside for comparison:

	CASE (A)	
	IDEAL GAS	REAL GAS
Inlet pressure[bar]	46.00	46.00
Inlet temperature[K]	300.00	300.00
Calculated mass flow[kg/s]	17.28	18.25
Inlet velocity[m/s]	8.27	8.09
Throat velocity[m/s]	424.65	421.48
Sonic velocity at throat[m/s]	424.65	421.48
Inlet Enthalpy[kcal/kg]	225.61	214.87
Delta Kin. Energy[kcal/kg]	21.54	21.22
Outlet Enthalpy[kcal/kg]	204.07	193.64
Throat pressure[bar]	27.32	25.73
Throat temperature[K]	254.27	250.83

It should be noted that these results are quite similar for this case because the behavior of the gas (methane) is close to ideal at the conditions specified. However, these results would very likely differ significantly if, for example, the inlet temperature was considerably lower and the pressure much higher in which case the gas behavior would be markedly non-ideal.

CASE (B) RESUTS

Case (B) was also solved using the Lee-Kesler equation of state to calculate non-ideal gas properties. This solution is provided below, with the ideal gas results quoted alongside for comparison:

	CASE (B)	
	IDEAL GAS	REAL GAS
Inlet pressure[bar]	1.5	1.5
Inlet temperature[K]	300.0	300.0
Calculated mass flow[kg/s]	0.524	0.525
Inlet velocity[m/s]	7.71	7.70
Throat velocity[m/s]	364.7	364.2
Sonic velocity at throat[m/s]	435.06	435.3
Inlet enthalpy[kcal/kg]	226.21	226.21
Delta Kin. Energy[kcal/kg]	15.88	15.85
Outlet Enthalpy[kcal/kg]	210.33	210.36
Throat pressure[bar]	1.01325	1.01325
Throat temperature[K]	266.18	266.25

As might be expected, the gas mass flow rate is far lower for Case (B) compared to Case (A). Also, for Case (B), it was found that the conditions at the throat are virtually identical for the ideal versus the non-ideal gas.

2. ADDITIONAL EXAMPLES

To illustrate the importance of the issues discussed in this paper, we show a few additional examples for the differences in our numerical results compared to others.

2.1 Gas Release From A Hole In A Vessel

A 0.1[in] hole forms in a tank containing nitrogen at 200[psig] and 80[°F]. Determine the mass flow rate through this leak (Crowl[4], Example 4-4, p. 143).

The solution for this problem using the Lee-Kesler EOS was 4.26e-2[lb/s], compared to 3.86e-2[lb/s] in the reference (gas velocity at the throat was sonic). This difference is over 10% and not negligible.

2.2 Flow Through A Nozzle

Air at p_0=7.0e5[Pa] and T_0=293[K] discharges to the atmosphere. Compute the discharge mass flux, G, the pressure, Mach number, and velocity at the exit[5].

Using the Lee-Kesler EOS, the results are as follows:

Quantity	Lee-Kesler (this article)	Ideal gas (isentropic)[5]	Difference
Mass flux[kg/(m^2.s)]	1589	1650	-3.68%
Exit temperature[K]	238.5	244	-5.5 [K]
Exit pressure[Pa]	212041	370000	-42.7%
Mach number	1.0	1.0	-
Sonic velocity[m/s]	338	313	+8%
Velocity[m/s]	338	313	+8%

Note that the nozzle is not predicted to be choked when the downstream pressure is above 2.12[bar] (kinetic energy effects must be included), as opposed to 3.7[bar] in the original reference.

If a relief valve is to be designed for this scenario, the size would be under-predicted somewhat using the isentropic formulas used in the original reference. However, this discrepancy could be significantly larger in cases where an isentropic expansion incorrectly predicts partial condensation of the gas at the throat. In that situation, the density of the two-phase mixture at the throat could be far lower than that of the uncondensed gas (for the isenthalpic case) and the valve could be under-sized significantly.

3. CONCLUDING REMARKS

It has been shown that calculation of the release rate from a relief valve (or a hole in a vessel) should be done using an isenthalpic (i.e., irreversible adiabatic) path for the flow process. Using an isentropic thermodynamic path for this problem violates the First Law of Thermodynamics, as expressed for steady flow in a stationary relief valve or a hole in a pipeline. This could lead to an incorrect design both when sizing relief valves and estimating release rate from an existing valve or hole in a vessel.

In addition, it is important to use the proper equation of state to estimate real gas properties reliably when compared to the ideal gas assumption, especially when gas conditions are far from ideal. The Bernoulli equation (for throat pressure), and the sonic velocity limit (when flow is choked at the throat) must also be included for a complete definition of the problem. If the upstream pressure is low enough and the throat is un-choked, the isenthalpic methodology - as described in this paper - should again be followed to ensure a theoretically sound and consistent basis. Iterative computations for non-ideal cases (real gases) generally are too intensive for hand calculations and require use of computerized chemical engineering software.

REFERENCES

1. Hougen, O.A., K. M. Watson: "Chemical Process Principles" (John Wiley & Sons, New York, and Chapman and Hall, London, 1st Edition, Part II, p. 551, 1947, and 2nd Edition with R.A. Ragatz, Vol. II, p. 653, 1959)
2. Elliott, J.R., and C. T. Lira: "Introductory Chemical Engineering Thermodynamics" (Pearson Education, Inc. publishing as Prentice Hall, pp. 68-69, 2nd Edition, 2012)
3. Klein, S. and G. Nellis: "Thermodynamics" (Cambridge Univ. Press, 1st Edition, p. 168, 2011)
4. Crowl, D. A., and Louvar, J.F.: "Chemical Process Safety" (Pearson Education, 3rd Edition, pp. 130-131, 2011)
5. Green, D. W., and R.H. Perry: "Perry's Chemical Engineers' Handbook" (McGraw-Hill, 8th Edition, Section 6, pp. 23-24, 2008)
6. Himmelblau, D.M., and J.B. Riggs: "Basic Principles and Calculations in Chemical Engineering" (Pearson Education, 8th Edition, Fig. 9.18, Item 3, p. 525, 2012)
7. Cutlip, M.B., and M. Shacham: "Problem Solving in Chemical and Biochemical Engineering with POLYMATH, Excel, and MATLAB" (Prentice Hall, 2nd Edition, 2007)
8. Mathcad 2001 Professional: "Mathcad User's Guide With Reference manual", (Mathsoft Engineering and Education, Inc., 2001)
9. Freund, S.M., M. Jones, and J.L. Starks: "Microsoft Excel 2013: Complete (Shelly Cashman Series)", (Course Technology, 1st Edition, 2013)
10. Lewis, G.N., and M. Randall: "Thermodynamics", (McGraw-Hill, 2nd Edition, Revised by K.E. Pitzer and L. Brewer, p. 48, 1961)

548434 Engineered Solutions for Effective Flare System Analysis

Relief and flare system documentation is an important part of Occupational Safety and Health Administration (OSHA) Process Safety Management (PSM) 29 CFR 1910.119. Flare system analysis (sizing or rating) can be performed using commercially available software. Effective flare system analysis requires reliable Process Safety Information (PSI) such as process and instrumentation diagrams (P&IDs), relief system evaluation, Heat and Material Balance (H&MB), relief valve data sheets, isometrics, and disposal system loads. Good engineering methodologies can be implemented to effectively design header systems and rate existing header for any change. Some methodologies include use of dynamic simulation to predict realistic relief loads, reboiler pinch analysis for applicable systems, and selection of global relief scenarios. This paper will describes some methodologies and a combination of instrumented systems for effective flare system analysis. Effective flare analysis helps save capital expenditure spent on flare headers and possible flare load reduction yields reduced radiation footprint on a facility.

548364 Flares Are the End of the Line; How to Permit Correctly

INTRODUCTION

-

Flare is one of the key equipment in any refinery or LNG facility. Many relief valves relieve to flare, and if the flare is not permitted corrected – and functional - the plant is down. You cannot operate the plant without an active flare or flare field. The permitting of a flare is one of the prime activities in the permitting process. But the permit is dependent on the type of flare. This presentation covers various Federal Regulations involved in the permitting process, whether general or specific to each type of flare.

FLARE TYPES AND THEIR DESIGN FUCTURES

-

There are mainly three types of flares. The most common and most economical one is an elevated flare. There are elevated flares as high as 660 ft. The can be classified many ways; self-supported, guy wire supported, derrick supported, demountable type etc. One of the most important is the type of burner, whether a pipe tip or "furnace" burners. The second type is a ground flare with rows of furnace burners arranged in arrays at grade and surrounded with a wind fence. The last type is an enclosed flare with furnace burner enclosed with a tall stack. This type can be large as 70 ft in diameter and 150 ft tall. The flare type on a given project is selected based on the plot availability, location of the plant, sparing philosophy etc.

PERMITTING REGULATIONS

This presentation will discuss cover various Federal regulations that are applicable to all types of flares and the various parameter that the regulators are needing when apply for the permit. The presentation will cover the required destruction efficiency of the flaring gases, exit velocity for an elevated flare, alternate means of emission limitation for ground flares. In particular, the tip velocity, burner type, backpressure, emission testing and smokeless capacity are specific to each type of flare.

WHAT YOU WILL LEARN

-

· This paper will provide sufficient information to ensure the correct type of flare is selected and permitted

· This paper will cover various flares types employed in refinery and in LNG plants

· The limitations of these flare

· Importance of flare in any facility

· Application Federal Regulations

· Parameter required in the permitting applications

· Ground flare testing to establish minimum destruction efficiency

· Alternate means of emission limitation requirements in lieu of testing

548232 A Case Study on Application of HAZOP for a Brewery

Abstract

Beer is the most consumed alcoholic beverage in the world, and most popular beverage after water and tea. It's an alcoholic beverage produced by the saccharification of starch, often derived from the malted grains such as barley and wheat, and the fermentation of the resulting sugar. The brewing process can be broken into four main constituents before the bottling line, Mashing, Lautering, Wort boiling and Fermenting. Breweries commonly rely on ammonia refrigeration to keep their supplies cold. However, some brewers fail to realize that ammonia is one of the dangerous chemicals covered under 29 CFR 1910.119, OSHA's regulation for preventing or minimizing the release of toxic, reactive, flammable or explosive chemicals. "Safety matters at all levels", even for companies that may not be PSM level facilities.

A HAZOP study identifies hazards and operability problems. The study involves investigating how the plant might deviate from the design intent. It is based on the principle that several experts with different backgrounds can interact and identify more problems and failure scenarios working together and combining their results. This paper presents a review of consequences identified in brewery HAZOP. It also outlines some of the recommendations and action items identified from the case study of a brewery.

Keywords: Process Safety Management (PSM), PHA, HAZOP, Brewery, Ammonia, OSHA.

548410 A Basic Review of Fundamental of Thermodynamics and Its Application in the Analysis of Consequences and Accident Investigation

Once the stage of hazard identification has been concluded in a risk analysis, the next stage corresponds to select and specify the risk scenarios that will be modeled their effects or damages associated. This stage correspond to the consequences analysis and usually takes place in three basic steps: a) source term definition, b) dispersion model and d) effect or damage model.

Depending on the risk scenario to be modeled, several thermodynamic fundamentals or principles are applied in each of these steps supporting or sustaining the behavior and expected results of the consequences analysis model. In this work we propose to review the main thermodynamic fundamentals that support each of the models applied in each of the three steps that define a consequence analysis. This review of basic thermodynamic principles defines a set of different thermodynamic basic principles to apply to representative and different cases of consequence analysis. These basic thermodynamic principles are applied in the consequence modeling of a pair of cases of accident inventions to corroborate their validity.

548331 Evaluating Safe Location Criteria for Vents and Relief Devices Discharging to Atmosphere

In some process technologies, it may not be feasible to dispose the effluents from safety valves and vents to a closed system and discharging directly to the atmosphere can be the only alternative. Some examples include High Pressure Polyethylene Processes, Olefins and Storage units. The discharge can be directed to the atmosphere if:

- Local regulations permits it;
- Does not harm the environment;
- Does not harm the neighborhood;
- And is directed to a Safe Location.

The criteria for defining what a safe location is can vary, but it must include at least a consequence assessment of the flammable, toxic and thermal radiation effects. Some guidance can be found on industry standards and engineering references such as API 521 and the CCPS Guidelines for Pressure Relief and Effluent Handling Systems

One aspect to discharge hydrocarbons to a Safe Location is to evaluate the dispersion of the discharged hydrocarbons and ensure that it is diluted bellow the lower flammable before reaching the ground level or elevated structures. These areas are typically with people circulation and ignition sources. Another aspect that must also be evaluated is the thermal radiation generated due to an ignition of the discharge affecting personnel or damaging equipment or other assets. Lastly, if there is a possibility of the flammable mixture reaching confined spaces, the explosion possibility must also be assessed.

API 521 gives some guidance on chapter 5.8 for estimating the dispersion and concentration profile from a typical relief device discharge. It also describes that if the vapor relief velocity is high (> 30 m/s), the air entrance rate is also high and the vapors typically would be diluted below lower flammable limit in a very short distance.

This phenomenon explains the standard recommendation of locating the discharge of the relief device at least 3 m above any adjacent equipment, building or structure located within 15 m horizontally from the discharge point.

Experience has shown, however, that for some special cases these typical recommendations does not apply, such as:

- Heavy gas releases;
- Cryogenic releases;
- Control valve releases;
- Discharge piping with unconventional design (rain cap, horizontal, etc.);
- Discharge location within a building wake.

Adding to the complexity of the analysis, the relief devices can also be fitted with steam or other purge gas injection that could positively affect the dispersion.

After some case studies, which included the analysis of High Pressure Polyethylene plants Vent Stacks and Olefins unit relief and venting devices, Braskem developed an internal engineering standard with guidance and criteria for evaluating existing projects and new designs in order to ensure they comply with safe location. The guidance includes:

- Relief cases selection;
- Simplified and rigorous dispersion analysis;
- Consequence assessment endpoints;
- Thermal radiation effects to assets;
- Liquid, vapor and two-phase releases;
- Process Hazard Analysis;
- Diluting/snuffing steam;
- New systems and retrofitting criteria;

Applying the criteria in the developed standard allows the evaluation if the relief is in a safe location or not. If the discharge is not in a safe location, a risk ranking and recommendations to change the system must be made. One kind of a typical recommendation is to install the relief in a higher location. However, for some cases, achieving Safe Location is not possible and changes in the system are needed, such as relocating to a closed system or design changes to eliminate the relief scenarios associated with that relief device.

The development of an internal engineering guideline ensures a standard approach throughout the company for evaluating the Safe Location of hydrocarbon discharges and improves the quality of the risk evaluation during the PHA cycles.

548407 Experimental Study of Electrostatic Hazard inside Scrubber Column Using Response Surface Methodology

Experimental Study of Electrostatic Hazard inside Scrubber Column using Response Surface Methodology

Jingyao Wang[a,b], Yue Sun[a,b], Xiaodan Gao[a,b], Benjamin Wilhite[a,b], M. Sam Mannan [a,b]

[a]Mary Kay O'Connor Process Safety Center, Artie McFerrin Department of Chemical Engineering, Texas A&M University, College Station, Texas 77843-3122, USA

[b] Artie McFerrin Department of Chemical Engineering, Texas A&M University, 3122 TAMU, College Station, Texas 77843, United States

Abstract

Electrostatic ignition has triggered significant fire/explosion incidents in liquid processing industries including chemical, oil/gas, food, pharmaceutical and transportation. In this work, the impact of operating flowrate and temperature on the electrostatic energy accumulation in scrubber columns was studied. An *in-situ* charge measurement apparatus combined with a particle scattering laser function was installed inside a pilot-scale scrubber column, which can measure the aerosol particle size distribution, concentration and charge simultaneously. Four electrostatic ignition scenarios were considered including discharge from a single aerosol droplet, aerosol clouds, ungrounded conductor charged by the aerosol and ungrounded conductor charged by the liquid. Response surface methodology was applied to construct a quadratic empirical model correlating the electrostatic energy response with the flowrate and temperature. The obtained electrostatic response shows that corona from a charged single droplet was too small for safety concern. However, the ungrounded conductor polarized by charged aerosol or liquid can accumulate significant energy that is higher than the minimum ignition energy of some flammable chemicals. Temperature has two-sided effect where high temperature generates more concentrated charged aerosol but also facilitates the charge relaxation in the liquid phase. In all scenarios, increasing gas/liquid flowrate can cause significant electrostatic energy accumulation and can increase the hazards of many incidents. To reduce the electrostatic hazards, it is recommended to reduce the gas/liquid interaction flowrate and eliminate the presence of ungrounded conductors in liquid process operations.

Keywords: Scrubber column design, Electrostatic hazards, Response surface methodology, Process safety, Hazard identification, Aerosol hazards.

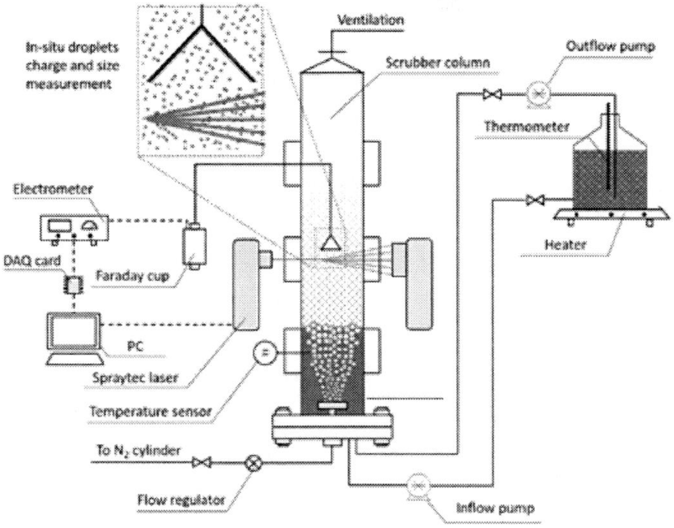

550852 Process Safety in Oil Refineries: Vacuum Distillation Column Hazop Study

Errors that can occur in the identification studies for hazards and risks which is one of the elements of a process safety management cause resulting major accident having undesirable long - term effects on human and environment. For preventing major accidents, identification and analysis of process hazards due to equipment failures, design or montage defects, improper maintenance, misuse, material defects, metal fatigue, corrosion, welding defects, human failure etc. are the main principles of process safety management. The aim of present study is to investigate possible accidental risks occurring in Vacuum Distillation Column of oil refining within contributing to the process safety frame.

In this cross - sectional study, semi quantitative HAZOP analysis was used for identifying accidental risks. By means of HAZOP studies 66 deviations, 238 deviation causes, 287 consequences, 651 barriers and 113 recommendation were identified. In the determination of accidental risk in the vacuum distillation column by applying semi quantitative HAZOP analysis method, effective results were obtained. The most common causes of deviations in the column have been found as control loop failure, leakage and mechanical failure. It is stated that the most common accidental risks are as a portion of 66.9 % operational deterioration, fire, explosion, release known as major accident risk. It is identified that the accidental risks mostly cause existence damage and the highest accidental risk occurs in the "vacuum column overhead system" and "bottom of the vacuum column system" 87.7 % of 113 recommendations are modification in the design, 5.3% is about modifications in the process safety management system.

551179 API Process Safety Site Assessment Program - Trends and Good Practices in Process Safety

The Process Safety Site Assessment Program is a collaborative industry effort managed by API and is focused on improving process safety in petroleum refining and petrochemical facilities. The program has conducted over 65 process safety assessments and 38 HF RP-751 assessments since its inception in 2012. The assessments are conducted by independent and credible third-party teams of industry-qualified process safety expert assessors using industry-developed protocols.

The presentation will review trends, both positive and low scoring, seen in the extensive benchmarking data collected. Good practices observed in the assessments will be identified and discussed. The emphasis will be to have the attendees leave with an understanding that typically some site(s) has developed a way to handle challenging areas of process safety.

545445 Emerging Trends in Process Safety Management Relating to Contractor Workforce Compliance

Following a Six Sigma Black Belt process improvement session conducted 2016 to 2018 for petrochemical complexes along the Gulf Coast the presentation identifies several emerging trends for Best Practices in industrial workforce compliance.

Safety Councils work for petrochemical plants and their contractors to facilitate innovative ways to reduce incident rates resulting in a 95% reduction in recordables. 2017 marked an evolution in Safety Council deliverables, slowly migrating away from site specific training to develop new workforce compliance and risk management processes. Working within the industry we developed several internationally recognized Best Practices for contractor compliance, risk management, labor cost management. ISTC collaborated with our partners bringing years of quantitative data to identify problems that are developing, have occurred in the workplace and formulating solutions, cost structures and most importantly compliance management to combat workforce issues. The presentation, best practices in contractor compliance highlights the journey from identifying the problems, and the associated issues with hiring, compliance, auditing, accident response and even Department of Homeland Security, CFATS inspections. Beyond the logistics of building solutions for a particular problem, the presentation reflects on how companies were able to share their issues among peers without violating confidentiality or company procedures to solve what was to be discovered as an industry trend. Using Six Sigma tools, the units were able to identify root problems, collaborate on a solution, model cost recovery and implement the solutions to the contractor workforce with a minimum of disruption to the business at hand.

Análisis Cuantitativo De Riesgo (QRA) Como Herramienta Para La Toma De Decisión Sobre La Infraestructura En Sistemas De Transporte De Fluidos Peligrosos

Daniel Mauricio Moncada Saavedra
CENIT -Transporte y Logística de Hidrocarburos
Cll 113 #7-80 Piso 13 Bogotá – Colombia
daniel.moncada@cenit-transporte.com

Maria Isabel Montoya Rodriguez
CENIT -Transporte y Logística de Hidrocarburos
Cll 113 #7-80 Piso 13 Bogotá – Colombia
maria.montoya@cenit-transporte.com

Prepared for Presentation at
American Institute of Chemical Engineers
2019 Spring Meeting and 15th Global Congress on Process Safety
New Orleans, LA
March 31 – April 3, 2019

Análisis Cuantitativo De Riesgo (QRA) Como Herramienta Para La Toma De Decisión Sobre La Infraestructura En Sistemas De Transporte De Fluidos Peligrosos"

Daniel Mauricio Moncada Saavedra
CENIT -Transporte y Logística de Hidrocarburos
Cll 113 #7-80 Piso 13 Bogotá – Colombia
daniel.moncada@cenit-transporte.com

Maria Isabel Montoya Rodriguez
CENIT -Transporte y Logística de Hidrocarburos
Cll 113 #7-80 Piso 13 Bogotá – Colombia
maria.montoya@cenit-transporte.com

Palabras clave:

Abstract

La pérdida de contención en activos utilizados para el transporte de hidrocarburos puede ocasionar impactos considerables a personas, medio ambiente, económicos y de reputación. Con el fin de mitigar la ocurrencia este tipo de eventos, se implementan estrategias sobre los activos, siendo la identificación y análisis de riesgo un elemento clave durante el proceso.

El presente análisis utilizó un enfoque cuantitativo, tanto para probabilidad como para consecuencia, con el fin de poder estimar el riesgo que genera la operación de una línea de transporte en un área de alta consecuencia clasificado como sitio identificado, planteando posibles alternativas de control y mitigación del riesgo.

Como resultado del estudio realizado se pudo establecer el nivel de riesgo actual asociado a la operación de la línea en el sitio de interés. Como actividades de mitigación para reducir el riesgo asociado se identificó: la reparación de los defectos, así como el realineamiento y reubicación de la tubería.

1 Introducción

Para el presente análisis se implementó un cálculo de la probabilidad de falla a partir del análisis de integridad estructural del sector de interés. Adicionalmente, se consideró el análisis de la susceptibilidad que se presenta a partir de las amenazas presentes, la cuales son identificadas a partir de las actividades de monitoreo e inspección desarrolladas sobre la infraestructura. El cálculo de la interacción entre las amenazas a la integridad; así como el resultado final de probabilidad de falla, se puede realizar mediante diferentes enfoques, el presente estudio utilizo cálculo de probabilidad de excedencia y aplicación de redes bayesianas para la estimación.

Para el cálculo de la consecuencia se buscó verificar la posible afectación a personas, producto de la ignición del producto transportado. Dentro de los sucesos finales se consideró: incendio de piscina, incendio por chorro de fuego, llamarada y la sobrepresión debida a la explosión. El cálculo de estos efectos se realizó en el sitio donde se presenta la perdida de contención, así como también para el caso en el que el hidrocarburo es transportado por un cuerpo de agua o se desplaza en tierra.

Una vez determinada tanto la probabilidad de que se desarrollen los sucesos finales identificados como la consecuencia que cada uno de ellos puede generar, se procedió a realizar el cálculo del riesgo expresando el resultado en términos de riesgo social e individual, comparando el resultado contra criterios de tolerabilidad referentes en la industria. Para la condición actual establecida se identificaron alternativas las cuales buscaban disminuir el nivel de riesgo, implementando actividades de mitigación del nivel de riesgo.

2 Análisis de Riesgo Cuantitativo

3 Frecuencia Final de Ocurrencia de los Sucesos Finales

En el sector bajo interés se tienen que la línea es susceptible a las siguientes amenazas:

3.1 Análisis de Condición.

En este aparte se relacionan los resultados de análisis de probabilidad de falla, obtenidos en el sector de interés, PK 43+000 al PK 45+000 del poliducto, para las amenazas de corrosión externa, corrosión interna, geotecnia y daños por terceros involuntarios:

3.1.1 Corrosión Externa (CE).

A continuación, se describe el estado de salud de las barreras de CE del poliducto en el sector solicitado (ver Ilustración 1), así como el análisis de probabilidad de excedencia de las anomalías ILI identificadas con la corrida del 2014. En la Ilustración 2 se observa la disponibilidad de la URPC y se reporta que la tubería cuenta con recubrimiento en buen estado. Bajo las condiciones y premisas consideradas para el cálculo y proyección de la POE (Probabilidad de excedencia; entre ellas una rata de corrosión de 2-3 mpy aproximadamente; no hay riesgo de excedencia de los criterios de profundidad y ERF (Factor Estimado de Reparación) de las anomalías ILI hasta el año 2021 en el sector 43+590 y 42+746 (Anomalías ILI con DR(Distancia de registro): 2646.76, 3490.08, 3492.38, 3492.71) (Ilustración 2).

Ilustración 1. Perfil de riesgo y salud de barreras CE.

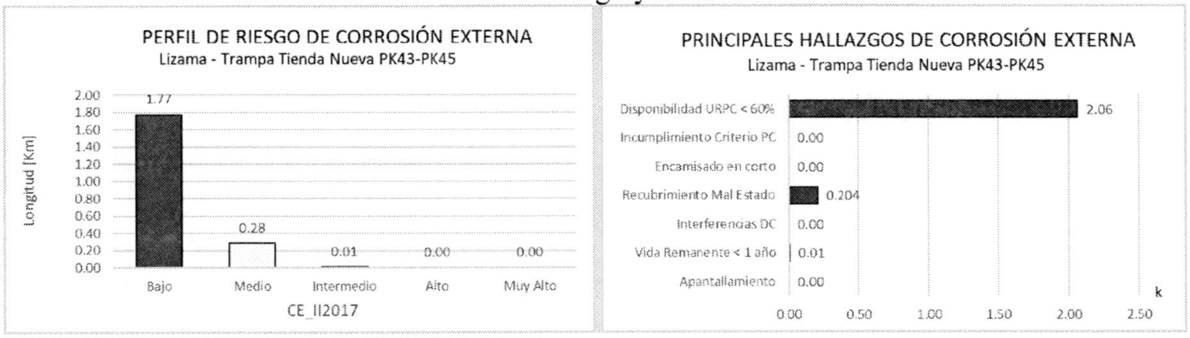

Ilustración 2. Análisis de POE de anomalías de CE en el tiempo.

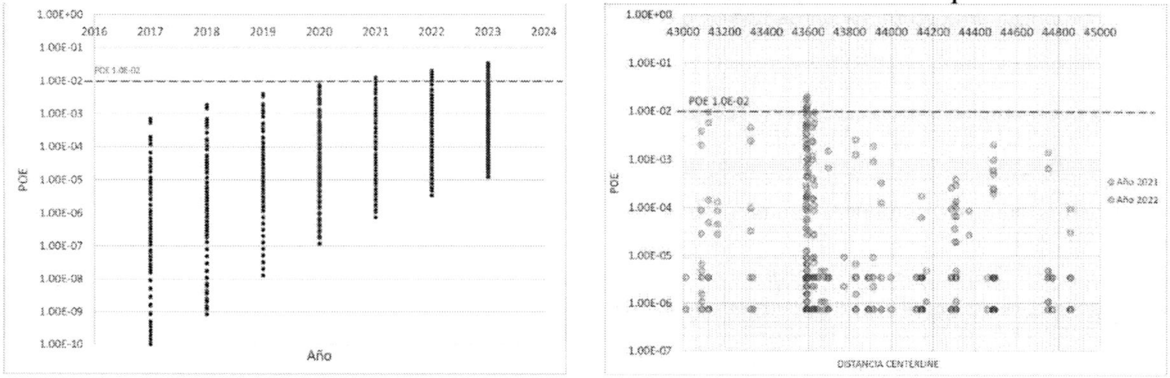

3.1.2 *Corrosión interna (CI).*

A continuación, se describe el estado de salud de las barreras de CI de la tubería en el sector solicitado (ver Ilustración 3), así como el análisis de probabilidad de excedencia de las anomalías ILI identificadas en la corrida del 2014. De acuerdo con los resultados de valoración de riesgo, se presenta un correcto estado de salud de las barreras de corrosión interna. Adicionalmente, de acuerdo con las condiciones y premisas consideradas para el cálculo y proyección de la POE, no hay riesgo de excedencia de los criterios de profundidad y ERF de las anomalías ILI en el mediano plazo (Ver Ilustración 4).

Ilustración 3. Perfil de riesgo y salud de barreras CI.

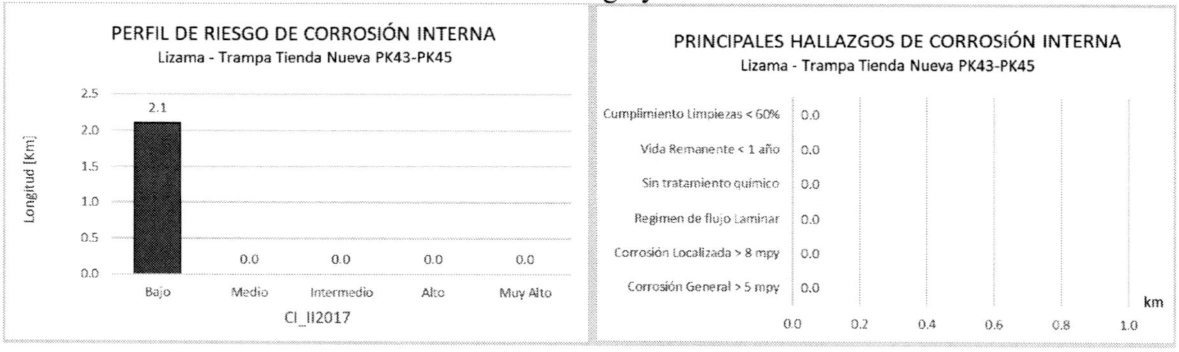

Ilustración 4. Análisis de POE de anomalías de CI en el tiempo.

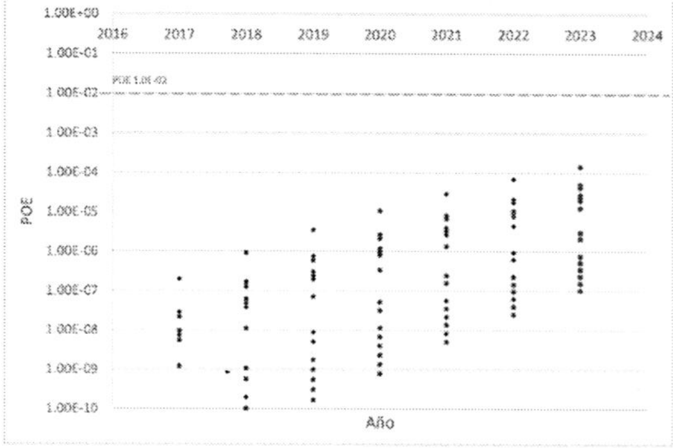

3.1.3 Geotecnia.

En el sector de interés se reporta hallazgo de geotecnia en la distancia centerline 44643 – 44703 (PK 44+400), donde se presenta proceso erosivo incipiente. Actualmente, no se identifica avance de la condición, razón por la cual, se continuará seguimiento a través de la inspección visual del derecho de vía (IVDDV).

Ilustración 5. Perfil de riesgo y salud de barreras de Geotecnia.

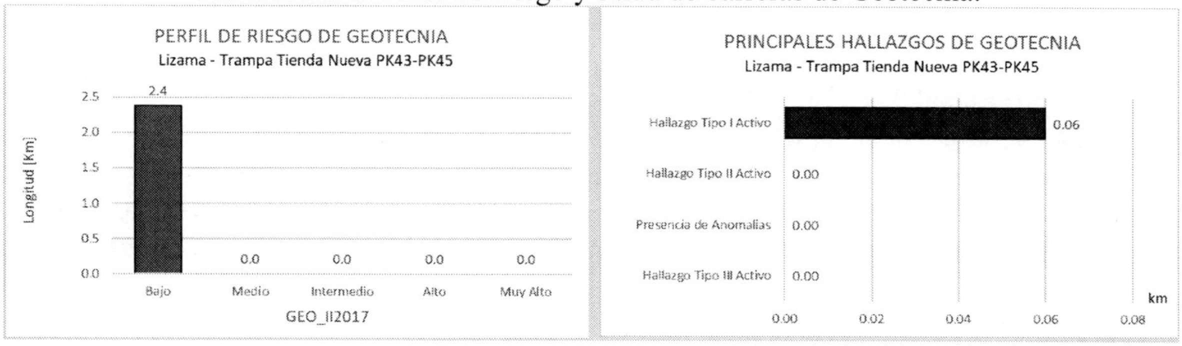

3.1.4 *Daños por Terceros Involuntarios (DPT).*

Debido a la condición de la tubería y a la gestión realizada por el operador para el control de la amenaza, ésta presenta una susceptibilidad baja de ser golpeada por un tercero sobre el derecho de vía. A continuación, se relaciona el perfil de riesgo de la amenaza de DPT en el sector analizado:

Ilustración 6. Perfil de riesgo y salud de barreras de Daños por Terceros Involuntarios.

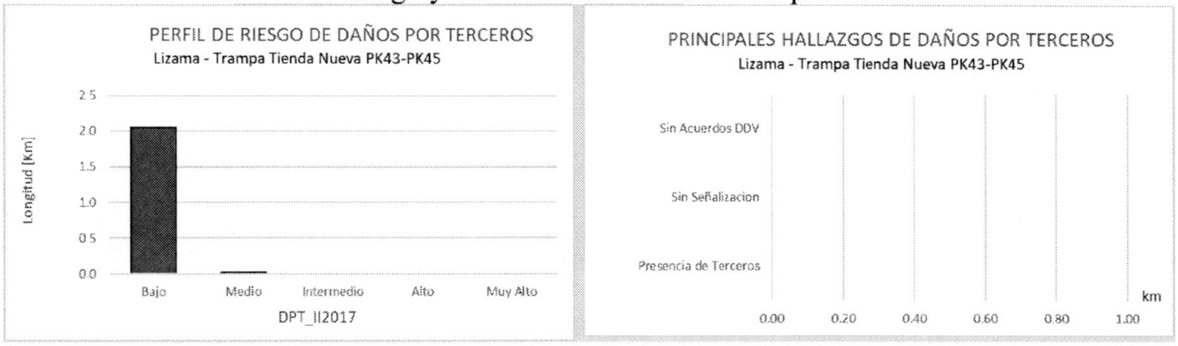

De acuerdo con el análisis ejecutado, el evento más probable obedece a una pérdida de contención generada por la amenaza de corrosión externa; donde el análisis estructural identifica que el modo de falla se daría por fuga. Sin embargo, las anomalías encontradas no presentan riesgo de excedencia de los criterios de profundidad y ERF hasta el año 2021.

Con el fin de poder estimar la probabilidad de falla de la línea en el sector de interés, se realizó el análisis estructural considerando principalmente defectos por perdida de espesor, bajo estas premisas se obtuvo un valor de probabilidad de excedencia de los criterios de profundidad y ERF de 7.09E-4. Del análisis de condición se concluye que el iniciador más probable en este sector es la falla de la tubería por fuga asociado a corrosión externa, para el cual a continuación se presenta el análisis de riesgo.

4 Análisis de consecuencia

A continuación, se presenta el análisis de consecuencia para los sitios en los cuales se ha identificado que se puede llegar a presentar una pérdida de contención.

4.1 Identificación de escenarios

4.1.1 Selección de sustancias.

En el poliducto se transportan las siguientes sustancias: Gasolina, GLP, Jet A1 y B2

4.1.2 Identificación de iniciadores.

Los iniciadores representan las situaciones iniciales de pérdidas de contención o exposición que se pueden presentar en los das diferentes equipos. Los iniciadores pueden ser de dos tipos, genéricos o específicos. Los iniciadores genéricos, involucran una pérdida de contención debida a causas no específicas del proceso u operación, como pueden ser corrosión, errores de construcción, fallas de la soldadura, etc. Los iniciadores específicos se caracterizan por tener causas concretas debido a las condiciones de proceso, el diseño, los materiales utilizados, etc.

A continuación, se describen los iniciadores tenidos en cuenta para cada uno de los sitios en los cuales se identificó la inestabilidad de taludes.

- **Sector PK 43+590.** Este sitio tiene la menor distancia desde la infraestructura a la escuela bajo análisis, 30 metros aproximadamente, sumado a esto en las cercanías de este se ha identificado la presencia de anomalías las cuales, bajo las condiciones actuales de operación y deterioro hacia el año 2021, pueden exceder los límites de profundidad critica a la falla.

- **Sector PK 43+805 – PK 44+110.** En este sitio se tiene que ante la posible pérdida de contención del producto se puede llegar a presentar la formación de una ruta de derrame en tierra, por la cual se desplazaría el producto hacia el interior de la Escuela.

4.1.3 Identificación de sucesos finales.

La identificación de los sucesos finales que pueden tener lugar después de que se da el iniciador, se debe realizar utilizando los árboles de eventos o sucesos. Dichos árboles permiten ver el comportamiento de la sustancia tras la descarga y estimar la probabilidad de ocurrencia de cada suceso, teniendo en cuenta los sucesos intermedios que condicionan la evolución del suceso final.

Esta identificación de escenarios se lista en Tabla 11 para los escenarios que involucran sustancias inflamables tales como Gasolina o Nafta. Los posibles sucesos finales identificados en el poliducto son: derrame continental, incendio de piscina, incendio de chorro y llamarada.

Tabla 1. Identificación de escenarios.

Escenarios	Descripción	Iniciadores	Sustancia	Tipo de pérdida
Afectación directa				
E1	Poliducto	L3	Gasolina	Incendio de piscina
				Incendio de chorro
				Llamarada
E2	Poliducto	L3	GLP	Incendio de piscina
				Incendio de chorro
				Llamarada

Escenarios	Descripción	Iniciadores	Sustancia	Tipo de pérdida
Afectación directa				
Afectación indirecta				
E3	Poliducto	L3	Gasolina	Incendio de piscina
				Llamarada

4.2 Estimación del área de afectación por los efectos que involucran producto

4.2.1 Condiciones para la simulación.

Para determinar las consecuencias, es necesario definir las condiciones propias del entorno de la infraestructura que servirán como base para la simulación de los efectos. Estos parámetros son presentados en la Tabla 2. Las estabilidades atmosféricas fueron determinadas a partir de la rosa de vientos basada en datos estadísticos y modelos meteorológicos, la cual es presentada en la **¡Error! No se encuentra el origen de la referencia.**.

Tabla 2. Condiciones atmosféricas

Parámetro	Valor	Estación de Referencia
Temperatura promedio (°C)	31	Colosal
Humedad relativa promedio (%)	83	
Rugosidad el terreno (m)	0,08	
Estabilidad atmosférica más probable y velocidad del viento (m/s)	2F	
Estabilidad atmosférica F y velocidad del viento (m/s)	2F, 5D	

4.2.2 Distancias de afectación directa

Las distancias de afectación representan las longitudes dentro de las cuales se pueden manifestar los efectos correspondientes a los niveles de afectación, para cada suceso final (incendios de piscina o llamaradas). Se denominan distancias de afectación directas, ya que los sucesos finales de que dependen se originan y desarrollan su efecto directamente en el sitio en el que se produce la pérdida de contención del producto

4.2.3 Estimación de Descargas.

Para la simulación de los posibles efectos producto de una pérdida de contención, es necesario conocer cuánto producto podría derramarse en cada punto de estudio sobre la infraestructura. El volumen derramado corresponde al estimado teniendo en cuenta el llenado de la tubería, las condiciones de operación y el tiempo de respuesta operativo, es decir, el tiempo de detección más el tiempo de actuación (generalmente el seccionamiento del sistema o parada de bombeo).

Tabla 3. Tiempos de actuación

Equipo	Tiempo de Detección (min)	Tiempo de Actuación (min)
	Fuga	**Fuga**
Poliducto	30	180

Tabla 4. Volumen y Tasa de descarga

Volumen (US bls)	Tasa de descarga (bls/hr)
59	22

4.2.4 *Simulación de Efectos de los Sucesos Finales.*

En este numeral se presentan los resultados de las distancias de afectación directa, obtenidas para cada uno de los niveles de afectación asociados a los sucesos finales. En la **¡Error! No se encuentra el origen de la referencia.** se presentan las distancias de afectación obtenidas para el evento incendio de piscina, es importante resaltar que este depende en gran medida del área de empozamiento que se forma una vez se produce la descarga del producto.

En la **¡Error! No se encuentra el origen de la referencia.** se presentan las distancias de afectación para incendio de chorro obtenidas durante las simulaciones, es importante resaltar que la distancia obtenida depende de las propiedades químicas de la sustancia y de las condiciones en las cuales se produce la descarga.

Tabla 5. Distancia de afectación incendio de piscina.

Equipo / Abscisado	Iniciador	Sustancia	Volumen Derrame	Niveles de Afectación por Incendio – Incendio de Piscina						
				1.6 Kw/m²	5.0 Kw/m²	7.3 Kw/m²	9.5 Kw/m²	14.5 Kw/m²	20.9 Kw/m²	37.5 Kw/m²
			(Bbl)	Distancia (m)						
K43+590	Rotura mínima	GLP	59	46,94	28,63	24,58	21,98	18,01	14,55	8,38
K43+590	Rotura mínima	Gasolina	59	94,63	49,69	35,17	27,20	20,98	20,98	-

Tabla 6. Distancia de afectación incendio JET FIRE.

Equipo / Abscisado	Iniciador	Sustancia	Volumen Derrame	Niveles de Afectación por Incendio – Chorro de fuego						
				1.6 Kw/m²	5.0 Kw/m²	7.3 Kw/m²	9.5 Kw/m²	14.5 Kw/m²	20.9 Kw/m²	37.5 Kw/m²
			(Bbl)	Distancia (m)						
K43+590	Rotura mínima	GLP	59	45,9	33,61	30,93	29,33	27,13	25,23	23,42
K43+590	Rotura mínima	Gasolina	59	44,43	31,20	28,46	26,82	24,58	22,96	20,84

En la **¡Error! No se encuentra el origen de la referencia.** se presentan las distancias de afectación por llamarada; las cuales están relacionadas con la volatilidad del producto junto con las condiciones climáticas de las zonas del evento.

Tabla 7. Distancia de afectación Llamarada.

Equipo / Abscisado	Iniciador	Sustancia	Volumen de Derrame	Niveles de Afectación por Llamarada más probable		Niveles de Afectación por Llamarada más estable	
				LII	LII/2	LII	LII/2
			(Bbl)	Distancia (m)		Distancia (m)	
K43+590	Rotura mínima	GLP	59	24,12	58,53	26,48	66,7
K43+590	Rotura mínima	Gasolina	59	-	-	-	-

4.2.5 *Distancias de Afectación Indirecta*

Las distancias de afectación indirecta representan las longitudes de potencial afectación por sucesos finales que se originan en una trayectoria o ruta de derrame (en tierra o en agua) o, a partir de estancamientos o acumulaciones de producto. En este numeral se presentan los resultados de las distancias de afectación obtenidas para cada uno de los niveles de afectación asociados a los sucesos finales que se pueden presentar en la ruta de derrame. Estos resultados corresponden a la radiación generada por piscina de fuego, evento que en los cuerpos de agua depende de la sección del cuerpo de agua, así como por el frente de llama generado por el volumen derramado.

En la Tabla **88 y ¡Error! No se encuentra el origen de la referencia.**9 se referencian los corredores de afectación generados por incendio de piscina sobre las rutas de derrame en tierra identificadas en el poliducto, en el sector comprendido entre los PK 43+000 - PK 45+000.

Tabla 8. Distancias de afectación indirectas POOL FIRE.

Abscisado	Medio de Transporte	Sustancia	Incendio						
			Incendio en piscina						
			1.6 Kw/m²	5.0 Kw/m²	7.3 Kw/m²	9 Kw/m²	14.5 Kw/m²	20.9 Kw/m²	37.5 Kw/m²
			Distancia (m)						
Km 43 - 45	Ruta en tierra	Gasolina	82,02	54,68	27,34	24,61	19,14	13,67	8,20

Tabla 9. Distancias de afectación indirecta FLASH FIRE

Abscisado	Medio de Transporte	Sustancia	Niveles de afectación por llamarada más probable		Niveles de afectación por llamarada más estable	
			LII	LII/2	LII	LII/2
			Distancia (m)		Distancia (m)	
Km 43- 45	Ruta en tierra	Gasolina	2,14	4,74	8,64	32,24

4.2.6 Áreas de Afectación

En sistemas de ductos, el área de afectación directa se define por un corredor medido desde el eje central del ducto. Este corredor está dado por la mayor distancia probable de afectación obtenida en la simulación de efectos para todas las sustancias analizadas. Por otro lado, el área de afectación indirecta está dada por varios corredores a lo largo de las posibles rutas de derrame tanto en suelo como en agua. En la Ilustración 77 se presenta el área de afectación identificada.

Ilustración 7. Área de afectación identificada.

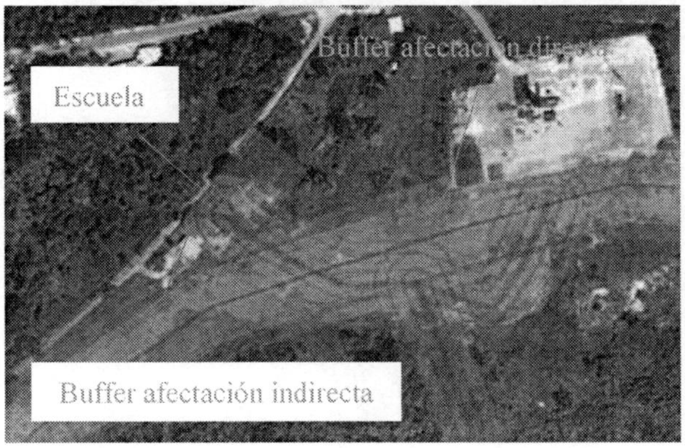

4.2.7 Identificación de Elementos Vulnerables

Para calcular el riesgo de la población respecto de la posible afectación producida por la pérdida de contención descrita, se requiere el número y localización de las personas, en la Tabla 100 se presenta la población identificada al interior de la escuela.

Tabla 10. Presencia de población Escuela

Presencia de Población							
Escuela - Dia				**Escuela - Noche**			
Niños	63	Profesores	6	Niños	0	Profesores	2

5 Evaluación del Riesgo.

A continuación, se presentan los resultados del análisis de riesgo el cual tuvo en cuenta la información presentada en los numerales anteriores asociada a las posibles causas de falla, los modos en que puede darse cada falla identificada, la frecuencia de ocurrencia de la falla, la probabilidad de ocurrencia de los sucesos finales, y finalmente, la estimación de la magnitud de las consecuencias de tales sucesos finales en el entorno en términos cuantitativos.

5.1 Riesgo Individual.

Los resultados de riesgo individual expresados en número de fatalidades por año (o año-1), sí permiten establecer zonas de aceptabilidad del riesgo según los diferentes criterios que se utilizan en la industria. En la Ilustración 8 se presentan los resultados de riesgo individual en el sector comprendido entre los PK 43+000 y 45+000.

Ilustración 8. Riesgo Individual.

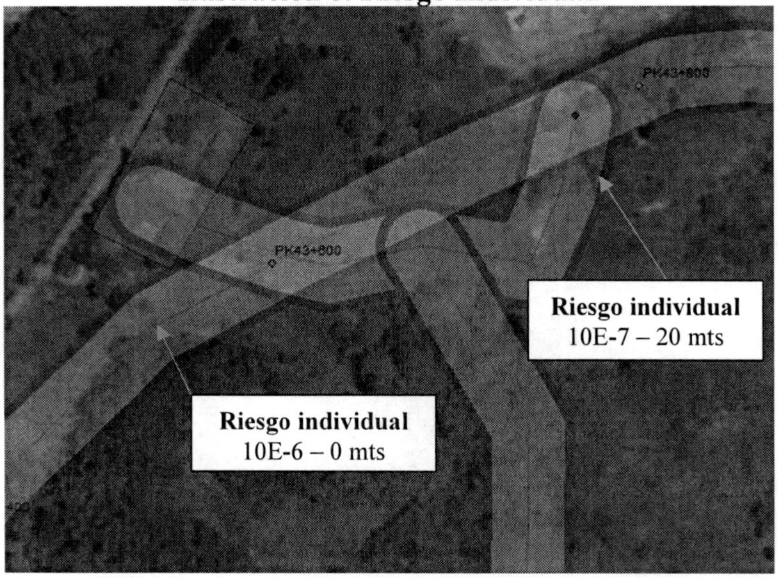

5.2 Riesgo Social

El riesgo social representa el número de fatalidades que pueden llegar a ocurrir en los sitios evaluados en el poliducto en el PK 43+000 – PK 45+000, al revisar la información contenida en la Ilustración 9 se puede apreciar como el riesgo se mantiene en la región tolerable, muy cercana al límite inferior ALARP. El evento dimensionante corresponde a una llamarada en la ruta de derrame, con el ducto transportando gasolina, la cual puede llegar a afectar a ocho (8) personas.

Ilustración 9. Riesgo social Escuela PK 43+000 – PK 45+000.

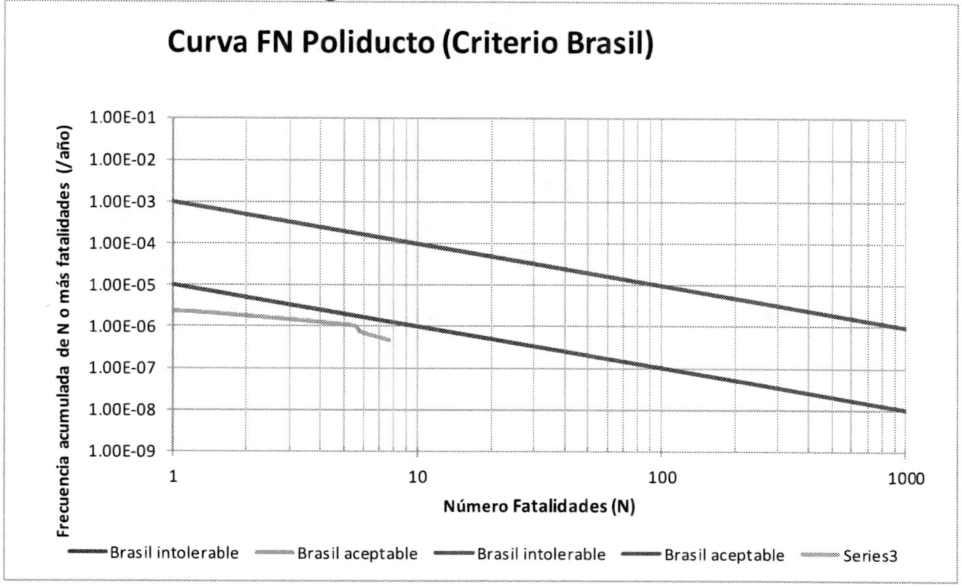

6 Alternativas de mitigación de Riesgo

Considerando los resultados del análisis de riesgo cuantitativo sobre el Poliducto para el sector PK 43+000 - PK 45+000, se tiene que ante una rotura total el evento dimensionaste corresponde a una llamarada con una distancia de afectación de 135 metros, para el caso de una fuga por corrosión el evento dimensionaste corresponde a una llamarada con una distancia de afectación de 32,24 metros, finalmente como resultado de una perdida de contención se puede llegar presentar la formación de una ruta de derrame en tierra la cual ingresaría a la escuela, con una distancia de afectación de 8.6 metros generada por una llamarada, tal como se presenta en la Ilustración 10.

Ilustración 70. Distancias de afectación

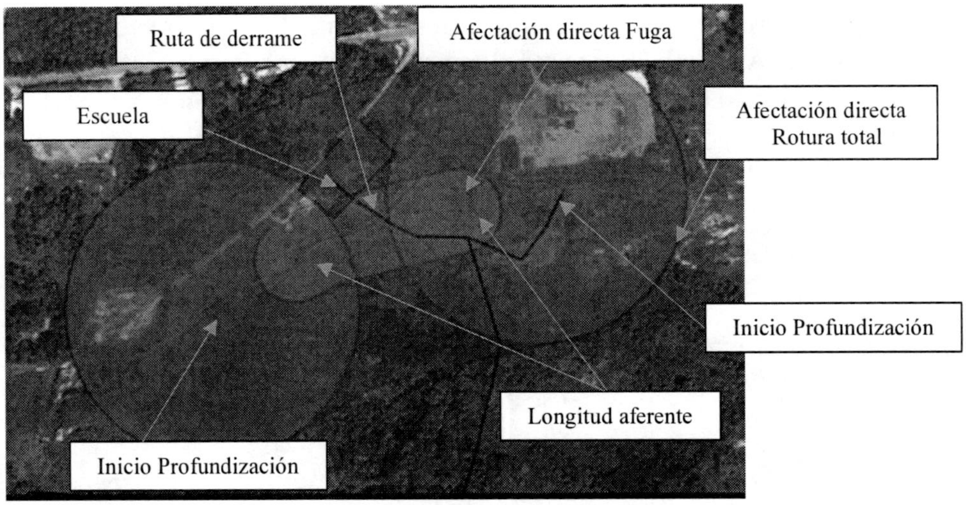

Teniendo en cuenta que el evento corresponde al presentado en la ruta de derrame, se hace necesario considerar la evaluación de alternativas de mitigación de riesgo. Con base en lo anterior, se consideró que la actividad de realineación y profundización cubra la longitud aferente de esta ruta, la cual es de 155 metros e inicia en la coordenada 74°33'35.466"O - 6°9'8.213"N y finaliza la coordenada 74°38'15.025"O - 6°9'8.737"N. Lo anterior, tiene como objeto de eliminar la ruta de derrame actual que conduce a la escuela en caso de una falla y perdida de contención, dado que el vertimiento se genere en sentido contrario al actual.

Sumado a lo anterior es necesario considerar una distancia de protección adicional para el dado caso en el que se llegue a presentar la rotura total de la línea, esta distancia se establece a partir del radio de impacto del evento llamarada, localizando los puntos sobre la línea que no presentan afectación a la escuela. A partir del análisis se estableció que esta distancia es de 316 metros e inicia en la coordenada 74°31'33,57146"O - 6°9'8,22442"N y finaliza en la coordenada 74°32'28,98241"O - 6°9'9,70756"N, esta distancia cubre la longitud aferente descrita anteriormente.

Dado que la alternativa de mitigación tiene como fin eliminar la afectación indirecta y el impacto directo sobre la escuela a continuación se presentan dos alternativas de trazados propuestos, que aseguran que las rutas de derrame en caso de una perdida de contención tendrán dirección opuesta a la ubicación de la escuela.

Ilustración 80. Opción 1 realineamiento entre PK 43 + 695 a PK 44 + 205.

Ilustración 91. Opción 2 realineamiento entre PK 43 + 625 a PK 44 + 460 de 752 metros.

En la ilustración 12 se presenta las dos alternativas propuestas y la distancia de separación del trazado actual de tubería, así como de la escuela.

Ilustración 102. Alternativas propuestas para mitigación de riesgo

7 Conclusiones

Del análisis realizado se tienen las siguientes conclusiones.

- El máximo número de fatalidades a generar es de ocho (8) y es generado por una llamarada en la ruta de derrame en tierra la cual, en el dado caso de que se presente el evento, ingresaría a las instalaciones de la escuela.
- La mayor susceptibilidad de la línea es por la amenaza de corrosión externa, especialmente en algunos sectores en los cuales el recubrimiento se encuentra en mal estado, esta condición junto con las anomalías encontradas en la línea, impactaron la frecuencia de falla establecida para el sector.
- En términos generales el nivel de riesgo se encuentra en la zona tolerable, muy cercana al límite inferior ALARP.

8 Recomendaciones.

- Es recomendable realizar actividades tendientes a disminuir el volumen disponible a desplazar en la ruta de derrame, para ello se puede realizar la profundización de la línea lo cual además de disminuir el efecto de los eventos por JET FIRE y FLASH FIRE disminuye la cantidad de producto disponible a desplazar al favorecer la evaporación de este.

- A partir del análisis realizado se tiene que la distancia a realinear y profundizar es:

  Opción 1 realineamiento entre PK 43 + 695 a PK 44 + 205 con una longitud de 516 metros
  Opción 2 realineamiento entre PK 43 + 625 a PK 44 + 460 con una longitud de 752 metros.

- Con ambas alternativas se elimina la afectación indirecta y la afectación directa, sobre la escuela, estas distancias cubren todos los posibles eventos que se pueden presentar y tienen afectación sobre la escuela en mención.

- Se pueden construir cortacorrientes los cuales eviten que la ruta de derrame ingrese o haga su tránsito en las cercanías de la escuela, junto con la construcción de estructuras de retención las cuales detengan el avance del producto mientras se realizan actividades de recuperación de este.

- Es recomendable rehabilitar el recubrimiento en aquellos sectores en los cuales se encuentra en mal estado, junto con la atención de aquellas anomalías que sean críticas.

9 Referencias

[1] Fingas, M. (2011). Oil Spill Science and Technology. Kidlington, UK: Gulf Professional Publishing Elsevier.

[2] D.O.T. Cornell Law School. Obtenido de 49 CFR 192.905 - How does an operator identify a high consequence area?: https://www.law.cornell.edu/cfr/text/49/192.905 Accesado 15 Enero, 2019

[3] D.O.T. Cornell Law School. Obtenido de 49 CFR 192.5 - Class locations: https://www.law.cornell.edu/cfr/text/49/192.5 Accesado 15 Enero, 2019

[4] Yellow Book. TNO. Methods for the calculation of Physical Effects Due to releases of hazardous materials (liquids and gases) – Third edition Second revised print 2005

[5] Purple Book. TNO. Guidelines for quantitative risk assessment – First edition 1999/2005

Referencias no citadas:

[6] ASME.ASME B31.8S. Managing System Integrity of Gas.New York, USA.2005.

[7] ASME.PIPELINE TRANSPORTATION SYSTEMS FOR LIQUID HYDROCARBONS AND OTHER LIQUIDS ASME B31.4-2002. New York, USA.2002.

[8] API.API Std 1160. Managing System Integrity for Hazardous Liquid Pipelines. Washington, USA.2008.132 Pags.

548036 Best Practice in MOC and Pssr Automation for Opex

Management of Change (MOC) and Pre Start-up Safety Review (PSSR) are still the most challenging elements of OSHA's Process Safety Management (PSM) standard and the EPA's Risk Management Program (RMP) rule. Now MOC and PSSR are requirements of the US Department of Interior's Bureau of Safety and Environmental Enforcement's Safety and Environmental Management Systems (SEMS). Even though the PSM standard has been around since 1992 and the industry has been managing change for several decades, we still can get it wrong, sometimes with disastrous results. The mindful or diligent efforts of many working in concert are necessary to ensure that change is identified, analyzed and executed in a quality way. But just when we get the process on paper and get the workforce up-to-speed, we have employee turnover, complacency sets in and sometimes there is regulatory change.

Many high-profile companies have addressed this challenge by creating and automating standard processes across their manufacturing or production facilities. MOC and PSSR are two of these major processes that have been successfully automated or **digitized**. Technology has been used to **automate** each step of the MOC workflow, enabling the collection of change ideas, formalizing the analysis of the ideas, and documenting the approval and execution of the change. Reports and dashboards can provide visibility of the health or current state of your MOC and PSSR processes. The use of **software** can enable the standardization of best practices.

On a well-meaning quest to standardize on a MOC/PSSR software solution and process, many companies are stifling creativity and innovation. Operational Excellence cannot be achieved in such ways. This paper discusses how the chemical process industry has defined the MOC and PSSR best management practices and how they should be automated for operational excellence.

Chemical Process Safety Requirements in the Fire Code

Ronald Melucci
Engineering Planning and Management, Inc.
959 Concord Street
Framingham, MA 01701
rbm@epm-inc.com

Prepared for Presentation at
American Institute of Chemical Engineers
2019 Spring Meeting and 15th Global Congress on Process Safety
New Orleans, LA
March 31-April 3, 2019

Ronald Melucci
Engineering Planning and Management, Inc.
959 Concord Street
Framingham, MA 01701
rbm@epm-inc.com
Keywords: NFPA Fire Code, International Fire Code, Process Safety

Chemical Process Safety Requirements of the Fire Code

Abstract

When most Chemical Process Safety professionals think about regulation and enforcement, they think about the applicable Occupational Safety and Health Administration (OSHA) and Environmental Protection Agency (EPA) regulations. However, the majority of US states adopt NFPA 1 or the International Fire Code (IFC) as their state fire code which contain requirements for chemical process safety. These requirements are applicable for processes of all sizes and levels of complexity below PSM and RMP thresholds and they address both fire and non-fire related chemical hazards. These requirements are enforced by municipal or state fire officials/marshals, who are becoming more aware of these requirements and chemical process safety in general. This paper will review some of the code requirements with a focus on requirements that may be commonly overlooked in the industry.

Introduction

The OSHA Process Safety Management (PSM) and EPA Risk Management Program (RMP) regulations, enforced by OSHA and EPA, respectively, are familiar to most safety professionals who work in industries that include processing of large quantities of chemicals. However, the applicable fire code requirements related to chemical process safety are often more restrictive than the federal regulations and are generally lesser known to such safety professionals.

The two model fire codes adopted by states in the US, are: *NFPA 1, Fire Code,* published by The National Fire Protection Association (NFPA), and the *International Fire Code* (IFC) published by the International Code Council (ICC). The states that do not adopt the IFC will typically use NFPA 1 and each state may also institute their own NFPA 1 amendments. Local amendments are almost always more restrictive than the model code verbiage, and many states have legislation that prohibits fire code amendments that reduce the level of safety/standard of care intended by the model fire code.

Both of these model fire codes contain many requirements that are similar to PSM and RMP regulations and contain references to them for additional guidance. Generally speaking, the model fire codes do not regulate all hazardous materials, but focus on hazards that are acutely dangerous to people or property. The PSM and RMP regulations address these hazards as well but additionally include materials that present a risk of chronic health or environmental hazards.

However, many states and municipalities have local fire code amendments that often include lower chemical quantity thresholds, such as those of the Commonwealth of Massachusetts as described in the following sections. The enforcement of these requirements is therefore not limited to OSHA and the EPA, as is commonly believed. The requirements are also enforceable by the local (municipal) fire authorities through the fire code references to these regulations.

Fire Code Requirements

The Commonwealth of Massachusetts currently adopts NFPA 1 – 2015 edition, with amendments, as 527 CMR 1.00: *Massachusetts Comprehensive Fire Safety Code*. Since 2006, 527 CMR has included chemical process safety requirements that are more restrictive than the federal regulations. The impetus for this was several accidents that occurred involving chemical process quantities below the OSHA and EPA thresholds. These Massachusetts (MA) amendments have been added to Chapter 60 of NFPA 1 – 2015 edition and include specific reference to OSHA PSM and EPA RMP regulations. Chemical processes with quantities above the PSM/RMP thresholds are defined in MA as "Category 5" processes. However, the 527 CMR amendments have also defined processes with lower quantities, i.e. Category 1, 2, 3 and 4, each with specific and unique requirements as described herein.

The NFPA 1 committee is currently reviewing proposals to adopt requirements with a "tiered" quantity threshold approach similar to MA. This is in recognition of the fact that severe hazards often exist, and severe accidents have occurred, in chemical processes with quantities below the PSM/RMP thresholds. It is likely that such a "tiered" approach will be incorporated into NFPA 1 and subsequently adopted by many US states. A preview of the applicable NFPA 1 and 527 CMR requirements are summarized herein.

Massachusetts Amendments to NFPA 1, Section 1.12.8.34.2 introduces the requirements as: *Process or Processing of Any Hazardous Material at any Facility* and it states the following:

(1) "A permit shall be required for the process or processing of any hazardous material at any facility identified in this *Code* as Category 2 through Category 5, as described in Chapter 60.
(2) The Authority Having Jurisdiction (AHJ) may require technical assistance in accordance with *1.15* to evaluate the adequacy of a Category 3 or Category 4 facility process safety conditions, programs, procedures, and practices undertaken at the facility, but only after a notice of denial has been properly served upon the person making application."

As part of the MA amendments to NFPA 1 – 2015 edition, Section 60.8, ***Hazardous Material Process or Processing,*** was added to 527 CMR. Section 60.8 applies to both new and existing facilities that process hazardous materials, with a list of exceptions such as:

- Motor vehicle service stations
- Refrigeration systems using a refrigerant other than ammonia or LPG,
- Certain wastewater treatment operations
- Storage of hazardous materials in atmospheric vessels.

Definitions:

NFPA 1 defines "**Hazardous Materials**" as "A chemical or substance that is classified as a physical hazard material or a health hazard material, whether the chemical or substance is in usable or waste condition."

OSHA and EPA define "**Process**" as "Any use, storage, manufacturing, handling, or the on-site movement of hazardous chemicals, or combination of these activities." The MA state fire code defines "Process" as "A sequence of operations in which the sequence can be inclusive of physical operations such as heating, cooling, mixing, distilling, compressing, and pressurizing, and chemical operations, such as polymerization, oxidation, reduction, and other chemical reaction processes. The sequence can involve but is not limited to: preparation, separation, combination, purification, or any actions that cause a change in state, energy content, or chemical composition."

MA Amendments to NFPA 1 Section 60.8 categorizes hazardous material processes into five (5) categories, as follows:

Table 1: Hazardous Material Threshold Quantities in the MA Amendments to NFPA 1

Category	**Quantity (gallons)**
Category 1	< 2.5
Category 2	2.5 – 60
Category 3	60 – 300 (or Group H)
Category 4	300 and not Category 5
Category 5	≥ the threshold quantities stated in 29 CFR 1910.119

The MA Amendments also add Section 60.8.5.1 to NFPA 1, which requires that facilities operating hazardous material processes must maintain the required documents and procedures for periodic inspection and review by the head of the Fire Department to remain in compliance with the state fire code. Requirements for each process category are as follows:

Table 2: Outline of Regulatory Requirements in the MA Amendments to NFPA 1

Category	Requirements
Category 1 (Vessel Capacity ≤ 2.5 gal)	• OSHA 1910.1450 (Occupational exposure to hazardous chemicals) • OSHA 1910.1200 (Hazard communication to employees) • Policy per NFPA 1, Ch. 66 (Flammable liquids) and Ch. 67 (Flammable solids) • Requirements of NFPA 1, Section 60.1 – 60.5
Category 2 (Vessel Capacity ≤ 60 gal)	• All requirements of Category 1 • Permit required per NFPA 1, Sections 1.12.8.34.2 and 60.8.1.1
Category 3 (Vessel Capacity ≤ 300 gal OR Group H Occupancy)	• All requirements of Category 1 & 2 • Category 3 Hazard Evaluation • Requirements of NFPA 1, Section 60.6
Category 4 (Vessel Capacity ≥ 300 gals AND not Category 5)	• All requirements of Category 1-3 • Limited Safety Program (PSM "lite")
Category 5 (Vessel Capacity ≥ the threshold quantities stated in 29 CFR 1910.119)	• All requirements of Category 1-4 • Process Safety Management (PSM) Program 29 CFR 1910.119 (OSHA) • Risk Management Program (RMP) 40 CFR Part 68 (EPA)

Category 1 – Requirement Details

Category 1 processes, as stated in Table 2 above, must possess documentation of compliance with OSHA 1910.1450, *Occupational Exposure to Hazardous Chemicals in Laboratories*, OSHA 1910.1200, *Hazard Communication,* maintain and implement a policy in compliance with NFPA 1, Ch. 66, *Flammable and Combustible Liquids*, and Ch. 67, *Flammable Solids*, demonstrate compliance with NFPA 1, Section 60.1.5.1, *Emergency Action Plan*, and Sections 60.5.1.4.3.2(5) through 60.5.1.4.3.2(9) and comply with permitting requirements in Sections 1.12 and 60.8.4 of the MA Amendments to NFPA 1.

OSHA 1910.1450 and 1910.1200

Compliance with OSHA 1910.1450 requires facilities to set and adhere to exposure limits, monitoring plans, hygiene plans, precautions, training, medical plans, and other plans, as found to be acceptable by the AHJ, to react to and minimize potential exposure to hazardous chemicals. OSHA 1910.1200 requires hazard communication methods for employees such as inventory documentation, hazard programs, labels, safety data sheets, and training.

NFPA Chapter 66 and 67

In addition to demonstrating compliance with OSHA standards, facilities meeting Category 1 requirements must implement and maintain a written policy in compliance NFPA 1, Ch. 66 and Ch. 67.Chapter 66 is divided into 29 sections, most of which contain multiple requirements. Section 66.6 contains the prescriptive requirements for fire and explosion prevention and risk control, and Section 66.6.4 goes on to list requirements related to a facility's hazards analysis, which is required to be reviewed to ensure that fire and explosion hazards are addressed by fire prevention, fire control, and emergency action plans. The AHJ should be consulted with for the development of the hazards analysis to ensure that the extent of the fire prevention and control is deemed acceptable or an engineering evaluation of the operation and application of fire protection and process engineering principles is soundly applied. The hazards analysis must include, but not be limited to, analysis of the following:

(1) Fire and explosion hazards
(2) Emergency relief from process vessels
(3) Applicable facility design requirements in Chapters 17, 18, 19, 28, and 29 of NFPA 30
(4) Applicable requirements for liquid handlings, transfer, and use, as covered in Chapters 17, 18, 19, 28, and 29 of NFPA 30
(5) Local conditions/exposures
(6) Emergency response capabilities of the local emergency services

Additional guidance for conducting a fire risk assessment can be found in NFPA 551, *Guide for the Evaluation of Fire Risk Assessments.* Facility management must repeat the hazards analysis when any of the following occur:

(1) Changes occur in the materials in process
(2) Changes occur in process equipment
(3) Changes occur in process control
(4) Changes occur in operating procedures or assignments

This is consistent with the requirements of OSHA 1910.119 Sections (e) and (l). Not all sections of Chapter 66 may be related to processes; however, Section 66.17 is for processing facilities. Section 66.17 includes prescriptive requirements for processing facilities such as location of process vessels and equipment (Section 66.17.4), containment, drainage, and spill control (Section 66.17.10), and management of operations hazards (Section 66.17.15). Later sections of Chapter 66 include additional requirements for dispensing, handling, transfer, and use of liquids (processes) (Section 66.18) and specific operations such as recirculating heat transfer system, vapor recovery and vapor processing system, solvent distillation units, and cooking oil storage tank systems in commercial kitchens (Section 66.19).

NFPA 1 Chapter 60

In addition to Chapters 66 and 67 of NFPA 1, Category 1 processes must also adhere to the requirements of 60.1 – 60.5 of NFPA 1 and the MA Amendments to NFPA 1. Depending on whether the quantities of materials at the facility exceed the MAQ or not, determines which requirements of Section 60.1 are applicable (Section 60.1.3).

Section 60.1.5 of NFPA 1 requires that an emergency action plan, consistent with the available equipment and personnel, shall be established to respond to fire and other emergencies in accordance with requirements set forth in NFPA 1. Section 10.8 of NFPA 1 includes additional requirements and guidance on generating emergency action plans, such as what must be included in the plans, purpose, responsibilities, etc. Section 60.4 of NFPA 1 contains requirements related to permissible storage and use locations and sets prescriptive requirements on the maximum allowable quantities for each individual control areas based on the area's occupancy type.

Section 60.5 contains the fundamental requirements for hazardous materials, including requirements for personnel training, ignition sources, systems, equipment, and processes (Section 60.5.1.4, 60.5.1.5, and 60.5.1.6 respectively). In Massachusetts, Section 60.5 includes multiple requirements that come from both NFPA 1 and the MA Amendments to NFPA 1. Despite Category 1 processes having a vessel capacity much less than Category 2-5 processes (Category 1 processes have a vessel capacity ≤ 2.5 gallons), there are numerous NFPA and state fire code requirements that must be followed. Therefore, the importance of fire protection engineers and their understanding of the NFPA and state fire codes cannot be understated for processing facilities. Each subsequent category process contains additional fire code fire protection/chemical process safety requirements of increasing complexity.

Category 2: Requirement Details

Category 2 processes must comply with all Category 1 requirements and additionally submit for a permit to the AHJ, as required per 1.12.8.34.2 and 60.8.1.1 of MA Amendments to NFPA 1. Permits submitted to the AHJ must comply with the requirements of Section 1.12 of NFPA 1 – 2015 edition. The nature of the operations and materials at the facility determine the permit requirements. Depending on the nature of the process, the AHJ may require technical assistance to evaluate the adequacy of facility process safety conditions, programs, procedures, and practices undertaken at the facility, but only after a notice of denial has been properly served upon the person making application.

Category 3: Requirement Details

While meeting all the requirements of Category 1 and 2 processes, Category 3 processes must also provide the following documentation and evaluations:

- Compliance with the requirements of Category 1 and 2 processes;
- Category 3 hazard evaluation; - per Section 60.8.2
- Implementation of process safety controls to mitigate hazards of normal and abnormal conditions identified in the hazard evaluation;
- Maintain documentation for review for a minimum of two (2) years;
- Third-party review (if required by the fire department/AHJ);
- Post-Incident Analysis per 60.8.6 (to be initiated within 48 hours of incident);

MA Amendment to NFPA 1, Section 60.8.2.2 provides requirements and details for the hazard evaluation that is required for Category 3 processes. **Hazard evaluations** are meant to identify hazards, including adjacent vessels that contain hazardous materials, and determine the required preventative, protective, and safety control measures in conformance with **r**ecognized **a**nd **g**enerally **a**ccepted **g**ood **e**ngineering (and safe work) **p**ractices (RAGAGEP) associated with a particular process or condition. Section 60.8.2.5 requires that a person with *specialized knowledge* beyond that of an average person, about risk assessment, process hazard analysis, and/or process safety management principles, for the process or processes being evaluated *conduct the necessary evaluations*. Section 60.8.2.5 further emphasizes the importance of both chemical process safety specialists and fire protection engineers for analysis of hazardous processing facilities.

In addition to generating a hazard evaluation, Category 3 processing facilities (vessel capacity ≤ 300 gal OR Group H Occupancy) must also meet the requirements of NFPA 1, Section 60.6, *Emergency Action Planning, Fire Risk Control and Chemical Hazard Requirements for Industrial Processes.* Where quantities of materials in use require compliance with Protection Level 1, Protection Level 2, Protection Level 3, or Protection Level 4 based on materials exceeding the Maximum Allowable Quantity (MAQ) in the following categories shall comply with the requirements of Chapter 7 of NFPA 400, *Hazardous Materials Code*:

(1) Unpackaged organic peroxide formulations that are capable of explosive decomposition in their unpackaged state
(2) Oxidizer Class 3 and Class 4: solids and liquids
(3) Pyrophoric solids, liquids, and gases
(4) Unstable reactive Class 3 and Class 4: solids, liquids, and gases
(5) Highly toxic solids, liquids, and gases
(6) Water-reactive liquids, Class 3

NFPA 400, Chapter 7, *Emergency Planning, Fire Risk Control, and Chemical Hazard Requirements for Industrial Processes,* contains requirements and guidance on topics such as:

- Process review and plan preparation
- Operating and maintenance procedures
- Safety reviews
- Incident investigation plans

Section 7.2.3 contains specific requirements for Emergency Action Planning that facilities must comply with. Some requirements of Section 7.2.3 include:

- Procedures for notifying and coordinating with the fire department and other applicable agencies
- Procedures and schedules for conducting drills
- Alternative measures for occupant safety
- Maintenance of fire protection equipment

Lastly, in lieu of complying with Chapter 60 in its entirety, occupancies containing high hazard Level 1 to high hazard Level 5 contents shall be permitted to comply with Chapter 10 of NFPA 400, as stated in NFPA 1, Section 60.7. MA Amendments to NFPA 1 further states that complying with NFPA 1, Section 60.7, is subject to an independent review in accordance with Section 1.15 and a copy, including its recommendations, shall be submitted to the Building Official.

Category 4: Requirement Details

Category 4 processing facilities (vessel capacity ≥ 300 gal) must meet:

All the requirements of Category 1, 2, and 3 processes

And also create and maintain a **Limited Safety Program (PSM "lite")**. The Limited Safety Program documents evaluation, policy and/or required procedures to ensure compliance with the following requirements of the MA Amendment to NFPA 1:

(1) Process information including, but not limited to, MSDS for the chemicals and products being processed, process chemistry, piping and instrumentation diagram, safety relief design, process control safety alarms and interlocks;

(2) Facility suitability including, but not limited to, the *Building Code* compliance, electrical hazard classification, ventilation design, fire alarm and fire protection, spill containment and control;

(3) A process hazard safety analysis including, but not limited to, effects in the event of failure, suitable administrative and engineering controls to minimize failure and to control unanticipated releases, and emergency responses to safeguard life and property;

(4) Written procedures, including routine operating and maintenance, as well as precautionary, shut-down and emergency response measures;

(5) A written training program for operating and maintenance personnel and outside contractors whose work or activity may affect process safety;

(6) A written records management protocol, which tracks any changes, including but not limited to changes to chemicals, equipment, operating procedures training program. Such records shall include the date of such change and the name of the manager responsible for such change; and an internal review at a maximum every three years.

These requirements are similar to those found in OSHA 29 CFR 1910.119, *Process Safety Management of Highly Hazardous Chemicals*.

Category 5: Requirement Details

Category 5 processing facilities must provide evaluation and documentation per 29 CFR 1910.119 OSHA, *Process Safety Management of Highly Hazardous Chemicals*, and 40 CFR Part 68 EPA, *Risk Management Program Rule*, all while meeting the requirements for Categories 1-4. OSHA's *Process Safety Management* (PSM) *of Highly Hazardous Chemicals* was issued in February 1992 and contains 14 elements that focus mainly on worker safety and catastrophic loss of containment. The Process Hazard Analysis (PHA) is the "heart" of the standard and is the most time consuming. The PHA must be completed every five (5) years and *is required to be completed by someone knowledgeable with the PHA methodology*. Additionally, compliance audits for the standard must be completed every three (3) years and may be completed in-house or by a third party (preferred). EPA's *Risk Management Program* (RMP) *Rule* consists of 12 elements and is focused on public safety and the environment. The program contains Chemical Accident Prevention Provisions (CAPP) and Accidental Release Prevention/Risk Management Plan Rule. The RMP must be submitted to the EPA every fire (5) years or if there are any changes.

Conclusion

Some fire code requirements for chemical process safety apply to processes with quantities much lower than the PSM/RPM thresholds. Failure to comply with these requirements can result in fire department citations, fines, remedial work and in extreme cases, shutdown of operations. Between 2013 and 2018, 214 OSHA PSM violations were recorded in Massachusetts alone, which resulted in over $485,000 in penalties. There are far more non-PSM facilities than PSM facilities in MA, so the magnitude of the problem can be seen by simple extrapolation. Failing to comply with the fire code and the federal requirements can also result in significant property loss, injuries and fatalities with much higher costs than the violation penalties. Compliance with the local fire code and federal regulations should be determined by experienced chemical process safety specialists and fire protection engineers to minimize the risk of significant loss.

Fire protection engineering and science is its own discipline and chemical process safety is its own discipline usually practiced by the chemical engineering field. Finding ways to create awareness and bridge the gap between the two disciplines, and sharing chemical process safety management concepts with those in the fire protection arena, will likely improve the oversight, understanding and enforcement of process safety resulting in better overall safety. There may be a way of joining these disciplines into one overarching safety category and therefore giving a better sense of how interrelated they really are.

546769 Poly-Butadiene Formation in Olefins Debutanizer Condenser

Abstract Title : Poly-butadiene formation in Olefins Debutanizer condenser

Discipline : Process Safety

Topic : Poly-butadiene formation in Debutanizer condenser

Author Name : Ali Bazroun

Participating team: Fahad AlShammari, and Eldridge Monette, Bertrand, Da, Cruz, Fahad AlSaad, Abdullah Al-Ismail

ABSTRACT

In one of Sabic olefins plant Plant, the condenser of the debutanizer tower was fouling. This was evident through the pressure increase in the tower and the cooling water control valve reaching the maximum opening (69%) limited by the mechanical stopper. The condenser process service is butadiene, which has the tendency of polymerizing and formation of popcorn, which could plug relief piping or accumulate in the condenser until the vessel shell is over stressed causing loss of containment and leading to fire or explosion.

The risk assessment team evaluated that the risk as significant, meaning that it could result into fatality and multiple lost time injury with a likelihood of 10E-3. With the propagating shell deformation, the likelihood could increase to L3 (10E-2) over a longer period. The risk could become C2-L3 and remain Significant. To ensure effective risk mitigation, a major plant modification was executed by constructing a new process line that bypasses the debutanizer tower and sends the material to the flare. The project execution was remarkable as it was completed in about a week. 15 tons of C4 is being flared per. This, however, allowed the olefins unit to mitigate the risk effectively without any interruption of the production. The change was not done before in the local industry, and it reflected how effective measures can be taken to mitigate risk through strong leadership and excellent engineering. Interim mitigations, moreover, were in place until the modification was completed: for example, inhibitor injection in the exchanger to inhibit the popcorn formation as soon as possible, and an emergency procedure in case loss of containment. Without the effective mitigation, Sabic operation team would have had to shutdown the olefins unit for over 2 months. This olefins unit feeds two poly ethylene that would shutdown consequently, causing significant losses over the mentioned period.

After isolating the exchanger, borescope test proved the formation of polymer. A root cause analysis was initiated, along with cleaning the tower and fabricating a new exchanger. The details of the RCA will be shared in the full paper.

Industrial process plant safety by collaboration of a Safety Instrumented System and a Dynamic Simulator

Tadaaki Ando
Yokogawa Corporation of America
12530 West Airport Blvd, Sugar Land, TX 77478, USA
Tadaaki.Ando@us.yokogawa.com
(Tadaaki.Andou@jp.yokogawa.com)

Yasuhiko Yamashiro
Yokogawa Electric Corporation
2-9-32 Nakacho, Musashino-shi, Tokyo, 180-8750 Japan
Yasuhiko.Yamashiro@jp.yokogawa.com

Prepared for Presentation at
American Institute of Chemical Engineers
2019 Spring Meeting and 15th Global Congress on Process Safety
New Orleans, LA
March 31 – April 3, 2019

Industrial process plant safety by collaboration of a Safety Instrumented System and a Dynamic Simulator

Tadaaki Ando
Yokogawa Corporation of America
12530 West Airport Blvd, Sugar Land, TX 77478, USA
Tadaaki.Ando@us.yokogawa.com
(Tadaaki.Andou@jp.yokogawa.com)

Yasuhiko Yamashiro
Yokogawa Electric Corporation

Keywords: Industrial process plant, SIS (Safety Instrumented System), Dynamic Simulator

Abstract

Safer HMI design for operators' quick and appropriate response is very important. Allowed maximum delay for operator's action from the alarm is not same all the cases because it is depends on many factors, i.e. ambient temperature, density of material, O2, and etc. So design of SIS includes HMI should be based "worst case". However, the 'worst case' is difficult to be found.

A "Dynamic Plant Simulator" can precisely simulate behaviors of the real process by visualizing how a hazard is diffused at various stages of a chemical reaction. And combination of DCS/SIS/HMI simulator with "Dynamic Plant Simulator", safety logic design is optimized about trip point and alarm threshold, and HMI design is modified safer manner.

The effective result of application of "cyber model by Dynamic Process Simulator" to a vinyl acetate monomer (VAM) plant and its value for operator training will be discussed.

1 Introduction

Industrial process plants in Japan have been seeking for the highest productivity and optimization of their processes, same with any process plant all over the world. These plants were renovated and expanded as time passed for the purpose, the plants themselves became complicated and troublesome as a result. Few years ago, severe process plant incidents occurred in Japan. The most probable causes of these severe incidents are out of "incorrect judgement and operation" and "shortage of prediction for technical aspects" during exceptional manual work that has been done by incidents. Understanding the hazardous situation by risk analysis and designing safety measures utilizing a SIS (Safety Instrumented System) are important and effective for resolving "incorrect judgement and operation" and "shortage of prediction of technical aspect". Safety measure includes human intervention via SIS man-machine interface which showing alarm, event, and process data.

The safety measure and its function are developed on the desk by brainstorming or so-called "hazard and operability study (HAZOP)." The verification of the safety function hardware could be established by reliability based analysis, SIL calculation in other word. However, the verification of safety logic is the issue because it has only been verified "on the desk". Especially the determination of the criteria for the emergency shutdown is the most important, but it is not possible to be confirmed with "try and error" on a running plant.

2 Safe HMI Design

Safer HMI design with considering different conditions and situations in the real process for operators' quick and appropriate response is very important. To design it, knowing the behavior of the process itself and related factors which affect the process is necessary.

For example, a reaction speed varies on the ambient temperature, density of material, O_2, catalyst, and etc., so allowed maximum delay for operator's action from the alarm is not same all the cases. So design of SIS includes HMI should consider "worst case". However, finding the worst-case scenario with how much is the allowed operator reaction is difficult because of many factors.

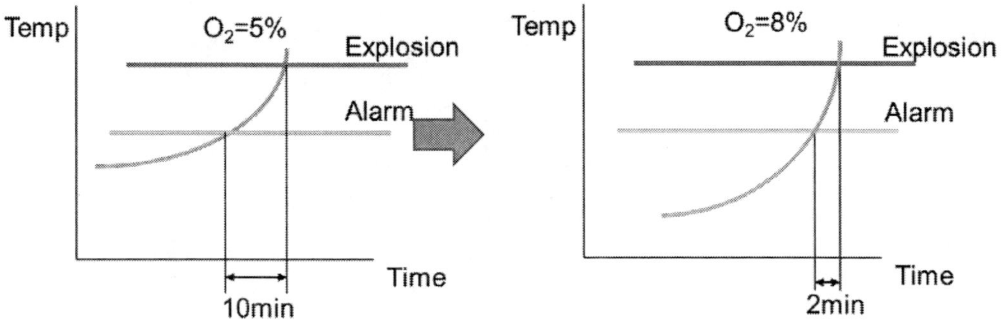

Fig.1 Example: Difference of O_2 density vs. allowed reaction time

Another example is the 'thermal runaway'. Even if feeding of material, heat, and catalyst were stopped by ESD because of a high temperature of a reactor, reaction continues due to a feed-forward loop of heating, means reaction generates heat to continue the reaction itself.

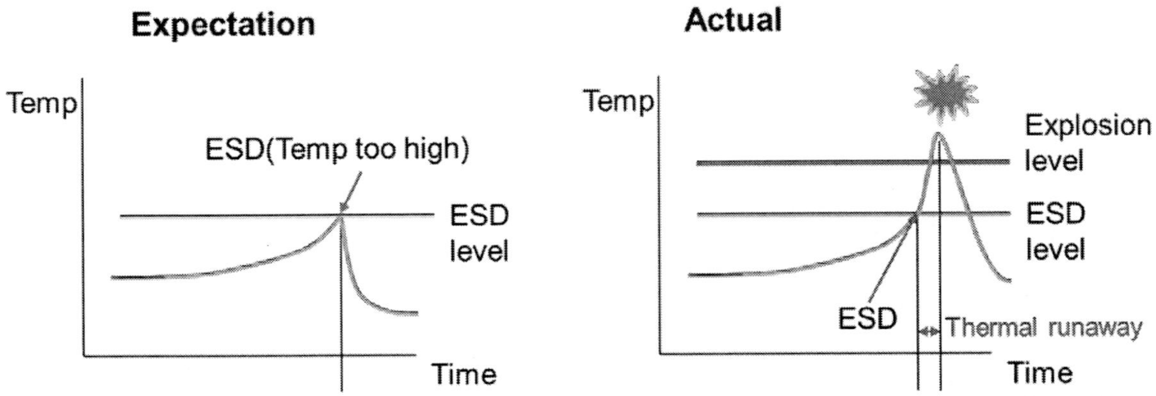

Fig.2 Even after ESD, thermal runaway leads the process to explosion

Thus, considered 'worst case' could not be a 'worst', or could be 'too much pessimistic'.

A "Dynamic Plant Simulator" can precisely simulate behaviors of the real process by visualizing how a hazard is diffused at various stages of a chemical reaction. And combination of DCS/SIS/HMI simulator with "Dynamic Plant Simulator", safety logic design is optimized about trip point and alarm threshold, and HMI design is modified safer manner.

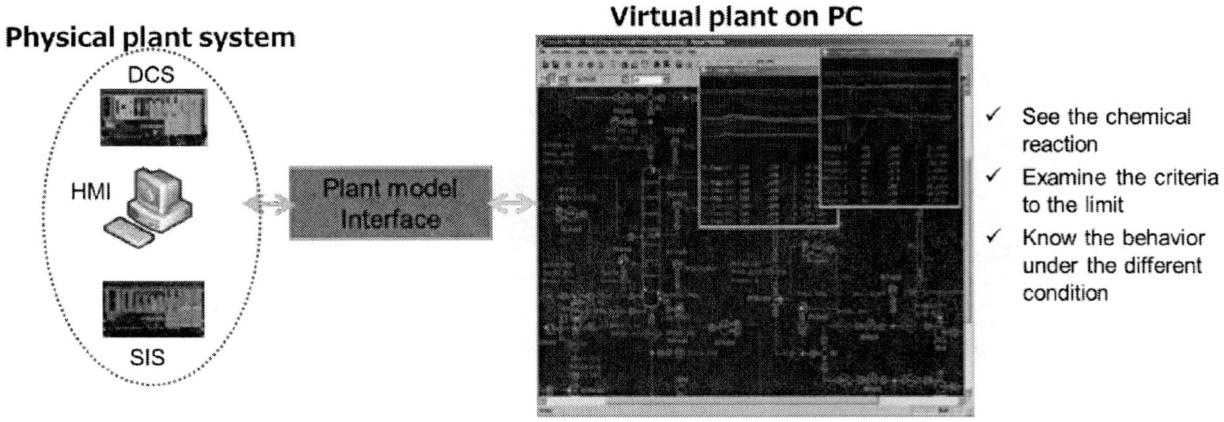

Fig.3 Dynamic Plant Simulator

3 "Dynamic Plant Simulator" at VAM (Vinyl Acetate Monomer) process

"Dynamic Plant Simulator" with "DCS/SIS/HMI simulator" was applied for several shutdown conditions of VAM plant. "Dynamic Plant Simulator" can simulate a variation of factors like ambient temperature and thermal runaway, because is based on a theory of chemical reaction. And "DCS/SIS/HMI simulator" is copying a real function and operation of DCS/SIS/HMI.

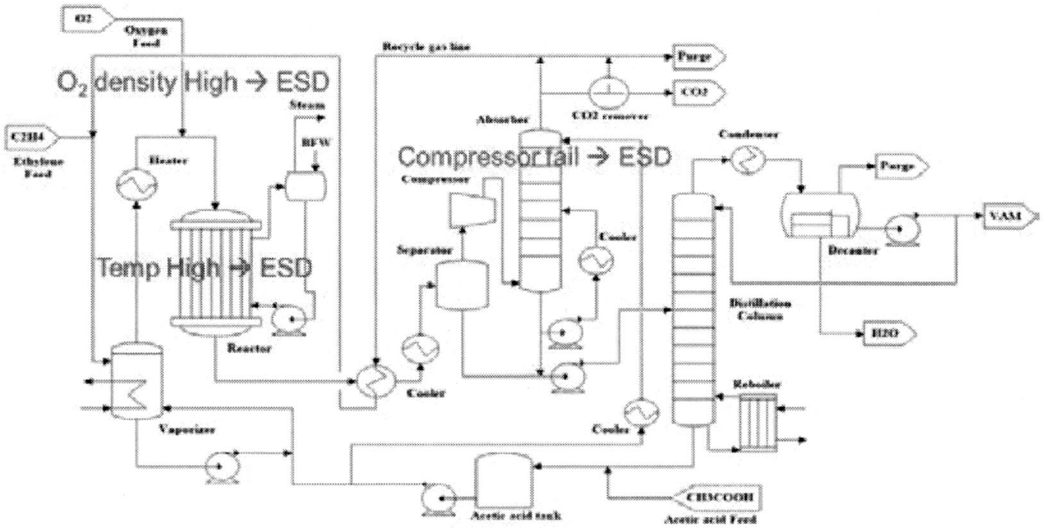

(Main Reaction) $C_2H_4 + CH_3COOH + 1/2O_2 \rightarrow CH_2{=}CHOCOCH_3 + H_2O$

(Side Reaction) $C_2H_4 + 3O_2 \rightarrow 2CO_2 + 2H_2O$

Fig.4 VAM process model

3.1 Shutdown by high temperature to avoid a damage to reactor

Reactor temperature must be monitored and controlled lower than 200 Degrees C (392 Degrees Fahrenheit) to avoid a damage of the reactor. If the temperature reached to a set point for emergency shutdown, emergency shutdown action stops feeding of Ethylene and O_2.

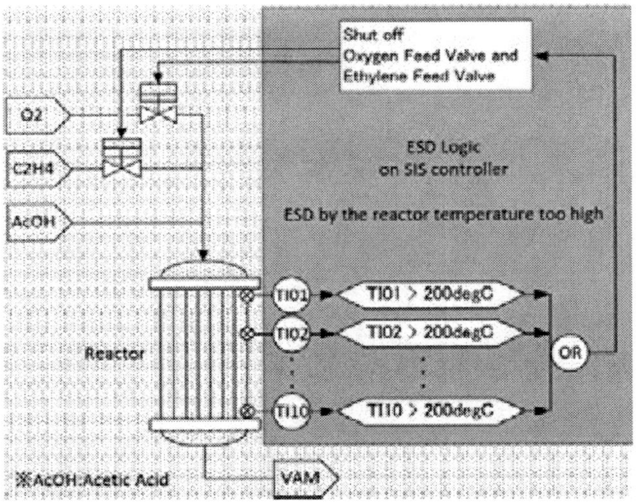

Fig.5 ESD logic for high temperature shutdown

When 190 Degrees C (374 Degrees Fahrenheit) is selected as the set point to shutdown, the result of simulation is, the reactor temperature must become over 200 Degrees C due to a runaway reaction in the reactor. It means reactor must be damaged. When 180 Degrees C (356 Degrees Fahrenheit) is selected, the simulation shows the reactor temperature can be kept lower than 200 Degrees C, so no damage to the reactor.

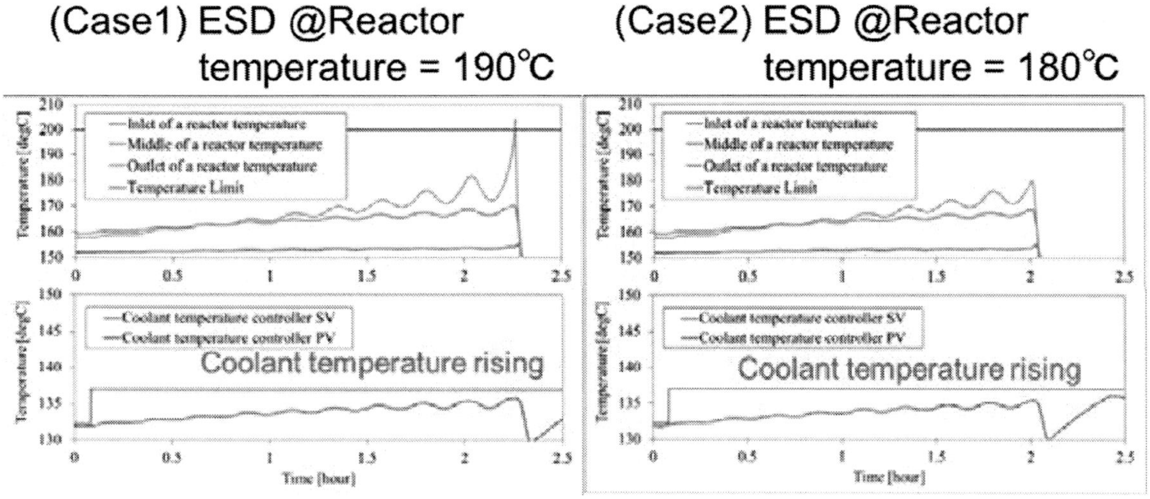

Fig.6 Simulation results

And simulation suggested the reactor temperature must be monitored in multipoint, middle of the reactor and top of the reactor. When the reaction temperature is lower than 160 Degrees C (320 Degrees Fahrenheit), temperature of a middle of the reactor is higher than other points, but when the reaction temperature is higher than 160 Degrees C, the temperature of top of the reactor is higher than other points.

3.2 Shutdown by high O_2 density to avoid explosion

If the density of O_2 in the reactor exceeds 8%, there is a risk of explosion. However, the O_2 density varies depends on other factors, i.e. feed pressure of ethylene to the reactor. Simulation has been done about O_2 density when the feed pressure of ethylene is suddenly changed. Simulation of some cases of dumping the feed pressure to lower side which increase the density of O_2 in the reactor were done.

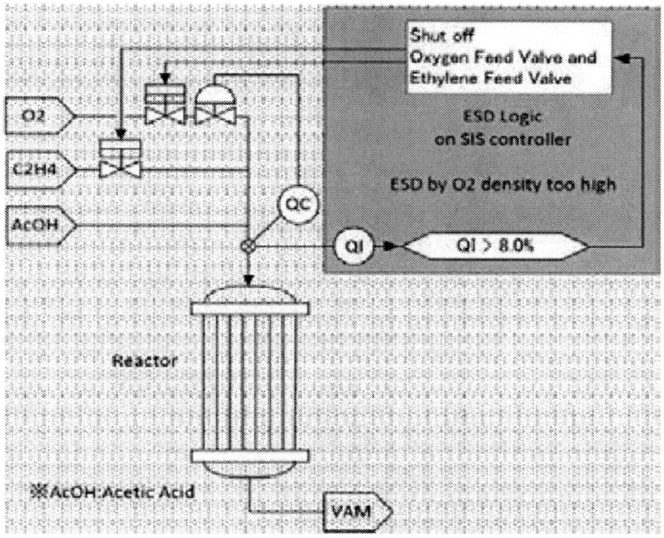

Fig.7 ESD logic for high O_2 density shutdown

When the feed pressure of ethylene dumped 10%, density of O_2 in the reactor doesn't reach to 8%. So there is no risk of explosion. However, the dumping of ethylene feed pressure is 20%, O_2 density will be higher than 8%, then an emergency shutdown must be activated. Shutdown does affect to business.

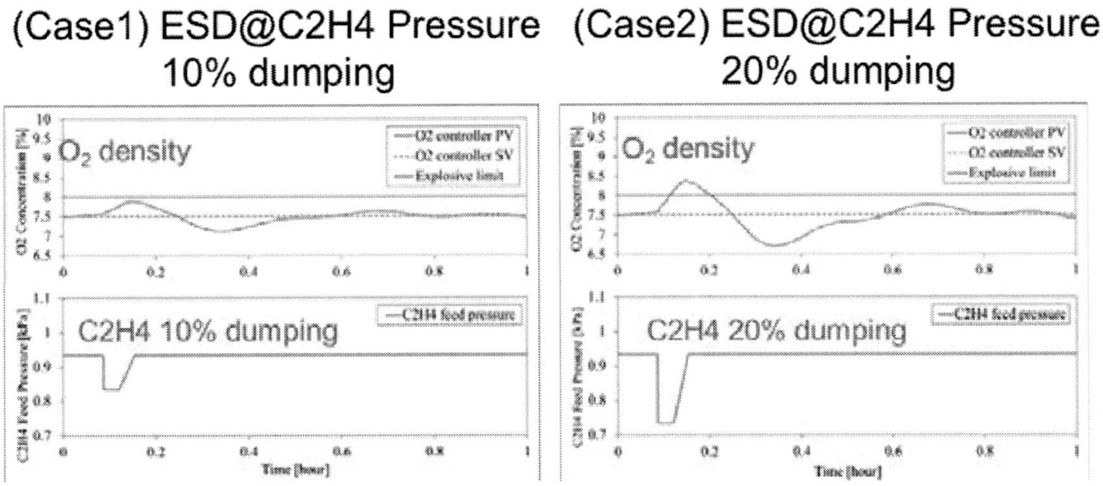

Fig.8 Simulation results

If there is a possibility of dumping ethylene feeding pressure, feeding O_2 in normal operation should be lower to avoid the risk of reactor explosion. On the other hand, lower O_2 density leads lower reaction speed, less production. Finding a best balance with safety and productivity is the key.

Using dynamic process simulator, many factors of process can be manipulated not only within normal operation range, but also close to abnormal, emergency situation or more un-realistic range, and results can be confirmed. This type of "try-and-error" approach is not possible on the real process, but possible in the "cyber" world to find a best balance of setting for maximize production with keeping or improving safety.

4 "Dynamic Plant Simulator" for Operator training

The training of operators is important in many reasons. Especially training for response to critical / emergency situation is indispensable, because operator rarely or never face such a situation and consequent secondary action to DCS/SIS or response to alarm flooding in a real plant. Therefore operator are impossible to familiarize with their required response in right procedure to critical / emergency situation. If operators are not capable or not knowing how to response such a situation, risk must propagate real hazard or accident. Furthermore, operators need to re-start the process after confirming the cause of emergency doesn't exist anymore and checking all the conditions of materials are steady, status of equipment are ready to re-start. This procedures are also not normal for operators.

"Dynamic Plant Simulator" realizes operator training with "equal to real" environment. Operators will be trained for correct and timely manner response until proficient. And operators will have more understanding for process itself by playing on the simulator.

5 Conclusion

The effectiveness of "Dynamic Process Simulator" for safer design of DCS/SIS/HMI is confirmed on VAM plant. The best balance of safety and productivity can be found and optimization of safety action is created by simulation. The created new design of DCS/SIS/HMI is possible to use an actual system directly. And operator training with "Dynamic Process Simulator" will increase a total safety of the plant.

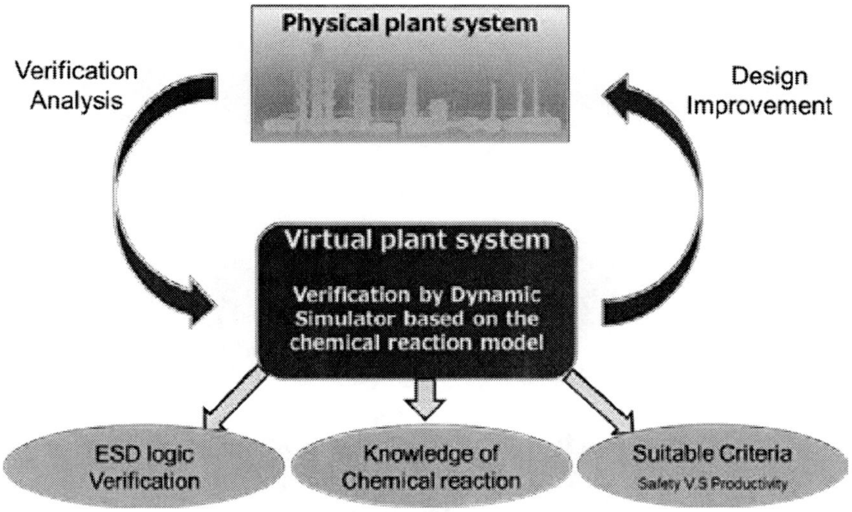

Fig.9:Virtual plant system realizes well balanced plant operation

Analysis of Chemical Effluent Discharge through the Emergency Relief System using Air-Water Properties

Basar Ozar
Jensen Hughes
600 W Fulton St. Suite 500
Chicago, IL 60661
bozar@jensenhughes.com

Jens Conzen
Jensen Hughes
600 W Fulton St. Suite 500
Chicago, IL 60661
jconzen@jensenhughes.com

Damian D. Stefanczyk
Jensen Hughes
3610 Commerce Dr. Suite 817
Baltimore, MD 21227
dstefanczyk@jensenhughes.com

Prepared for Presentation at
American Institute of Chemical Engineers
2019 Spring Meeting and 15th Global Congress on Process Safety
New Orleans, LA
March 31 – April 3, 2019

Analysis of Chemical Effluent Discharge through the Emergency Relief System using Air-Water Properties

Basar Ozar
Jensen Hughes
600 W Fulton St. Suite 500
Chicago, IL 60661
bozar@jensenhughes.com

Jens Conzen
Jensen Hughes
600 W Fulton St. Suite 500
Chicago, IL 60661
jconzen@jensenhughes.com

Damian D. Stefanczyk
Jensen Hughes
3610 Commerce Dr. Suite 817
Baltimore, MD 21227
dstefanczyk@jensenhughes.com

Keywords: two-phase flow, hydrodynamic load, emergency relief system, chemical effluent

Abstract

A sudden opening of a pressure relief device (typically a pressure relief valve or a rupture disk) during an overpressure event will cause the chemical effluent to accelerate through the emergency relief system (ERS) piping. The rapid change of momentum of the effluent creates transient hydrodynamic loads on the piping. ASME B31 in the United States and European Directive 2014/68/EU requires these transient loads to be considered along with other applicable loads (i.e., dead weight, thermal) in the system design. Additionally, the pressure distribution along the piping may be of importance, particularly when other vessels are connected to the same ERS. The back pressures generated on these vessels should be evaluated and compared with the vessels' MAWP (maximum allowable working pressure) as well as the set pressure of any isolation devices. As such, a detailed time history of the effluent pressure and hydrodynamic loads are required to perform a proper evaluation. The effluent may be composed of proprietary chemical mixtures containing various materials. Therefore, the exact material properties of the effluent during the transient might not always be known. In this study, a previously published methodology is used for calculating the dynamics of the effluent and the resultant hydrodynamic loads for a typical ERS by using the RELAP5 (Reactor Excursion and Leak Analysis Program) software. Air-water properties are assumed for the effluent. Key output parameters, such as pressure drop and hydrodynamic loads, are then compared with established hand calculation methods using the actual effluent properties. Given the mechanical nature of the hydrodynamic load event, it was found that the use of air-water properties provides a reasonable representation of the important characteristics of the actual effluent discharge in mechanical design space.

1 Introduction

Pressure build up in a system may occur as a result of a process upset, instrument failure or fire. These upset scenarios can result in either, or both, vaporization due to exothermic runaway reactions or generation of gases due to the decomposition process, which are of particular interest in terms of system and public safety. The extent of the potential hazards caused by such incidents are generally determined by the rate and amount of the energy being generated, the chemical substance makeup, and the quantity of the effluent.

The vessels or systems, which contain such chemicals (e.g. chemical reactors or storage tanks), are typically protected by an ERS that prevents from over-pressurization of the equipment as a consequence of the vaporization of chemicals and/or generation of gases. The process for designing of an ERS was previously discussed by Conzen and Wood (Ref. [1]) and Ozar (Ref. [2]). The design process consists of a process hazard analysis (PHA), defining credible and limiting upset scenarios, adiabatic calorimetry testing, vent sizing calculations, fluid transient analysis, pipe stress analysis, and a dispersion analysis. The central task in the design process is the fluid transient analysis (Refs. [1] and [2]), which is the main focus of this study.

The ERS is typically equipped with either a rupture disk or a pressure relief valve (PRV). Some systems may have both pieces of equipment. The sudden opening of the relief device during an overpressure event will typically result in a two-phase discharge and cause the chemical effluent to accelerate through the ERS piping. The rapid change of momentum of the effluent creates transient hydrodynamic loads on the piping. ASME B31 (Ref. [3]) in the United States and European Directive 2014/68/EU (Ref. [4]) requires these transient loads to be considered along with other applicable loads (i.e., dead weight, thermal) in the system design. Additionally, the pressure distribution along the piping may be of importance, particularly when other vessels are connected to the same ERS. The back pressures generated on these vessels should be evaluated and compared with the vessels' MAWP as well as the set pressure of any isolation devices. As such, a detailed time history of the effluent pressure and hydrodynamic loads are required to perform a proper evaluation.

The two-phase behavior of the chemical effluent has been studied extensively. Following earlier developments by Epstein et al. (Ref. [5]) and Grolmes and Leung (Ref. [6]), a simplified correlation was suggested by Leung (Ref. [7]) to predict flashing choked flow in a frictionless duct or nozzle. The correlating parameter, ω, was based on an isenthalpic (constant enthalpy) assumption.

Leung and Grolmes (Ref. [8]) integrated the momentum equation for homogenous two-phase flow in a horizontal duct using "ω" as a closure relationship for two-phase mixture specific density. Leung and Epstein (Ref. [9]) extended the derivation to an inclined duct configuration. Leung (Ref. [10]) provided the derivation of the reaction forces during a two-phase discharge from ERS piping.

Computational methods have also been used by various industries to investigate two-phase flow behavior in pipes and the hydrodynamic loads due to a sudden acceleration or deceleration of the fluid and the propagation of pressure waves. RELAP5 is a tool for analyzing small-break loss of coolant accidents and system transients in nuclear power plants. The program has the capability to model one dimensional single and two-phase fluid flow and heat transfer problems. The software has undergone extensive review and validation by the US Nuclear Regulatory Commission (Ref. [11]). It has also been benchmarked (Refs. [12], [13] and [14]) for predicting pressure wave propagation and hydrodynamic loads in piping systems. Refs. [15] and [16] provide a discussion about some of the nuclear power plant applications. However, RELAP5 is limited to non-condensable gas-water and steam-water two-phase flow applications.

This study provides a discussion about using the analytical methods developed by Leung and Grolmes (Ref. [8]) and Leung (Ref. [10]) to demonstrate the similarity between the two-phase mixture of common industrial chemical effluents and air/steam-water two-phase flow. Computational and analytical calculations for three different discharge scenarios are also provided. The comparisons between the computational and analytical results are used to demonstrate the applicability of using air/steam-water properties to simulate chemical effluent discharge to reasonably predict the hydrodynamic loads and pressure distribution in the ERS following the actuation of a pressure relief device.

2 Theoretical Background

The governing steady-state conservation equations for one-dimensional homogenous two-phase flow in a horizontal duct with a constant flow area are (Ref. [8]):

Mass

$$G = \text{constant} \tag{1}$$

Momentum

$$vdP + G^2\left(vdv + \frac{4fv^2dL}{2D}\right) = 0 \tag{2}$$

Energy

$$h_0 = h + \frac{G^2v^2}{2} \tag{3}$$

G, v, P, f, L, D and h are the mass flux, specific volume, pressure, two-phase friction factor, length of the duct, hydraulic diameter of the duct, and specific enthalpy, respectively. The specific enthalpy and specific volume are expressed in terms of quality (x) and phasic properties, i.e., $h_0 = h_{f0} + x_o h_{fg0}$, $h = h_f + xh_{fg}$, and $v = v_f + xv_{fg}$. Subscript 0 denotes stagnation conditions and fg the difference between vapor and liquid properties.

Epstein et al. (Ref. [5]) and Grolmes and Leung (Ref. [6]) suggested a simple relation between the two-phase specific volume and pressure as given by:

$$\frac{v}{v_0} - 1 = \omega\left(\frac{P_0}{P} - 1\right), \tag{4}$$

where ω is a non-dimensional parameter. Leung (Ref. [7]) correlated ω for flashing choked flow in a frictionless duct or nozzle. The correlating parameter was shown to be in the following simplified form:

$$\omega = \frac{x_0 v_{fg0}}{v_0} + \frac{C_{f0} T_0 P_0}{v_0} \left(\frac{v_{fg0}}{h_{fg0}}\right)^2 \quad (5)$$

C_{f0} is the heat capacity of the liquid. The last term on the right-hand side is zero for non-flashing gas-liquid two-phase flow and the non-dimensional parameter, ω, is provided by:

$$\omega = \frac{x_0 v_{fg0}}{v_0} \quad (6)$$

Leung and Grolmes (Ref. [8]) considered a two-phase horizontal discharge from a large reservoir (Figure 1). They integrated Equation 2 over the length, L, by substituting Equations 4 and 5 and using the isenthalpic (adiabatic and reversible) flow approximation. The momentum equations after integration yields to:

$$N \equiv 4f\frac{L}{D} = \frac{2}{G^{*2}}\left[\frac{\eta_1 - \eta_2}{1-\omega} + \frac{\omega}{(1-\omega)^2} ln\left[\frac{(1-\omega)\eta_2 + \omega}{(1-\omega)\eta_1 + \omega}\right]\right] - 2ln\left[\frac{(1-\omega)\eta_2 + \omega}{(1-\omega)\eta_1 + \omega}\left(\frac{\eta_1}{\eta_2}\right)\right] \quad (7)$$

The non-dimensional parameters are defined by:

$$\eta_1 \equiv \frac{P_1}{P_0}, \eta_2 \equiv \frac{P_2}{P_0}, \eta_3 \equiv \frac{P_3}{P_0}, \; G^* \equiv G/\sqrt{\frac{P_0}{v_0}}, N \equiv 4f\frac{L}{D} \quad (8)$$

η, $G^*$and f are the dimensionless pressure, dimensionless mass flux and two-phase friction factor, respectively. Leung and Grolmes (Ref. [8]) recommended a range between 0.003 and 0.005 for the friction factor.

G^* and η_1 are related via Bernoulli's equation for compressible fluid (Ref. [8]):

$$G^* = \frac{\sqrt{-2[\omega ln\eta_1 + (1-\eta_1)\omega - (1-\eta_1)]}}{\omega\left(\frac{1-\eta_1}{\eta_1}\right) + 1} \quad (9)$$

The exit condition can be either subcritical, in which case $\eta_2 = \eta_3$, or critical where the local choking statement would require (Leung and Grolmes, Ref. [8]):

$$G^* = \frac{\eta_2}{\sqrt{\omega}} \quad (10)$$

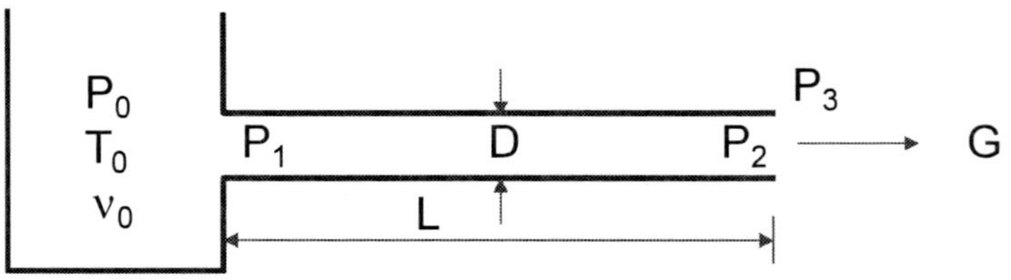

Figure 1. Two-phase horizontal discharge from a large reservoir.

With $\eta_1 = \eta_2 = \eta$, the above two equations satisfy the following transcendental equation (Leung and Grolmes, Ref. [8]):

$$\eta^2 + (\omega^2 - 2\omega)(1-\eta)^2 + 2\omega^2 ln\eta + 2\omega^2(1-\eta) = 0 \quad (11)$$

Similarly, the generalized thrust expression (T_R) is provided by Leung (Ref. [10]):

$$\frac{T_R}{P_0 A_e} = \frac{G^2}{P_0 \rho_0}\left(\frac{v_2}{v_0}\right) + \left(\frac{P_2}{P_0} - \frac{P_3}{P_0}\right) \quad (12)$$

The first term represents the fluid momentum expulsion rate and the second term represents the exit plane pressure (imbalance) force. For the case of unchoked exit, the exit pressure imbalance term vanishes, and the trust is only dependent on the exit momentum rate (Ref. [10]):

$$\frac{T_R}{P_0 A_e} = \frac{2[-\omega ln\eta_2 - (\omega-1)(1-\eta_2)]}{\omega\left(\frac{1}{\eta_2}-1\right)+1} \quad (13)$$

This approximate analysis shows that for two-phase flow in ducts (e.g. piping systems) ω is the sole independent variable and hence is the scaling parameter.

3 Transient Fluid Analysis using RELAP5

3.1 Description of the benchmark scenarios

A chemical reactor with a volume of 2.7 m^3 is considered for the benchmark scenarios. The reactor vessel is rated for 3 bars-g. The reactor is equipped with an emergency relief nozzle and a rupture disk of 41.1 mm inner diameter (DN40, 3.6 mm thickness). The nozzle is located near the top of the vessel. The piping downstream of the rupture disk has an inner diameter of 80.9 mm (4 mm thickness) and is 20 m long. It is a straight and horizontal section of piping.

Three different relief scenarios are used for the benchmarks. A summary of the postulated scenarios is provided below:

- Scenario 1: Polyurethane (PUR) and toluene diisocyanate mixture at 60 °C.
- Scenario 2: N-methyl pyrrolidone, formic acid, trimellitic anhydride and methylene diphenyl diisocyanate (MDI) at 90 °C.
- Scenario 3: Acetone under-fire scenario with vessel initially filled to 20% of its volume.

The chemical mixtures for Scenarios 1 and 2 are gas (CO_2, carbon dioxide) generating reactions. Alternatively, the scenarios are also called gassy reactions. A foamy mixture is expected inside the reactor vessel for these scenarios. Therefore, the chemical effluent discharging through the ERS is a two-phase gas-liquid mixture.

Scenario 3 is the boiling of a solvent due to a fire load applied on the vessel. The level of the boiling mixture is low, and liquid acetone is not entrained into the ERS. Only acetone vapor discharges through the ERS for this scenario.

3.2 Description of the computer model

The calculation was performed using the RELAP5/MOD3.3 (Patch 05) computer code, which is a light water reactor transient analysis code developed for the United States Nuclear Regulatory Commission (US NRC). It is suitable for the simulation of a wide variety of hydraulic and thermal transients in both nuclear and nonnuclear systems involving mixtures of steam, water, non-condensable gases and solutes under single phase and two-phase conditions (Ref. [17]).

A RELAP5 model was developed for the simulation of the venting scenarios from the reactor vessel into the ambient. The model was used to determine the hydrodynamic loads resulting from the sudden opening of the rupture disk. The mass flow rate of the two-phase effluent and the pressure distribution along the ERS piping were also calculated. Figure 2 provides an overview of the nodalization diagram of the model.

Time dependent volumes (TDVs) were used for the pressure boundary conditions and the thermo-physical condition of the fluid at those boundaries. Pipe components (P) were used to model the piping. Single junctions (SJs) connected the components such as pipes with pipes or pipes with TDVs. The valve junctions (V) was similar to the SJs but with the additional ability of the junction to open and close.

The length-to-diameter ratios (L/D) were set close to 1.0 for each pipe node and the time step was selected to be below the acoustic Courant limit. This is recommended for simulations involving propagation of acoustic waves and more generally for models where rapid fluid momentum changes are anticipated. The rupture disc was assumed to open linearly over a period of 10 msec, which is consistent with typical data (Ref. [18]). A steady-state calculation was performed for the first 5 seconds of each simulation to initialize the model. The transient portion of the simulation was initiated at 5 seconds, when each rupture disc started to open and was fully open at 5.01 sec. No heat transfer through the pipe walls was considered for the hydrodynamic fluid analysis. This is often a reasonable assumption for industrial processes due to the short duration of the relief event.

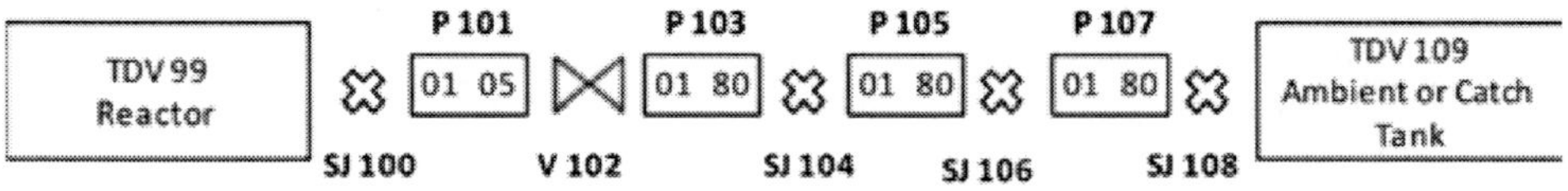

Figure 2. Nodalization diagram for the RELAP5 model.

The hydrodynamic loads (also commonly referred to as fluid force) on the modeled piping were computed through command files by using the RELAP5 output data. The generalized force equation in one-dimensional form can be resolved by a pipe segment bounded by two elbows as shown in Equation (14) (Ref. [19]). Note that the sign convention for the force is considered positive in the nominal direction of the flow as defined within the physical system.

$$F_W = -\frac{d}{dt}\int \rho A V dx \quad (14)$$

F_W is the hydrodynamic load, ρ is the fluid density, A if the inner flow area of the pipe, V is the fluid velocity, and x is the distance along the pipe. For pipe segments with an open end an additional force component needs to be considered, which is regarded as the thrust load. Examples of open ended pipe segments are pipes that connect to pressure vessels, large manifolds, or a catch tank. The total thrust load is calculated as shown in Equation 15, which is made up of the momentum flux (due to fluid momentum expulsion) and pressure terms:

$$F_T = -(P_E - P_A)A - \dot{m}_E V_E, \tag{15}$$

where P_E, P_A, $\dot{m}_E$ and V_E are the pressure at the pipe exit, pressure of the reservoir the open piping is connected to, and the mass flow rate and fluid velocity through the pipe exit, respectively.

3.3 Boundary conditions for the RELAP5 model

A summary of the upstream boundary (reactor vessel) conditions for the RELAP5 models of the three scenarios are provided in Table 1. 10% over-pressure (10% greater than the set pressure) was used for Scenarios 1 and 2, which is typical for relief system sizing of reactive chemicals. 21% over-pressure was used for Scenario 3 per API 520's (Ref. [20]) recommendation for under-fire scenarios. Two-phase gas liquid discharge was assumed for Scenario 1 and 2, where foamy mixtures inside the reactor vessel were established. Single-phase gas discharge was assumed for Scenario 3. The gas-liquid mixture of chemical effluent was simulated by using the air-water properties. The quality values of the air-water mixtures were adjusted so that the ω value of the gas-liquid mixture of the chemical and that of the air-water mixture would match. The quality and the ω value of chemical mixtures were determined based on the initial fill ratio and the homogenous two-phase mixture assumptions. A single-phase acetone vapor release scenario (Scenario 3) was simulated by using air. The determination of the choking condition at the rupture disk was made based on the analytical model, which was described in Section 2. The downstream boundary conditions were set to be at typical atmospheric conditions (air at 20 °C and 1 bar-a).

Table 1. Upstream Boundary Conditions for the RELAP5 Model

Upstream Boundary Conditions	**Scenario 1**	**Scenario 2**	**Scenario 3**
Pressure	4.4 bar-a	4.4 bar-a	4.8 bar-a
Flow regime	Two-phase gas-liquid	Two-phase gas-liquid	Single phase vapor
Simulant	Air-water	Air-water	Air
ω	0.085	0.336	Not calculated
Choking at the burst disk	No	Yes	Yes

4 Results and discussion

The time history of the mass flow rate at the rupture disk calculated using RELAP5 for Scenario 1 is provided in Figure 3. The flow quickly accelerates following the opening of the rupture disk. The effluent experiences the frictional and form losses only for the sections of the piping, which it has already occupied at a given time. Therefore, the transient mass flow rate is slightly greater than the steady-state mass flow rate while the effluent is filling up the ERS piping. Once the entire ERS piping is filled, the mass flow rate achieves a steady-state value.

Figure 4 shows the pressure-time histories downstream of the rupture disk and at the end of the ERS piping for Scenario 1. The pressure downstream of the rupture disk gradually increases to its steady-state value as the effluent fills the piping. The pressure at the exit of the piping remains near atmospheric pressure (downstream boundary condition).

Figure 5 shows the dynamic load calculated on the ERS piping. The total load and its constituents are provided. There is a spike for the wave load as the rupture disk opens. This spike is caused by the depressurization wave traveling from the rupture disk to the reactor vessel. Typically, this section of the piping is filled with gas or vapor for which the speed of sound is well known. Therefore, the initial spike is captured accurately and the speed of sound of the two-phase mixture (which is typically unknown) does not have a significant role on the results. The wave load continues to be the dominant mechanism of the load as the effluent continues to fill up the ERS piping. Once the piping is filled up, the flow momentum term replaces the wave load. The flow momentum is due to the discharge of the effluent from the ERS piping into the ambient or a catch tank and it is the first term on the right-hand side of Equation 15. The pressure force is zero since the exit of the ERS piping is at the same pressure as the ambient; there is not choking at the exit of the piping.

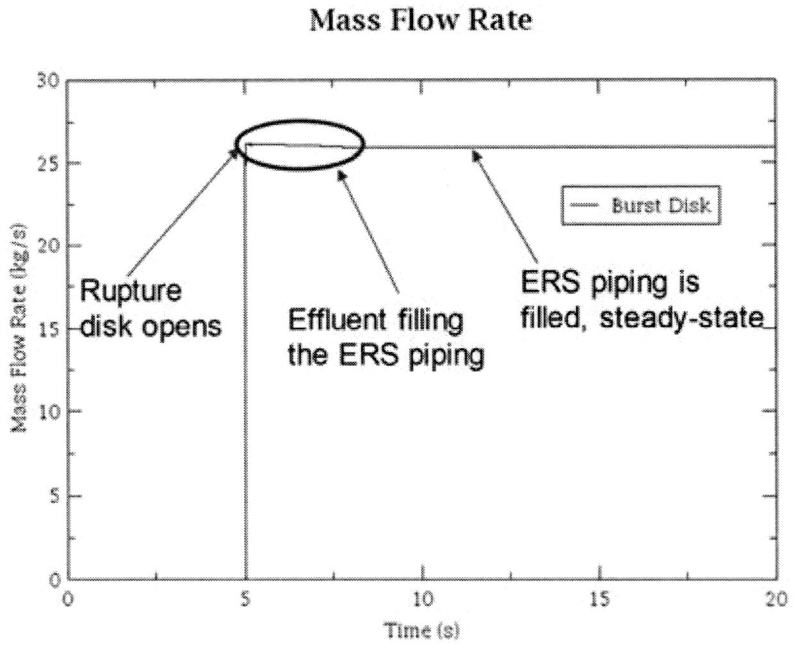

Figure 3. Mass flow rate for Scenario 1.

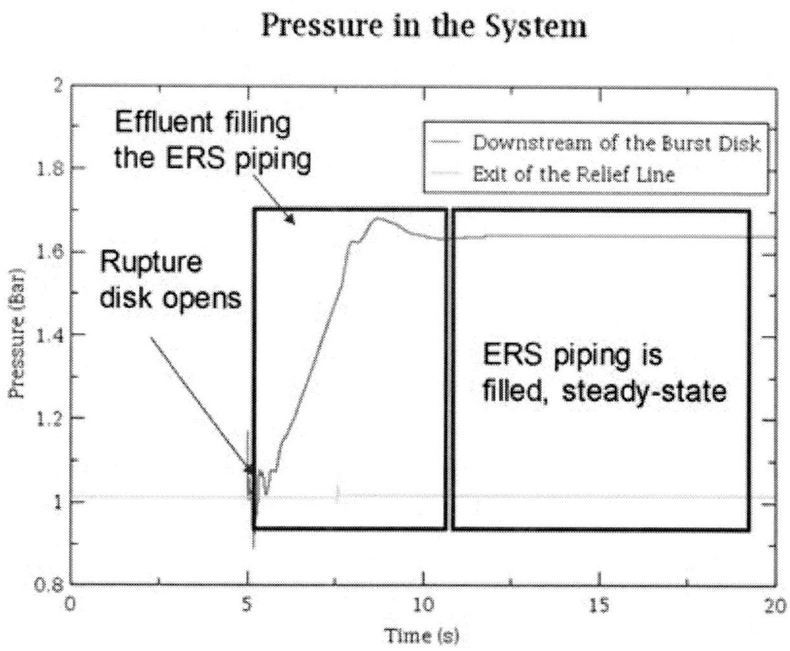

Figure 4. Pressure in the ERS piping for Scenario 1.

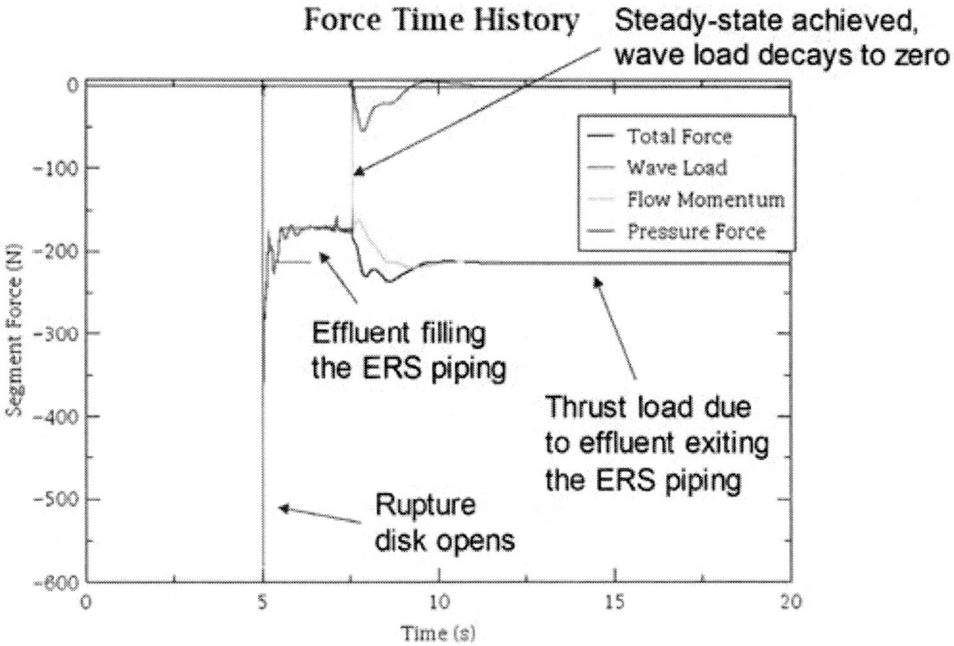

Figure 5. Dynamic loads for Scenario 1.

It is interesting to note that the wave load (while the ERS piping is filling up) and the load due to flow momentum (at steady-state) are almost similar. This is explained using the mathematical relations for the wave and thrust loads. The wave load for an effluent while filling up an empty piping (using Equation 14) is mathematically same as the flow momentum term in Equation 15.

The steady-state pressure, mass flow and thrust load values were compared between RELAP5 calculations and the analytical model, which was provided in Section 2. Actual chemical properties were used for the theoretical model. The comparisons between the two methods are provided in Table 2. The results demonstrate that the two-phase chemical effluent can be simulated with air/steam-water properties as long as the ω values are matched for these mixtures.

Table 2. Comparison between calculations using air-water (RELAP5) and actual two-phase chemical properties

Parameter	Scenario 1	Scenario 2	Scenario 3
Pressure downstream of the burst disk (bar-a)	1.6\1.7	1.6\1.7	1.9\1.6
Mass flow rate (kg/s)	25.9\26.1	16.3\16.4	1.2\1.2
Steady-state thrust load (N)	212.5\191.4	176.3\199.9	219.3\235.0

Black: RELAP5 results Red: Theoretical model

In the chemical manufacturing industries, the effluent can be composed of proprietary chemical mixtures containing various materials. Therefore, it can be challenging to know the exact material properties at any given time during an emergency relief event. The method presented in this study requires only the stagnation properties of the chemicals at the venting conditions for predicting key parameters, such as the pressure distribution and wave loads during the transient. These properties are typically established by adiabatic calorimetric testing or similar characterization of the chemicals. The method presented in this study may also be used with other commercially available software, which can predict transient fluid dynamics and wave loads.

5 Conclusions

In this study, a previously published methodology is used for calculating the dynamics of the effluent and the resultant hydrodynamic loads for a typical ERS by using the RELAP5 software and assuming air-water properties for the effluent. Key output parameters, such as pressure drop and hydrodynamic loads, are then compared with established analytical methods using the actual effluent properties. The following conclusions were observed:

- The fluid transient analysis is a key aspect of the design of the emergency relief systems. It provides the wave loads required for the dynamic pipe stress analysis. It also provides key parameters such as the pressure distribution, mass and volumetric flow rates through the system, and wave loads. These key parameters are often used to determine back pressure effects on other vessels, for chemical dispersion analysis and dynamic pipe stress analysis.

- The analytical approach shows that there is a similarity between the air/steam-water and the gas/vapor-liquid chemical mixtures if the "ω" values of the two mixtures are matched. This allows for an emergency relief event to be simulated with air/steam-water properties.

- Similar pressure distribution in the piping system and loads can be achieved through the similitude of "ω". Therefore, "ω" of the air/steam-water mixture should be matched with the "ω" of the chemical effluent at the stagnation conditions. The quality of the air/steam-water mixture can be adjusted until a similar "ω" is achieved.

- The sonic velocity of the effluent is typically not known. Also, it is practically difficult to match the sonic velocity of the air/steam-water mixture with the sonic velocity of the chemical effluent, which is a major requirement for tracking pressure waves and wave loads. This study demonstrates that the emergency relief scenarios are pipe filling events, which only require the momentum flux to be calculated for a reasonable prediction of the initial wave load. Therefore, the sonic velocity of the effluent is not required to capture the initial wave load.

- The effluent can be composed of proprietary chemical mixtures containing various materials. Therefore, it can be challenging to know the exact material properties at any given time during the transient. The method presented in this study may be used with most commercially available software and air/steam-water properties for a reasonable representation of the important characteristics of the actual effluent discharge in mechanical design space.

- The study needs to be extended further to identify applicability and limiting conditions for this method.

6 References

[1] Conzen, J., Wood. R.G., (2015), "Modeling and Computation of Reaction Forces on Relief Piping During Depressurization", AIChE 11th Global Congress of Process Safety.

[2] Ozar, B., (2018), "Analysis of Chemical Effluent Discharge through the Emergency Relief System using Stream-Water Properties," DIERS Fall Meeting, 15-17 October, Houston, USA.

[3] ASME. *ASME Code for Pressure Piping, B31.* The American Society of Mechanical Engineers, New York, NY, 2016.

[4] "Directive 2014/68/EU of the European Parliament and of the Council," Official Journal of the European Communities, May 2014.

[5] Epstein, M., Henry, R.E., Midvidy, W., Pauls, R., (1983), "One-dimensional Modeling of Two-phase Jet Expansion and Impingement," Thermal-hydraulics of Nuclear Reactors, 2nd International Topical Meeting Nuclear Reactor Thermal-hydraulics, Santa Barbara, CA.

[6] Grolmes, M.A., Leung, J.C., (1984), "Scaling Considerations for Two-phase Critical Flow," Multi-phase Flow Heat Transfer III, Part A: Fundamentals, T.N. Veziroglu and A.E. Bergles, eds., 549, Elsevier, Amsterdam.

[7] Leung, J.C., (1986), "A Generalized Correlation for One-Component Homogenous Equilibrium Flashing Choked Flow," American Institute of Chemical Engineers, Vol. 23, pp. 1743-1746, October.

[8] Leung, J.C., Grolmes, M.A., (1987), "The Discharge of Two-phase Flashing Flow in a Horizontal Duct," American Institute of Chemical Engineers Journal, Vol. (33), p. 524.

[9] Leung J.C., Epstein, M., (1990) "The Discharge of Two-phase Flashing Flow from an Inclined Duct," Transactions of the ASME, Vol. 112, pp. 524-527, May.

[10] Leung, J.C., (1992), "Reaction Forces During Two-phase Discharges," American Society of Mechanical Engineers Journal of Fluids Engineering, Vol. (114), p. 689-692.

[11] Information Systems Laboratories, (2016), "RELAP5/MOD3.3 (Patch03) code manual volume I to VIII, NUREG/CR-5535/Rev P5-Vol I to VIII", prepared for USNRC.

[12] International Agreement Report, (2007), "Simulation of the Propagation of Pressure Waves in Piping Systems with RELAP5/MOD 3.2.2: Comparison of Computed and Measured Results", TÜV SÜD - NUREG/IA-0206, June.

[13] Sokolowski, L., Koszela, Z., (2012), "RELAP5 Capability to Predict Pressure Wave Propagation Phenomena in Single- and Two-Phase Flow Conditions", Journal of Power Technologies 92 (3), page 150 – 165.

[14] Miller, J.S., Ramsden, K.B., (2005), "Validation of a Thermal Hydraulic Computer Code to Perform Two-Phase Multi-Component Force Calculations for Structural Evaluations", Proceedings of 13th International Conference on Nuclear Engineering (ICONE13), 16-20 May 2005, Beijing, China.

[15] Ozar, B., Henry, C.E., Harvill, R., Norton, D.A., (2012), "Analysis for Low Pressure Cooling Injection System Suction Hydrodynamics for a Boiling Water Reactor," Proceedings of 20th International Conference on Nuclear Engineering (ICONE20), July 30-August 3, 2012, Anaheim, CA.

[16] Henry, C.E., Lim, J., Ozar, B., (2012), "Investigation of Hydrodynamic Loads Associated with Pyrotechnic Valve Actuation," Proceedings of 20th International Conference on Nuclear Engineering (ICONE20), July 30-August 3, 2012, Anaheim, CA.

[17] Riemke, R.A., Davis, C.B., Schultz, R.R., (2009), "Recent Hydrodynamics Improvements to the RELAP5-3D Code", Proceedings of the 17th International Conference on Nuclear Engineering, ICONE17-75380, July 12-16.

[18] Health and Safety Executive, "Offshore Technology Report 2002/023: Testing and Analysis of Relief Device Opening Times," Pipeline Simulation and Integrity Ltd., Yorkshire/UK.

[19] Moody, F.J., (1990), "Introduction to Unsteady Thermofluid Mechanics", John Wiley and Sons, New York, Section 7.4, p. 365-366.

[20] API Standard 520, (2008), "Sizing, Selection, and Installation of Pressure-relieving Devices in Refineries," 8th edition.

547437 Evaluation of the Thermal Stability of Styrene in Contact with Impurities

Safe handling and storage of reactive monomers have long been an issue for polymer manufactures. The unsaturated double bond may self-react to initiate unintended polymerization reactions under many scenarios, for example, completely depletion of inhibitors, exposure to fire or external heat, and in contact with impurities. Usually, a trace amount of incompatible species is able to trigger unintended reactions and lead to catastrophic consequences. Thus, understanding how the impurities promote self-reaction is expected to provide valuable guidance to safe monomer-handling practices. Our goal in this study is to quantify the contamination effects of some most common contamination chemicals on styrene self-polymerization. Thermal analysis tools including DSC and ARSST are applied to examine the exothermic characteristics of styrene mixed with contaminating substances at different concentration levels and mixing ratios. From the preliminary results, we observed the onset of self-polymerization shifted to a lower temperature when in contacted with water and also alkaline. A trace amount of concentrated strong acid initiated unwanted exothermic activity even at room temperature. Self-heating rate was also elevated when diluted acid was present. Additional research is needed to understand what is the impact in of other types and concentrations of contaminants such as rust, metal ions. Our study demonstrates significant implications in the prevention of runaway incidents during transportation and storage of styrene.

549038 A Mixed-Integer Programming Framework for Placement of Fire Detectors in Chemical Processing Facilities

In chemical processing plants, gas and fire detectors provide an important layer of protection for personnel. However, given the large variety of chemical and physical sensors and number of factors influencing their capabilities to successfully detect hazards, the decision of where to place sensors is challenging. Gas detectors (e.g. ultrasonic, catalytic bead) must be in contact with gas leaks to be effective, while fire detectors (e.g. point infrared, open path infrared, optical) require a visual path to the fire, free of obstructions, for successful detection. Numerical optimization techniques provide an opportunity to overcome these challenges and determine effective sensor placement.

The optimal gas detector placement problem was formulated by [1] as a stochastic program based on the p-median facility location problem and relied on scenarios generated from CFD simulations within a petrochemical facility. Their results showed that coupling coverage with expected detection time using a scenario-based method outperformed the existing coverage-only approaches and has inspired formulations based on MINLP approaches for non-uniform failure probabilities [2].

But while gas detectors must be placed in the path of gas leaks for effective detection, fire detectors must be placed within unobstructed visual distance of flames. The optimal placement of visual fire detectors in processing plants can be particularly difficult due to the large number of physical obstructions (e.g. piping, valves, tanks).

This work focuses on an optimization formulation for placement of fire (visual) detectors. The scenario-based, p-median approach demonstrated by [1] for gas detectors is translated to a maximum coverage model for fire detectors in a 3D space with consideration to failure probabilities. We show that this approach allows for the placement of fire detectors for a fixed sensor budget while under uncertainty. We demonstrate the effectiveness of this formulation on several case studies. We further discuss how the framework is readily scalable for larger problem sizes and amenable to various objectives.

We would like to thank the Purdue Process Safety and Assurance Center for their support in this work.

References:

[1] Legg, S. W., Benavides-Serrano, A. J., Siirola, J. D., Watson, J. P., Davis, S. G., Bratteteig, A., and Laird, C. D. (2012). A stochastic programming approach for gas detector placement using CFD-based dispersion simulations. Computers and Chemical Engineering, 47:194–201.

[2] Liu, J., and Laird, C.D., A Global Stochastic Programming Approach for the Optimal Placement of Gas Detectors with Nonuniform Unavailabilities, to appear in Journal of Loss Prevention in the Process Industries, 2017.

546855 Advance Alarm Management Applied to DCS Improvement in a Petrochemical Plant: A Case of Study

Alarms use in process industry is fundamental for avoiding process accidents in which people, the environment or process equipment could be damage. Currently, Carbon Black production plants are equipped with distributed control systems (DCS) for setting alarms values and to provide indication of abnormal events to the control rooms operators. However, the basic control system could not meet the needs against the different production conditions and the presence of improper alarms could send a bad signal to control rooms and it could be the start of a process safety event.

In the present study, an alarm system assessment was performed in a petrochemical plant, located at Mamonal industrial park, in Cartagena – Colombia; based on data analysis, it was observed that when process conditions changes occured, the DCS was not be able to adjust alarms settings automatically, according to the load fluctuation and/or process conditions. In order to solve this problem, in first place the DCS performance was studied and it was reported an alarms list, which provided information regarding to the alarm flooding phenomena, alarm chattering and others alarm system issues.

On the other hand, the information earned and alarm rationalization methodology were used for establishing an action plan in order to improve DCS performance. Results showed the utility of these tools and it was identified the need to perform betters in the alarm system logic to avoid flooding alarm phenomena and to make operations safer.

548361 Overcoming Motivational Biases in Hazops

Hazard and Operability Studies (HAZOPs) are one method of assessing the risk in a facility. Based on the risk ranking of a scenario, recommendations may be required to reduce the risk. Often, these recommendations require additional capital expenditure and/or operational resources. In aging plants where resources are not readily available or in greenfield projects where there is a tight timeline to start up, there is sometimes a bias to risk rank scenarios such that there is no need to add additional safeguards. Motivational biases can outweigh common sense such that scenarios are not ranked according to their real risk. We have developed a method to overcome some of these motivational biases with the use of a polling system to engage all participants in the HAZOP. Participants are asked to vote anonymously on the severity and likelihood for a given scenario, then the results are discussed. During this discussion time, arguments for all the different risk rankings are discussed, including those voted on by the minority in the room. This facilitates an open discourse about the real risks for a given scenario rather than shifting the attention to the most vocal person in the room, who may be strongly influenced by a desire to reduce spending or minimize the effort required for maintenance. Following this discussion a second poll is performed to reach a consensus. We have found this to be an effective method for reducing motivational biases and ensuring that the right decisions are made based on the most logical argument instead of based on budget or time constraints. Tracking polling results increases the accountability of participants and makes them less likely to skew the risk ranking. The requirement for consensus ensures that all opinions are heard and considered, even if they are not popular. Together these make the polling system an effective method for overcoming motivational biases in HAZOPs and ensuring that safe decisions are made.

546515 Integrating Culture into Process Safety Assurance

Purpose: PETRONAS Chemicals Group Berhad (PCG) has embarked on a group-wide culture transformation, introducing the concept of the Results Pyramid® and a Culture of Accountability® i.e. a workplace culture where people take accountability to think and act in the manner necessary to achieve results.

Methodology: This paper will discuss two such activities anchoring on the Results Pyramid® that have been deployed across several plants and units within PCG.

The first activity is an enhancement of existing assurance programmes, where culture subject matter experts are assigned to observe plant leadership teams and front-liners during interviews and site verifications conducted by the technical assessors. The observations are synthesised into belief statements, which are presented together with the technical findings.

The second activity is Power Up, a session designed to incorporate powerful and specific feedback exchange between various teams, culminating in a specific list of actions to improve delivery of each team's activities in order to achieve the shared key results. This workshop can be done either as a follow-up to an assurance activity, in preparation for a project/turnaround activity or as a part of a periodic performance review.

Results: Examples of observations, feedback exchanges and specific actions from these activities will be shared, together with the impact on process safety leading and lagging indicators. Recommended practices and lessons learnt for continuous improvement will also be included.

Conclusion: Applying the Results Pyramid® concept to the design and implementation of Process Safety-related programmes has helped to identify the underlying leadership, mindset and behaviour issues impacting their success. By doing so, more specific interventions can be put in place to ensure full compliance to safety requirements and maximise effectiveness of these programmes.

548442 Emergency Plans Integration By Diferent Companies

In Brazil, different companies with pollution potential must have an emergency plan. However, each company analyzes its risks and contingency measures separately, based on the Conama 398 law. With the establishment of the Area Plan, companies from the same territory with pollution potential caused by oil in waters began to analyze how to combat a major accident quickly and effective together. This work shows how the integration of the individual contingency plans of each company and the cooperative result of the area plan in territory of significant scope and considerable environmental sensitivity.

No Brasil, diferentes empresas com potencial de poluição devem ter um plano de emergência. No entanto cada empresa analisa seus riscos e medidas de contingência separadamente, com base na lei Conama 398. Com a instituição do Plano de Área, empresas do mesmo território com potencial de poluição causadas por óleo em águas passaram a analisar como atender conjuntamente de forma rápida e eficaz a um grande acidente em seu território de abrangência. Este trabalho mostra como foi estruturada e organizada a integração dos planos de emergência individuais de cada empresa e o resultado cooperativo do plano de área em território de abrangência significativa e considerável sensibilidade ambiental.

548111 Moving Back to Flammable Refrigerants - Facility Siting and Personnel Safety Considerations

"Everything old is new again – Paul Allen"

Not just a catchy tune from the 1970's, this jingle summarizes a current technological shift in the world of refrigeration. Historically, refrigeration cycles used toxic and/or flammable materials as the working fluid. As the technology matured and made its way into transportation and consumer applications, the working fluid also evolved. Halogenated hydrocarbons, such as chlorofluorocarbons (CFCs), were developed as replacements to conventional refrigerants such as ammonia and propane. CFCs, which may have toxic effects, are desired for their tunable working range and for being non-flammable. However, the environmental drawbacks to the use of CFCs has been well researched and documented. As we embark into the future, where climate change and environmental quality are at the forefront of concern to consumers and industry alike, we are seeing a move back to flammable refrigerants. Everything old is, indeed, new again.

New construction and expansion projects can certainly consider and plan for the use of flammable refrigerants in commercial and industry facility design and siting, but what about retrofit applications? What are the risks and hazards associated with converting a CFC refrigeration system, designed and sited for the use of non-flammable refrigerants, to a system now utilizing flammable refrigerants? Beyond the need for additional fire protection systems, what risks are posed to the facility and personnel? In this paper, we seek to answer these questions by comparing the hazards and risks of mid-scale refrigeration systems using either non-flammable or flammable refrigerants. We explore how facility siting constraints may change with respect to the use of flammable refrigerants, and perform a quantitative risk assessment (QRA) for an example retrofit project in order to assess the potential for increased risk that should be considered in such projects.

548210 Making Sense of Risk Results

Risk based studies are very useful, but at the same time, the results are extremely complicated. While we have a basic set of results that can be used to illustrate the risks to people, these are relatively simple for risk practitioners and sometimes too complicated for site managers.

FN curves show the risk to all people from all sources, and individual risk contours illustrate the risk from all scenarios to people across specific locations, but how do we translate these results into actions. Generally, the action is to relocate people to lower risk areas, or upgrade buildings to mitigate the impacts identified. However, this is not always feasible. What if the site footprint is limited; there are public receptors who cannot be relocated; or building upgrades are cost-prohibitive or infeasible?

We must better quantify and evaluate the **sources** of risk rather than focusing on mitigating the impacts from the events. We need to evaluate the type of risks, the potential initiating events, and then we can assess the potential methods to effectively mitigate the risks. Existing risk result presentations are often very poor at illustrating the sources of risk. To effectively identify the sources of risk requires that we carefully dissect the risk results, while maintaining their validity and intent.

This paper will focus on how we find these risk drivers and make enough sense of them to create actionable recommendations that protect people.

551205 Ammonia Dispersion Modeling for Ammonia Storage with Multi-Scenario Failure System

PT Pupuk Kalimantan Timur is one of the largest fertilizer companies in Southeast Asia where located in Bontang City, East Kalimantan Province, Indonesia. The total production capacity of PT Pupuk Kalimantan Timur are 2.7 million Tons / year Ammonia, 3.43 million Tons / year Urea and 350 thousand tons / year NPK.

To accommodate ammonia production, PT Pupuk Kalimantan Timur currently operates 3 ammonia tanks with a capacity of 2 x 25,000 tons and 1 x 50,000 tons. In order to manage the hazard risk contained in the ammonia tank, it is necessary to do a modeling to map the potential and ammonia leakage potential areas and the impact of their exposure to the workforce and the community around the factory area.

The results of this modeling are expected to be used by Pupuk Kalimantan Timur's management in carrying out corrective actions for occupational health and safety performance, especially in controlling the potential hazard of ammonia, especially in ammonia tanks.

In this paper modeling is done using the AERMOD® Modeling System software. AERMOD Modeling System was chosen because it can be used on:

1. U.S. EPA regulatory Gaussian plume air dispersion model.
2. For near-field impacts (50 km max).
3. Prefered model for complex terrain.
4. Improved treatment of urban effects.

For modeling, we choose the AS-T-101 ammonia tank which has the largest capacity, which is 50,000 tons. Ammonia distribution models will be simulated based on weather conditions and ammonia leak source scenarios that can occur in the AS-T-101 ammonia storage tank system. Ammonia leakage sources will be displayed in 5 scenarios, i.e:

1. 1 Vent opens
2. 1 PSV poping
3. 2 PSV poping
4. 3 PSV poping
5. 3 PSV and 1 vent poping

The AERMOD Modeling System program is used to make ammonia distribution models for each event scenario assuming an exposure radius of up to 10 km. Based on the results of the calculation and modeling of the potential distribution of ammonia from AS-T-101 using the AERMOD Modeling System, the highest exposure of ammonia is in the 5th scenario, which is up to 36611 μg / M3.

546560 De-Mystifying the Concepts of Hazardous Area Classification

Purpose

The purpose of this topic is to address and clarify major items of confusion associated with hazardous area classification assessments and compliance auditing in classified locations. The ultimate objective is developing documentation that provides a framework for the control of electrical energy ignition sources in environments made hazardous by the potential presence of flammable vapors.

Background

In February of 1992 Occupational Safety and Health Administration (OSHA) 29 CFR 1910.119 Process Safety Management of Highly Hazardous Chemicals became law in the United States and confusion reigned supreme. Over the next several years industry professionals began the process of sorting through the responsibilities, regulations and impending deadlines that had been placed squarely in their laps. Acronyms such as PSM (Process Safety Management), PSI (Process Safety Information), RAGAGEP (Recognized and Generally Accepted Good Engineering Practices), MOC (Management of Change), and PHA (Process Hazards Analysis) became commonplace in the vocabulary of process safety professionals nationwide.

Employers now were required to meet stringent deadlines regarding the creation and management of a plethora of process information associated with equipment in the process covered by this law, i.e. any covered process. This information would be used to support the exhaustive efforts needed to create a baseline PHA for each covered process by May of 1997. In addition OSHA required all PHA's be placed on a five (5) year life cycle requiring their review and revalidation. Along with the revalidation of PHA's came the task of also revalidating all process safety information on the same five (5) year cycle.

As a result, every willing and able bodied employee was thrust into the quagmire of supporting, participating in, facilitating, or conducting PHA's. The required information used to support PHA's included the materials of construction, Piping and Instrument Diagrams (P&ID's), Relief System design and design basis, ventilation system design, material and energy balances, safety shutdown systems, and last but not least **electrical classification**. Yes, "electrical classification" was and still is legally considered an integral part of required process safety information.

This abstract addresses the following and provides an interactive discussion regarding the three (3) major myths surrounding electrical classification:

- assigning a classification to a location
- proper use of the classification information
- keeping the documentation evergreen

  The following topics will be addressed and clarified:

- What Is Hazardous / Electrical Classification?
- The Process of a Hazardous Area Classification Assessment
- The Concept of Electrical vs Hazardous Area Classification
- The Compliance Assessment (where does it fit in the process of controlling electrical energy ignition sources)
- Common Compliance Assessment Issues

546108 Maintaining Safety Excellence By the Implementation of Process Safety Inspection in Badak LNG

Badak LNG as one of the most experienced LNG Producers in the world, has undergone safety culture change throughout its operation. The goal of this change is to achieve excellent safety performances, indicated by the lower rate of Lost Time Injury Frequency (LTIF), lower number of Loss of Primary Containment (LOPC) Tier-1, High Potential Incident and multiple Train trip. This change was triggered by the occurrence of a fatal incident in 2005 as well as lost time injury incident in 2006. Since then, Badak LNG has integrated Process Safety Management (PSM) into its Company Management System, strengthened the roles and responsibilities of Managements, and pursued for interdependent safety behaviour of each personnel. One of PSM implementations is through process safety inspection.

Process safety inspection is an activity to assess the current equipment integrity based on visual observations compared with equipment's standard condition. The standard condition checklist for process safety inspection in Badak LNG is based on Physical Barrier Assessment tools, owned by Det Norske Veritas (DNV). The assessed items are divided into 5 categories: Inherent Safety, Hazard Prevention Measure, Hazard Control, Mitigations, and Support System. The overall score is determined by total number of standard items in all categories compared to the total number of verified items in all categories. The minimum acceptable score is at 80%.

During the year of 2012-2017, the average score of process safety inspections in Badak LNG is 90%. As a result of these inspections, Badak LNG had also identified the need of improvement in some items such as blinds, pipework and flanges, passive fire protection, fire & gas detection system and lock-open/lock-close block valves. In overall, by the implementation of PSM, Badak LNG has aligned on the right path to maintain its safety excellence achievement in LNG business, such as zero LTIF since 2007, zero multiple Train trip since 2012, zero LOPC Tier-1 since 2016, and zero High Potential Incident in 2017.

548411 Using HAZOP/LOPA to Create an Effective Mechanical Integrity Program

Many people view the conduct of a HAZOP/LOPA to address regulatory requirements as a chore, and stop there. However, the implementation of a quality HAZOP/LOPA has the potential to provide a framework for addressing numerous safety and operational optimization issues at plants, including the formulation/refinement of the Mechanical Integrity Program. The purpose of this paper is to focus on the mechanical integrity program, illustrate how a quality HAZOP/LOPA can support the effective implementation of some of the new requirements for California Refineries, and optimize key elements of an effective Mechanical Integrity Program, e.g.:

- Inspection/testing methods

- Testing intervals

- Maintenance outage periods

- Repair prioritization and allowable outage

- Identification of low priority equipment

544678 "*Identifying Critical Process Parameter to Ensure Asset Integrity" or "Pulse of Mechanical Integrity Management System"*

The management of equipment begins with an awareness that ageing is not about how old the equipment is, but is about what is known about its condition and the factors that influence the onset, evolution and mitigation of its degradation. The presentation provides an overview of Integrity Operating Window (IOW) parameters of any process that are key to understand the asset degradation and prevent Process Safety Incidents.

IOW are the situations in the process industries where a process parameters (temp., pressure, dosing rate, conc. etc...) when gets out of operating windows could result in significant loss of primary containment leading to potential severe consequences.

Arrazi - SABIC developed an IOW system that covers IOW identification, associated risk, mitigation (interim – long term), stewardship/management of IOWs and elimination.

The presentation shall cover Insight concept of the Topic, major Process Safety accidents related to breeching IOWs, actual example, IOW Management Process, its influence on PM plans etc.

three key bullet points that we can include in the agenda:

- Knowing operating limits with associated mechanical integrity or process safety risk to ensuring proper actions to be in operating envelope
- Developing the engineering leadership to systematically guide the operators, superintendents and engineers to be in safe zones
- Dynamic inspection frequency based on plant conditions and sharing some industrial major incidents for this critical item

The Practice of Process Safety Analysis and Management System Based on Safety Reliability
(THE NEW TITLE: The PHAMS system in SINOPEC)

Jiang Xue
SINOPEC Research Institute of Safety Engineering,
QingDao,China
Jiangxue.qday@sinopec.com

The PHAMS system in SINOPEC

Jiang Xue
SINOPEC Research Institute of Safety Engineering,
QingDao,China
Jiangxue.qday@sinopec.com

Keywords: Safety Barriers Reliability, PHAMS, HALOPA, SIL, Safety Management Online.

Abstract

The big data technique has been in trend for the process safety assessment. In China, it's the very need to transfer the qualitative safety management to quantitative safety management, which should rely on the professional risk analysis technique, as well as the standard safety reliability database. Based on the above, it's very important to integrate several professional risk analysis technique, as HAZOP, LOPA, SIL etc. to realize the integral risk identification and evaluation, grading and management plan making, supporting by the proprietary SINOPEC risk metrics and grading standards, as well as safety equipment reliability database,which is developed as the Process Safety Analysis and Management System(PHAMS) for SINOPEC, to support all SINOPEC branches to make dynamic risk analysis and management online

1 Introduction

With the smart factory's development, the static, periodic risk assessment and management can not satisfy the identification and real-time safety control of plant's risk events. The PHAMS which establishes the plant-wide dynamic risk analysis, can effectively solve the risk rating and monitoring of the unit and realize the plant-wide or corporate-wide management optimization.

The key data for HALOPA, SIL and Management module in PHAMS is the dependent protection barriers, which we rely on RBI correction, Personnel exposure probability and ignition probability to optimize the six mainly safety barriers' failure probability, simultaneously, do use of the Bayesian Correction Modelling to verify the instant failure probability on site, as well as the SIL identification, which can validly save the grading cost on time and personnel, and all of these are towards to the deeper research on dynamic risk analysis for PHAMS.

2 HALOPA Analysis System

2.1 SINOPEC RISK MARTIX

According to international common matrix criteria, as well as statistics from the SINOPEC accident data and the maintenance data, the frequency of risk occurrence and severity classes are coupled to the 8×7 matrix by a linear exponential model, and be assigned risk index to each risk level.

SINOPEC RISK MARTIX		Occurrence Frequency							
		1	2	3	4	5	6	7	8
		≤10^{-6}/y	10^{-6}-10^{-5}/y	10^{-5}-10^{-4}/y	10^{-4}-10^{-3}/y	10^{-3}-10^{-2}/y	10^{-2}-10^{-1}/y	10^{-1}-1/y	>1/y
Severity Classes	A	1	1	2	3	5	7	10	15
	B	2	2	3	5	7	10	15	23
	C	2	3	5	7	11	16	23	35
	D	5	8	12	17	25	37	55	81
	E	7	10	15	22	32	46	68	100
	F	10	15	20	30	43	64	94	138
	G	15	20	29	43	63	93	136	200

Figure 1 SINOPEC Risk Matrixp[1]

Based on this matrix, SINOPEC has established four-level risk list for establishing targeted management of branch network. For example, the executive board of the group is in charge of the red-level risk list, the branch directors is in charge of the orange-level risk list, etc.

2.2 HALOPA Analysis Modelling

2.2.1 System risk analysis based on equipment operation cycle

The HALOPA analysis model can be used to identify the hazard of the device or process unit, clarifying the initial event, process deviation, independent protective layer and consequence events, combined with the event tree analysis method to analysis the reduction influence of the occurrence frequency of initial events. The protection layer performs a failure frequency evaluation role to determine the final frequency of the on-site device risk events during each operating cycle under the coupling reduction of the existing protective layers at each level [2].

An important assumption of the model is that each protection layer is independent, that is, the failure of a single protection layer will not affect the risk reduction impact of other existing protection layers [3]. Each time a protective layer takes effect, the frequency of accidents is reduced by the risk reduction of the protective layer.

When the frequency of the initial event does not exceed 1 time/year or is less than twice the test frequency of the first protective layer, the evaluation of the consequence frequency can be referred to the following formula [4]:

$$f_I^C = f_I \times P_{enable} \times \prod_{j=1}^{J} PFD_{Ij} \qquad (1)$$

where

$\boldsymbol{f_I}$=Occurrence frequency of initial event I annually；

Penable=Moderation factors by enabling conditions and other conditional probabilities.

PFDIj=Failure probability of dependent protection layer j of initial event I.

When the frequency of the initial event exceeds 1 time/year or exceeds the test frequency of the first protective layer twice, the first protection layer failure shall be used as the initial event, and the calculation formula of the consequence frequency is as follows:

$$f_I^C = f_{\text{IPL\_first}} \times P_{enable} \times \prod_{j=2}^{J} PFD_{Ij} \qquad (2)$$

where

$\mathbf{f_{IPL\_first}}$= The failure probability of 1st dependent protection layer

The HALOPA analysis model in PHAMS designed a modified calculation model for equipment failure frequency and personnel exposure probability, and realized BPCS, SIS, physical protection, leakage storage and collection, fire and explosion suppression for equipment operation cycle and device in service. The system's six major types of protection layer failure frequency have the correction factors of risk reduction, and realize system risk assessment based on the device production cycle. For example, according to the system risk assessment of a diesel hydrogenation unit based on the PHAMS system, on-site data collection of various protective layers and key correction conditions under one cycle of the driving operation to date, online preliminary analysis of the risk overall value and grading of the feeding unit The results are as follows:

Figure 2 HALOPA Analytical Function Online

2.2.2 Dynamic risk assessment based on real-time data of safety control systems
According to the protection layer frequency analysis model, the failure probability of the protection layer lowering the frequency of a certain initial event can be regarded as θ , and the a priori risk probability distribution function is:

$$f_0(\theta) \propto \theta^a(1-\theta)^b \quad (3)$$

Its likelihood function $L(E|\theta)$ is

$$L(E|\theta) \propto \theta^m(1-\theta)^{n-m} \quad (4)$$

where

a, b=the prior distribution parameter of the failure probability of the protective layer;
m=The number of the posterior cumulative event occurred when the protective layer is valid;
n=The number of the posterior cumulative event occurred when the protective layer is failed.

With the posterior probability distribution function corrected by dynamic risk analysis as$f_1(\theta|E)$, the posterior probability distribution can be obtained by:

$$f_1(\theta|E) = \frac{f_0(\theta)L(E|\theta)}{\int f_0(\theta)L(E|\theta)d\theta} \propto \theta^a(1-\theta)^b\theta^m(1-\theta)^{n-m} \propto \theta^{a+m}(1-\theta)^{b+n-m} \quad (5)$$

Since m and n are parameters of the age change and the posterior probability distribution function seems to be continuous, the dynamic failure frequency of the protective layer at a certain moment can be obtained by the following formula[4-6].

$$P_i(\theta) = \int_{i-1}^{i} f(\theta)d\theta = (1-F_i(\theta|E)) - (1-F_{i-1}(\theta|E)) = F_{i-1}(\theta|E) - F_i(\theta|E) \quad (6)$$

In where $F_i(\theta|E)$ is $f_i(\theta|E)$ the cumulative failure frequency function as

$$F_i(\theta|E) = \int_0^i f(\theta|E)d\theta \quad (7)$$

And then bring in the following formula to get the dynamic risk frequency of the device or process unit:

$$f_I^C = f_I \times P_{enable} \times \prod_{j=1}^{J} P(\theta)_{Ij} \quad (8)$$

and

$$f_I^C = f_{IPL\_first} \times P_{enable} \times \prod_{j=2}^{J} P(\theta)_{Ij} \quad (9)$$

Where

$P(\theta)_{Ij}$=Posterior failure probability value of the jth independent protection layer

Based on this, the PHAMS system realizes the real-time update of the protection layer risk reduction frequency, and intelligently analysis the overall risk value and grading of the unit to plant.

3 SIL Intelligence Identification Online System

3.1 SIL Identification Modelling

The analysis data of the SIS protection layer of the HALOPA evaluation system in PHAM is updated and interconnected by means of the safety reliability database. After the HALOPA dangerous event using the same safety instrument loop is automatically brought into the SIL identification, the PHAMS system takes the same kind of consequences. C, to calculate without the cumulative frequency F^C before the SIF

$$F^C = \sum_{i=1}^{n} f_i^C \qquad (10)$$

where

$$f_I^C = f_I \times P_{enable} \times \beta \times \prod_{j=1}^{J} PFD_{IJ} \times \prod P_{condition} \qquad (11)$$

$$\beta = \frac{2^N - 1}{N} \qquad (12)$$

where

f_I^C= The occurrence frequency of the dangerous event I with consequence C

β= the common factors caused by several SIFs

N=the number of SIFs for the dangerous event I

Based on the modelling, the PHAMS system can intelligently group common risk events and assess the risk reduction factor RRF of SIF in each group [7] :

$$\text{RRF} \geq \frac{F^C}{TMEL} \qquad (13)$$

and

$$1 \times R_{\text{Factor}} < RRF_{SIL1} \leq 10 \times R_{\text{Factor}} \qquad (14)$$

$$10 \times R_{\text{Factor}} < RRF_{SIL2} \leq 100 \times R_{\text{Factor}} \qquad (15)$$

$$100 \times R_{\text{Factor}} < RRF_{SIL3} \leq 1000 \times R_{\text{Factor}} \qquad (16)$$

where

R_{Factor}= risk reduction factor between1 to 10.

Based on the SIL identification results of each group, the optimal solution is intelligently selected as the RRF of this safety instrument loop.

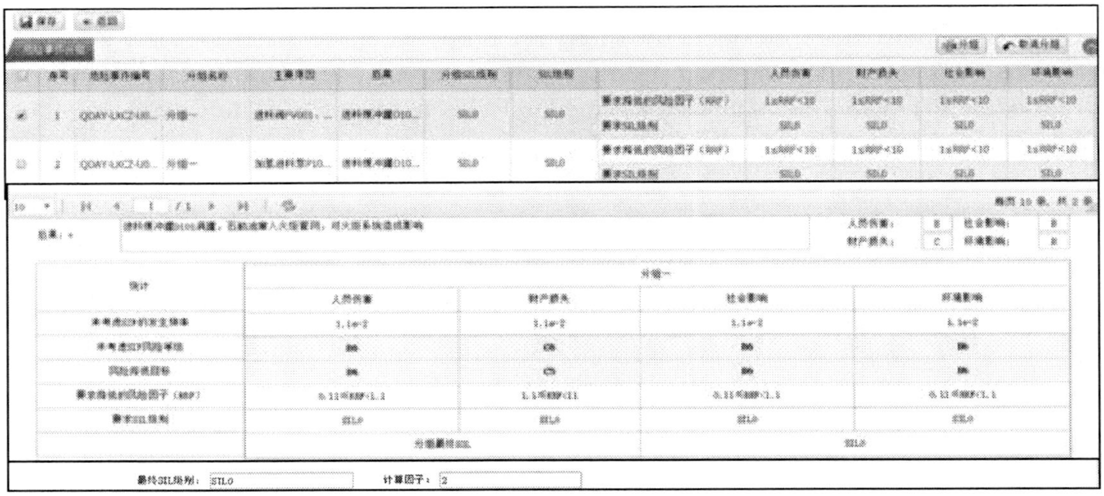

Figure 3 SIL Intelligence Identification Online

3.2 The SIL Identification Based on SINOPEC Risk Matrix in PHAMS

With the aid of data warehouse technology, PHAMS system can compare HALOPA assessment data which involved with SIS protection layers of the initial situation, the consequences classes and non SIS protective layer of data, and automatically take the residual risk of the events into the risk matrix. Then based on SINOPEC risk matrix, the SIL identification module can intelligently couple and assess the residual risk and RRF.

according to Sinopec's safety management focusing on high severity impact, PHAMS sets an independent TMEL requirements (as shown in the following table), and can automatically selects the most serious impact event as the final SIS identification data of this event. .

Table 1 TMEL of SIL Intelligence Identification[8-10]

Severity Classes	*Health & Safety (r/y)*	*Property Damage (r/y)*	*Social Impact (r/y)*
A	$\leq 1\times10^{-1}$	≤1	≤1
B	$\leq 1\times10^{-2}$	$\leq 1\times10^{-1}$	$\leq 1\times10^{-1}$
C	$\leq 1\times10^{-3}$	$\leq 1\times10^{-2}$	$\leq 1\times10^{-2}$
D	$\leq 1\times10^{-5}$	$\leq 1\times10^{-4}$	$\leq 1\times10^{-4}$
E	$\leq 1\times10^{-6}$	$\leq 1\times10^{-5}$	$\leq 1\times10^{-5}$
F	$\leq 1\times10^{-7}$	$\leq 1\times10^{-6}$	$\leq 1\times10^{-6}$
G	$\leq 1\times10^{-7}$	$\leq 1\times10^{-6}$	$\leq 1\times10^{-6}$

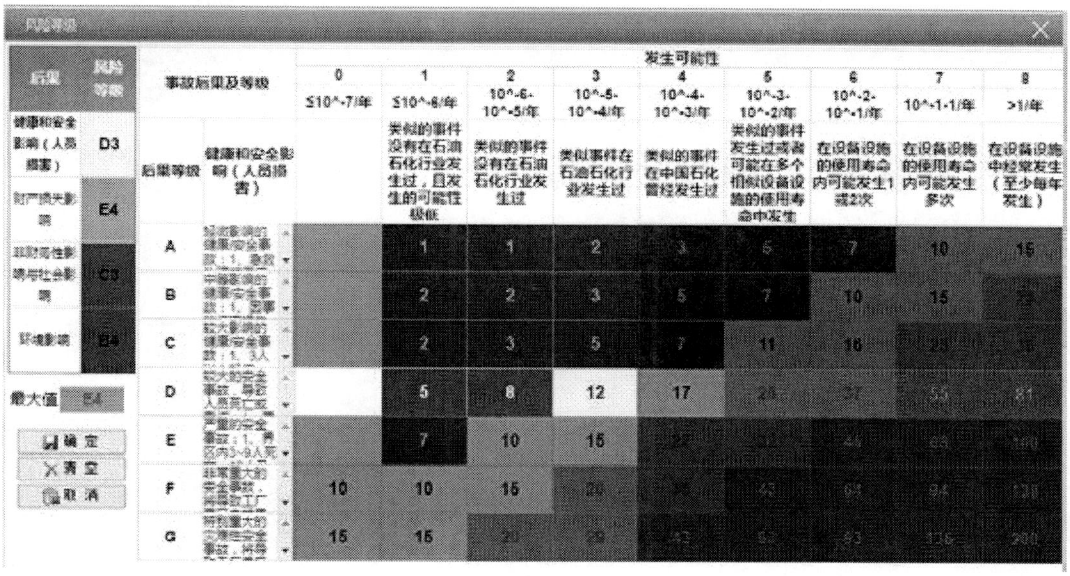

Figure 4 SIL Intelligence Identification on SINOPEC Risk Matrix

On the basis of SIL identification, the PHAMS system can also implement SIL verification based on multi-factor model, and evaluate the structural constraints, PFDavg and MTTF respectively. The verification result data is consistent with the accuracy requirements of IEC61508 and IEC61511.

4 Risk Management and Decision Support System

With the powerful safety and reliability database and expert experience database of PHAMS system, as well as the data mining of the result data of HALOPA and SIL system, the Risk Management and Decision Support System can implement the cumulative risk frequency calculation of independent devices and plants in construction or in service, to assess the overall risk based on different impact severities through the linear comparison [11-12].

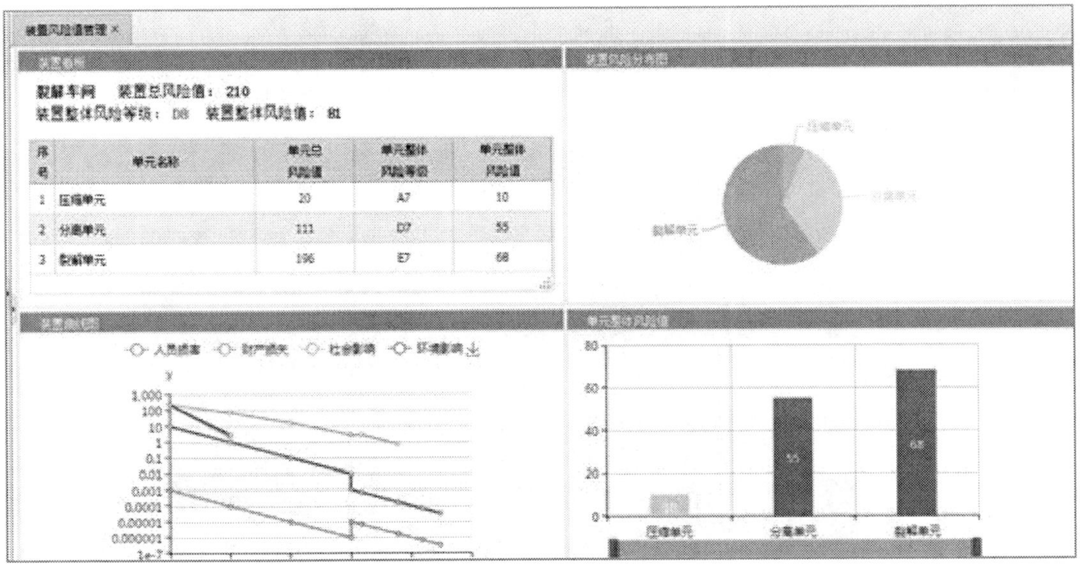

Figure 5 Overall Risk Rating and Risk Event Distribution Analysis of Ethylene Cracking Facility

5 Conclusion

1. PHAMS system sets up 9 categories of safety reliability database, expert experience data and Sinopec risk matrix, etc, which can support unified SINOPEC risk assessment and management standards, risk analysis template, process data storage and data mining for PHAMS system , as well as the interconnection and real-time update of data of each system.
2. HALOPA system implements plant overall risk analysis based on protection layer data, and performs data coupling evaluation from static analysis based on equipment cycle and dynamic risk correction to realize real-time evaluation of risk assessment technology and device operating conditions. The dynamic risk analysis research will be further strengthened to realize the system risk prediction function.
3. The SIL identificaiton and verification system can realize the evaluation of safety instrument function of in service and in construction. In the long run, it helps to improve the safety management level of the plant's overall safety instrument function
4. The risk management and decision support system can realize the overall risk assessment and risk distribution based on various evaluation results, through in-depth mining and statistical analysis, the PHAMS can realize the big data analysis management of Sinopec branches' risk management.

6 References

[1] The Bureau of Safe Production Supervision and Administration of SINOPEC. Guide for Safety Risk Identification and Assessment of SINOPEC[S].2018

[2] AQ/T 3054-2015. Guide for Application of Protective Layer Analysis Methods. Chapter 8： Scenery Frequency Calculation [S].

[3] Khan,Seyed Javad Hashemi,Nicola Paltrinieri,Paul Amyotte,Valerio Cozzani,Genserik Reniers.

Dynamic risk management: a contemporary approach to process safety management[J]. Faisal Current Opinion in Chemical Engineering.

[4] CCPS. Risk-based Process Safety [M]. 2013.

[5] Maryam Kalantarnia,Faisal Khan,Kelly Hawboldt.Dynamic risk assessment using failure assessment and Bayesian theory[J].Journal of Loss Prevention in the Process Industries . 2009 (5)

[6] Rakesh Ranjan,Sonam Singh,Satyanshu K. Upadhyay. A Bayes analysis of a competing risk model based on gamma and exponential failures[J] . Reliability Engineering and System Safety . 2015

[7]Zugravu-Soilita N. How does Foreign Direct Investment Affect Pollution? Toward a Better Understanding of the Direct and Conditional Effects[J]. Environmental & Resource Economics, 2017:1-46.

[8] Khakzad N, Yu H, Paltrinieri N, et al. Chapter 5 – Reactive Approaches of Probability Update Based on Bayesian Methods[J]. Dynamic Risk Analysis in the Chemical & Petroleum Industry, 2016:51-61.

[9] Villa V, Paltrinieri N, Khan F, et al. Towards dynamic risk analysis: A review of the risk assessment approach and its limitations in the chemical process industry[J]. Safety Science, 2016, 89:77-93.

[10] Mark Adrian van Staalduinen,Faisal Khan,Veeresh Gadag. SVAPP methodology: A predictive security vulnerability assessment modeling method[J]. Journal of Loss Prevention in the Process Industries.

[11] Podofillini L, Dang V N. Conventional and dynamic safety analysis: Comparison on a chemical batch reactor[J]. Reliability Engineering & System Safety, 2012, 106(4):146-159.

[12] R A F. Problems with risk analysis in the chemical industry. A detailed examination of the theoretical and practical problems faced by the risk analyst in the study of a chemical plant[J]. Plant/operations Progress, 1983, 2(3):185–190.

549550 Process Hazard Analysis

Process Safety Management System starts with managing the process related risks and from the first step of establishing a new process unit that is needed to evaluate the risks and define the necessary measurements. Process Hazard Analysis methodologies are the systematic tools which help to evaluate the risks in a process. For different needs there are several methodologies like DOW FETI, HAZOP, What if, HAZID, Fault Tree and Event Tree Analysis, Scenario Modelling, Facility Siting etc. Some of these methodologies are semi quantitative some are quantitative methodologies. In order to define and prevent major industrial accidents in our refinery we use DOW FETI, HAZOP, FTA and ETA and Scenario modelling. Moreover for some kind of projects we may also apply What if and HAZID analysis instead of HAZOP depending on the complexity. Facility Siting is another tool for risk related studies which is used to evaluate the risks for buildings in order to protect people inside the buildings. The workflow for a full unit PHA study starts with DOW FETI calculation which is used for risk ranking of al static equipment inside the unit. DOW FETI calculation is done in accordance with the equipment temperature, pressure, hold-up amounts, components and also reaction types if there is any. Then risk assessment process continues with HAZOP studies which is for scenario identification. In HAZOP, first step is chopping the unit into small parts to be evaluated which is called nodes. After that all nodes are evaluated using P&IDs and PFDs by keywords like temperature, pressure, flow, level etc. HAZOP helps to understand what kind of scenarios are possible inside the unit and these critical scenarios are taken through FTA and ETA studies. HAZOP is a semi quantitative risk assessment method where FTA and ETA are quantitative methods. FTA is applied for critical scenarios in order to find out the probability of top event then ETA is then applied in order to find out the probability of scenarios like fire, explosion, toxic dispersion etc. caused by this top event. FTA and ETA analysis all use the statistical PFD and failure rate values of all initiating events and barriers defined in HAZOP for the related scenario. Scenario Modelling is the following step which is used for understanding the effect of the probable scenario. The most probable scenarios coming from ETA is modelled to see the effect zones. These effect zones are used to plan Emergency Response Studies for all these potential major accidents.

548517 Site-Wide Asset Integrity Analytics & Benchmarking

All too often, reliability engineers focus on the equipment that is deemed as most critical or covered by some regulatory reference, while sometimes forgetting about those areas where something as simple as water can cause major issues. Non-destructive testing technicians and API inspectors alike can help facilities gather and compile data to perform predictive analytics and even benchmark sites with similar assets. This data can be used to identify assets where others may be seeing issues. This also allows facilities to compare reliability engineering effectiveness. This presentation will share some sample results of non-destructive testing and inspections of both regulatory required and non-required assets in the fuel ethanol production sector as well as an example of how that data can be assembled and shared across any sector to educate and learn from each other. Afterall, there are no secrets in safety!

548330 Reliability, Availability, and Maintainability (RAM). What Does It Provide?What Should You be Aware of before You Conduct a RAM Study?

Reliability, Availability and Maintainability (RAM) studies can significantly add value to your Process Safety Management (PSM) activities, especially with regards to Mechanical Integrity (MI). RAM studies are an excellent way to determine which equipment limits you from achieving optimum availability and throughput. Using industry failure rate data or even site-specific data can help you identify areas for improvement and determine how best to maintain your equipment. But RAM studies can go beyond purely theoretical studies by using the experience and knowledge gained from operating and maintaining similar facilities to develop a pragmatic approach to solving mechanical integrity-related issues. This paper will cover the pros and cons of site versus industry data and will present a case example of utilizing experience to better define the proposed problems and solutions. If done correctly, RAM studies can be more than statistics; they can be roadmaps to process reliability improvements.

546707 Driving Asset Integrity and Process Safety Consistency across a Multinational Organization

Mexichem is in the process of implementing world-class process safety management (PSM) and Asset Integrity Management (AIM) programs globally for their facilities. This is a major undertaking as Mexichem, to a large extent, has grown through acquisitions. Therefore, each plant has a safety culture and management system unique to that asset. As we all know, it's difficult to drive global consistency when dealing with different types of operations in various regions throughout the world under differing regulatory regimes. This is especially difficult when there is no common understanding of process safety. One of the first steps Mexichem has taken is to develop a "fit-for-purpose" management system with an emphasis on developing a common understanding of process safety and ensuring the culture is robust enough to support the system when it is implemented.

To aid Mexichem in this overwhelming task, Risktec Solutions were engaged to develop and deliver a bespoke set of interactive training workshops on a variety of PSM / AIM topics. The first of which was related to the identification and management of safety critical equipment (SCE). This presentation will outline the plan and methodology Mexichem and Risktec have used to transfer such knowledge and discuss some of the successes and challenges along with way.

548216 Alarm Floods and Their Connection to Plant Incidents and Near Misses

Investigations of many catastrophic plant events have shown alarm flooding to have been significant distractions for the operators just prior to the incident. Alarm floods have also been determined as a cause or contributing factor to smaller incidents and near misses.

In fact, many of the more recent findings by U.S. Chemical Safety and Hazard Investigative Board (CSB) include alarm management as one of the key issues in the investigative reports.

Even after several years of trying, many plants still struggle with controlling alarm floods. Many also have problems generating consistent and accurate alarm metric reports for the entire plant. Now that the ISA 18.2 Alarm Performance Metrics are considered "generally accepted good engineering practices," we should make haste to demonstrate where each of our plants fall (particularly the PSM plants) versus these metrics before the regulating authority moves in this area. If we are not able to prove we meet the metrics (particularly for alarm flooding) under all operating conditions, we should remediate.

Static rationalization can reduce your average number of alarms, but without controlling the alarm floods, there is no help for the operator when he needs it the most. For most, we must begin the process by recognizing that alarm floods occur most of all during a change of process state.

The means of controlling alarm floods has been available to us for many years. Unfortunately, only a select few have implemented the solution and are achieving the desired results today. These plants have achieved equal to or better than ISA 18.2 since before it was published.

This session will cover the justification for true alarm management from the safety and economic perspective with several practical steps that everyone should take if they are not currently meeting the ISA 18.2 specifications during all operating conditions.

548318 Do Incident Data & Lessons Learned Really Facilitate Process Improvement?

Regulations are a set of guidelines for the industry for protecting the people and environment, when it comes to process safety systems. While some companies tend to find that complying with regulatory requirements are always a challenge, on the other hand some tend to go overboard in their bid to stay compliant. In spite of the route taken, there are incidents – major, minor and near misses.

Other than the accumulation of this data, the investigations also result in lessons learned. Over the past four decades, lot of incident data and lessons learned has accumulated, been assimilated, and also analyzed. Complementing that are the audits, system reviews, hazard identification exercises, changes to regulations, and knowledge sharing forums.

It is a very efficient and effective mechanism to leverage these vast sources of current information developed, and implement its learnings, in the operations of existing processes. While the core concepts of safety are universal across the various industrial sectors, its application may differ. But once again, the lessons learned databases are effective sources of making this leveraging process more efficient.

This paper will discuss the some case studies of how the concept of lessons learned and the knowledge of the incident data has been leveraged by the authors in various manners ensuring success from one industry to another – example (1) Learnings from an incident due to gaps in the Emergency Response Plan, bringing about success at another chemical manufacturing site; (2) Incident data from a steel plant, when analyzed and applied to a chemical plant resulted in improved of effectiveness.

546673 Structured & Scrutinized Approach Towards HSE Excellence; A Success Story

Author: Usman Ali, Process Safety Engineer

Co-Author: M Ahsan Sarfraz, Unit Manager HSE

This paper studies the practices & programs that support reporting of Near misses, minor unsafe conditions, incidents and quantitative risk assessment strategies in our organization for analyzing and learning from past events of operational failures, injuries, safety and environmental incidents.

A near miss is a leading indicator to an incident that, if scrutinized appropriately, can prevent injuries and accidents. Some persons may be tempted to write off near misses as "no harm, no damage" situations, but organizations who track near misses, determine how and why they occurred, and take corrective action can prevent similar or more serious incidents from occurring in the future. Fatima Fertilizers Limited believes that, "We don't look at what happened; we look at what could have happened".

Keeping in mind the importance of near miss reporting, FFL initiates drive and implement systems for

1. Capturing and increasing near miss reporting at whole site.
2. Logging data for analysis and using this valuable data for identifying gaps in existing process safety system.
3. Implementing sustainable improvement plans and safety campaigns (Safe hands, Safe Routines, Job Cycle Checks, Safe Drive etc) in our process safety systems for making our site a safer workplace.

These actions has improved the site performance which is reflected as Site has now over 45 Safe Million Man Hours,

Total Record-able Injury Rate (TRIR) is below 0.2 since last 03 years and No Loss time injury since last 06 years.

548162 Events Report Database Analysis for SHE/PSM Programs Improvements

Incident databases contain records of events that have occurred in a petrochemical facility during recent years. The most of chemical industries report databases are filled with accident valuable data (frequency and severity). Harnessing data mining to explore incident databases, it was used data mining to analyze the incidents database of the studied plant. Inferent approaches are ineffective for low frequency, high consequence events because of their rarity. Given this information limitation, this work uses database analyses to forecast incident frequencies, their relevant causes, equipment involved, and their possible consequences, in specific petrochemical plants.

Databases information analysis, also help to avoid future accidents, thereby reducing the risk. More specifically, this study presents dynamic analyses of incidents in a petrochemical plant ubicated in Mamonal industrial park, in Cartagena- Colombia. Differences in the chemical-specific predictions with the overall predictions are demonstrated. Furthermore, value theory is used for consequence of rare events by formulating distributions for events over a threshold value. Finally, the accidents number is used to estimate the capital at risk within an industry utilizing the frequency and loss-severity distributions.

In conclusion, it was recommended to update de maintenance program (predictive, preventive and corrective) in cases of abnormal changes in production rates, comparing the current production levels, with the analyzed plant production forecast. Also, it was suggested to improve the report culture, due to gaps founded regarding to events classification.

549483 The Process Safety Analysis and Management System in Sinopec

The big data technique has been in trend for the process safety assessment. In china, it's the very need to transfer the qualitative safety management to quantitative safety management, which should rely on the professional risk analysis technique, as well as the standard safety reliability database. Based on the above, it's very important to integrate several professional risk analysis technique, as HAZOP, LOPA, SIL,BOWTIE etc. to realize the integral risk identification and evaluation, grading and management plan making, supporting by the proprietary SINOPEC risk metrics and grading standards, as well as safety equipment reliability database,which is developed as the Process Safety Analysis and Management System for SINOPEC, to support all SINOPEC branches to make dynamic risk analysis and management online

The key data for HAZOP, SIL and Management module in PHAMS is the dependent protection barriers, which we do focusing on the failure probability, at the same time, we do use of the Bayesian Correction Modelling to verify the failure probability of safety equipent and protection barriers on site, which can validly save the grading cost on time and personnel, all of this is towards to the deeper research on dynamic risk analysis for PHAMS.

Thanks to the PHAMS, we has set up and implemented several standards on HAZOP, LOPA and SIL for SINOPEC.

552170 Analysis and Improvement of Safety Integrity Level Considering Human Factors

Safety Instrumented System (SIS) is an irreplaceable active protection layer that prevents the propagation of process accidents and plays an important role in reducing the risk of accidents. In addition to the sensor response triggering the SIS, a human emergency button is often provided in practice to ensure that the SIS is triggered by personnel in the event of the sensor part failure. Due to human factors added, the emergency button will improve the reliability of the SIS. However, there is no research to study the human reliability of the SIS upon sensor part failure in a continuous operation chemical plant, and how human factors can improve and affect the overall reliability of the SIS still remain undiscovered. In this paper, a verification methodology was proposed which integrated with Cognitive Reliability and Error Analysis Method (CREAM) and Reliability Block Diagram (RBD) to emphasize human reliability analysis of the SIS verification. The operator behavior is subdivided into several steps and the Human Error Possibility (HEP) is calculated by a modified CREAM. Then the RBD method is used to verify the SIS reliability considering human factors. An SIS system of one reacting furnace was taken as a case study using this method to show how human factors can improve the reliability of the SIS. The proposed methodology allows SIS verification considering human factors to improve safety integrity level in real practice thus help stakeholders saving costs and making more decisions on SIS implementation.

551948 Do we *REALLY* need more Process Safety Leaders? The Criticality of Effective Followership

Process safety, as is the case with many other disciplines, emphasizes the need for strong leadership and commitment. Leadership from the top is a critical element in any process safety program. However, it is a necessary, but not a sufficient condition for success.

In this presentation, the need for competent followership will be discussed and illustrated.

Industrial-Academic Collaboration to Teach Chemical Process Safety

John R. Schlup, PhD
Professor
Kansas State University
jrsch@ksu.edu

Jennifer L. Anthony, PhD
Associate Professor
Kansas State University
anthonyj@ksu.edu

Dr. Keith L. Hohn, PhD
William H. Honstead Professor of Chemical Engineering
Kansas State University
hohn@ksu.edu

Chris Aiken
EHS Senior Process Safety Director
Cargill, Inc.
Chris_aiken@Cargill.com

Christopher Frampton
Integrated Project Manager
Cargill, Inc
Chris_Frampton@cargill.com

Prepared for Presentation at American Institute of Chemical Engineers
2019 Spring Meeting and 15th Global Congress on Process Safety
New Orleans, LA
March 31 – April 4, 2019

Abstract

Chemical process safety is historically an overlooked topic within chemical engineering education. In response to a reactive chemical explosion at T2 Laboratories that killed four employees and injured 32 others including employees of surrounding businesses, the Chemical Safety Board emphasized the critical need to improve chemical process safety and hazard recognition education within undergraduate chemical engineering.

As part of the AIChE Foundation's 'Doing a World of Good' campaign [1], and in conjunction with the Center for Chemical Process Safety's (CCPS), industry and academia have come together on a major global initiative to improve and accelerate process safety education at the university level. Experience has shown that presenting process safety concepts as part of the chemical engineering curriculum proves most effective, something the Accreditation Board for Engineering and Technology (ABET) also recently adopted as a requirement in 2012.

Cargill, Inc. and the chemical engineering faculty at Kansas State University believe the education of future engineers is a critical success factor in the pursuit of zero harm to people, communities and our environment. One significant way to achieve exceptional process safety performance in our industry is through collaboration between faculty and industry to incorporate process safety concepts into the engineering curriculum and design courses. The collaboration between Kansas State University and Cargill, Inc. started in a CCPS Process Safety Faculty Workshop hosted by Cargill in 2016 where chemical engineering faculty members were shown how the concepts of process safety are put into industrial practice and to highlight the role that process safety plays in the design and operation of a chemical process industry (CPI) facility. One workshop outcome was a deeper partnership that would enhance process safety education in a KSU capstone design course. In this partnership, each student design team was assigned a process safety coach from Cargill. This coach led sessions to help the student designers understand how to apply fundamental concepts of process safety and build awareness that the choices they make in the design of a process impacts the safe operation of a facility. The coaching sessions included:

- Identification of process safety hazards during the design of a process
- Considerations for inherently safer design
- Identification, evaluation and mitigation of risk
- Understanding of types of codes, standards, and Recognized and Generally Accepted Good Engineering Practices (RAGAGEPs)
- Process safety education materials from AIChE Safety and Chemical Engineering Education Program (SAChE) and the Center for Chemical Process Safety

1 Introduction

Designing and operating chemical processes in a safe manner is a key job requirement for practicing chemical engineers. Chemical engineers working in industry spend extensive amounts of time reviewing the safety of the facilities in which they work and are frequently asked to review the potential hazards of processes or changes in the operation of those processes. For this reason, employers of chemical engineers expect chemical engineering students to acquire a strong background in chemical process safety during their undergraduate studies.
Safety education in general is an essential but historically overlooked topic, particularly in chemical engineering. Initiatives have been implemented within the chemical engineering community to enhance process safety education. The Safety and Chemical Engineering Education (SAChE) program was initiated in 1992 to provide teaching materials and programs to enhance the teaching of process safety at the undergraduate and graduate level[1]. This program is a cooperative effort between the Center for Chemical Process Safety (CCPS) and universities. Initiatives implemented by SAChE at the undergraduate level include an on-line certificate that students can earn by completing several modules on safety and the SAChE Student Design Competition for Safety in Design that recognizes students who do an outstanding job considering process safety in their submission to the American Institute of Chemical Engineers (AIChE) Annual Student Design Competition.

However, in 2007, a reactive chemical explosion at T2 Laboratories killed four employees and injured 4 more T2 employees and 28 employees of surrounding businesses[2]. Within their final report, the Chemical Safety Board emphasized the critical need to improve chemical process safety and hazard recognition education within undergraduate chemical engineering programs. The final CSB report recommended that the ABET and the American Institute of Chemical Engineers (AIChE) ensure reactive hazard awareness is included within ChE curriculum (U.S. Chemical Safety and Hazard Investigation Board, 2009). In response, AIChE and ABET worked to specifically address these concerns into the program criteria for chemical, biochemical, and biomolecular engineering programs. Many programs, Kansas State University included, have also added a safety class to the curriculum, and industry has begun offering process safety workshops for faculty.

The above initiatives have undoubtedly improved student understanding of process safety, but they typically do not directly engage industry process safety experts in educating students. Industry engagement is an effective tool for training students in relevant professional skills. Given the importance of process safety in the daily job of practicing chemical engineers, engaging practicing chemical engineers in educating the next generation of chemical engineers has significant potential.

A partnership between the chemical engineering department at Kansas State University and Cargill has been established over the past two years to enhance process safety education in a capstone design course. In this partnership, each student design team was assigned a process safety coach from Cargill. This coach provided guidance on how the teams should incorporate process safety in their designs. The coaching sessions also provide a practical application of an established tool (Hazards and Operability (HAZOP) analysis) to analyze process hazards of a new design. This paper reports on the impact of this collaboration by analyzing student survey data before and after the semester, and student performance data on the safety component of their design reports. This data shows that the industrial-academic collaboration positively impacts students, most significantly by enhancing their ability to apply HAZOP analysis to their design.
Improving the chemical processing industry's process safety performance is paramount. Cargill, Inc. and the chemical engineering faculty at Kansas State University believe the education of our future engineers is a critical success factor in the pursuit of zero harm to people, communities and our environment. One significant way to achieve exceptional process safety performance is through collaboration between faculty and industry to incorporate process safety concepts into the engineering curriculum and design courses.

2 Process Safety Knowledge of Student Designers

Process safety has previously been implemented into the KSU chemical engineering program in several ways. First, student design teams in the two-semester design sequence (ChE 570 and ChE 571) have been required to address process safety in the three different design projects they complete. Their treatment of process safety is evaluated based on a four-point rubric. To help the students properly address safety in their design reports, topics such as different process hazards analysis methods (HAZOP, What-If Analysis, Fault Tree Analysis), pressure relief valve sizing, and layers of protection analysis are often discussed in these courses. In addition, a required course in process safety was added as a curriculum requirement in 2015 to provide a more in-depth treatment of process safety. Course topics include toxicology, industrial hygiene, toxic release and dispersion models, fires and explosions, safety relief systems, and HAZOP and risk assessment. This course was initially offered as a 1-hour course that students took during the first part of the senior spring semester, which is the same timeframe as their second design class and the Cargill coaching. It should be noted that in 2018, this course was expanded to three credit hours and moved to the fall semester, concurrent with their first design class. This change means the current Spring 2019 class of students who are just beginning to interact with the coaches from Cargill have already taken a course in process safety. However, all the discussion in this paper is for students who took the original 1 credit safety class concurrent with their final semester of design.

3 Process Safety Coaching Objectives

The purpose of the process safety coaching sessions during their capstone design project is to help chemical engineering students apply fundamental concepts of process safety, build their awareness that the choices they make in the design of a process impacts the safe operation of a facility, and emphasize the degree of importance industry places on process safety.

An experienced Cargill engineer leads three coaching sessions with each student design team and supports the team throughout the semester to understand and address process safety risks associated with their design. These coaching sessions help the student design team understand the process hazards, process safety risks and how their choices early in the design can affect the operating risk of the process they designed.

Upon completion of the process safety coaching the student designers should have a strong understanding of:

- Identification and assessment of process safety hazards
- Inherently safer design considerations
- Identification, evaluation and mitigation of risk
- Understanding of types of codes, standards, and Recognized and Generally Accepted Good Engineering Practices (RAGAGEPs)
- Resources available for process safety education including materials within the AIChE Safety and Chemical Engineering Education Program (SAChE) and CCPS guidelines and concept books, etc.

3.1 Session 1: Introduction

The purpose of session 1 is to review with the student design teams the purpose of the coaching program, individual sessions, review the timeline and plan, set expectations for the design teams, and communicate what the teams can expect of the coach. This alignment is critical to set the foundation for successful interaction between the coaches and student designers. Also included is a brief refresher on process safety concepts the teams have already encountered in the ChE 570 course such as risk, hazards, inherently safer design and process hazard analysis. Session 1 typically lasts 30-45 minutes and is scheduled during normal class periods within the first 2 weeks of the course. Session 1 helps establish a safe atmosphere where student designers feel empowered to engage openly, ask questions and risk making mistakes with experienced engineers of potential future employers.

3.2 Session 2: Hazards and Inherently Safer Design

The purpose of session 2 is to help the student design teams understand the process safety hazards associated with their process, concepts of inherently safer design, existence of codes and Recognized and Generally Accepted Good Engineering Practices (RAGAGEP) and awareness of other resources available to the chemical process industry.

At this point most students have only a basic level of understanding, so the coach assesses the knowledge level of the team and determines the amount of information to provide so that the team is not overwhelmed. Even within a design team, students have different knowledge levels based on their experiences, such as industry internships and coops, research experience and previous coursework. The goal is not to make students process safety experts in one semester but rather to ask thought provoking questions specific to their design project and guide the teams through their discussions. At times the coaches may provide specific answers to questions but in general keep the discussion to guidance. For example, the coach might ask about vacuum in a tank and let them think and discuss this for a couple of minutes. Then the coach might need to share specific examples to explain how vacuum can occur in a tank, how atmospheric tanks collapse under vacuum, and guide the teams toward a potential design solution to first remove this risk, and if not possible, manage this risk to an acceptable level.

In preparation, one week before session 2 the student design teams send their assigned coach an email detailing the different process options, list of chemicals, and anticipated process hazards. The students also watch two videos produced by the US Chemical Safety Board highlighting inherently safer design and understanding chemical reactivity. The coach in turn reviews the submitted design information, identifies further potential hazards and inherently safer design considerations for discussion and identifies applicable codes, standards, and RAGAGEPs.

The session is a one-hour teleconference with the following agenda:

1) Introductions
2) Zero Harm Contact - Discuss what the student designers learned from the 2 videos

3) Identification of the hazards
    a. Hazards associated with the materials
    b. Hazards associated with equipment
4) Inherently safer design
    a. Process selection
    b. Equipment design
    c. Facility siting
5) Codes, standards, and Recognized and Generally Accepted Good Engineering Practices (RAGAGEPs)
6) Other industry resources available to the student designers (i.e. product stewardship bulletins)

3.3 Session 3: Process Hazard Analysis

The purpose of session 3 is to help the student design team understand how the choices made in the design of their process impacts the safe operation of a facility. More specifically the students will conduct a "partial, qualitative" risk assessment on a single unit operation utilizing the Hazard and Operability study (HAZOP). The goal is to have the student designers think about what can go wrong (process loses control), what can happen (end consequence) and what safeguards are in place (what brings it back to control or minimizes the consequence).

The focus is not for the coach to teach how to perform a HAZOP, but for the students to realize that things malfunction, errors are made, and processes do not only operate at steady-state conditions but have deviations. This session also challenges the students' understanding of how the unit operation works and drives a deeper understanding of the unit operations' physics and chemistry.

In this third session the goal is to increase the student designer's awareness that:

1) The process will operate at conditions other than steady state
2) Operational upsets occur and must be anticipated
3) Designs must be robust enough to be able to withstand deviations
    a. A single human error or valve failure should not directly cause a catastrophic failure
    b. How to include layers of protection in their design
4) People (operators & maintenance) are part of the process and directly affect it
5) The design must also take into account transient operations (startup, shutdown)

Similar to previous sessions, the coach must evaluate the team and each individual's level of knowledge and adjust the approach. Initially the coach may need to provide an example scenario, safeguards, and recommendations. Then as the session progresses the coach shifts to asking probing questions to guide the team's discussions as required. The teams will almost certainly require technical help at times, but the coach must be cognizant not to complete the HAZOP themselves but rather walk the team through the analysis process, allowing them the time to consider and discuss amongst themselves. Given the high-level goals of session 3, the HAZOP does not incorporate risk ranking matrices or consider matters of independence.

Midway through the project the design team submits a process flow sheet and in preparation for session 3 the coaches and faculty select a single unit operation to perform the process hazard analysis on. The design team then defines in detail the specific design requirements for the selected unit operation including sizing, instrumentation, safety devices, and control strategy. The piping and instrumentation diagram, along with other design information, is sent to the process safety coach one week prior to session 3 so the coach can prepare to facilitate the process hazard analysis. To prepare the teams for scenario development during the HAZOP, the students are also asked to watch two videos from the US Chemical Safety Board (Inherently Safer: The Future of Risk Reduction and Runaway: Explosion at T2 Laboratories) providing examples of process deviations that were initiating causes to catastrophic incidents. For example, the T2 Laboratories incident was caused by failure of a reactor cooling system which did not have a backup.

The session is a three-hour video conference, with screen sharing ability, with the following agenda:

1) Zero Harm Contact – discussion of what the student designers learned from the 2 videos
2) PHA overview
3) Conduct the PHA
    a. Select 2-3 trigger words to focus on (i.e. pressure – high, low)
    b. Look to develop 10-15 scenarios
    c. Focus on building a good scenario and recording already designed safeguards (if any)
    d. Make recommendations to modify design or add additional safeguards
    e. Leave 20 minutes to discuss siting issues – i.e. control room, truck traffic
        i. The siting plan is unlikely done, but the discussion increases awareness
        ii. Where should the control room be
        iii. Discuss area electrical classification or building ventilation if appropriate
4) Last 10 minutes ask for feedback how the session went

4 Results

KSU faculty have been pleased with the results of this collaboration with Cargill. A few key benefits have been noted. First, the students get an in-depth experience in process hazard analysis and inherently safer design that would be difficult to reproduce in a purely classroom academic setting. In addition, the experience is tailored to their unique interests, since they are addressing the process hazards associated with their specific design. Another strength of the collaboration with Cargill is that the students gain an industrial perspective on process hazards analysis. The students generally reacted positively, recognizing that they were learning from professionals who were applying the same skills in their jobs. Anecdotally, student comments indicate that these interactions with the Cargill coaches emphasize an increased appreciation for how critical process safety considerations are in industry.

Qualitatively, the KSU design instructors noted improvements in the students' performance in recognizing and designing for the process hazards in the design project. This is demonstrated quantitatively when student scores on different design reports were compared. Students complete three different design reports in their senior year: a team report based on their design project in ChE 570 Chemical Engineering Systems Design I, a team report associated with the AICHE design competition, and a team report for a design project on the "local" design project (projects selected by the course instructors). The rubric for evaluating their designs includes two categories with metrics specifically related to process safety. These categories and evaluation criteria from the rubric are:

Consideration of Safety

> 4 (Exceptional): An initial attempt at HAZOP or similar analysis (beyond an initial "what-if" analysis) has been performed on at least one unit of the design. Implications of process safety specifically incorporated into the design; for example, use of HAZOP considerations in the P&ID diagram. Considerations of appropriate standards and regulations have been included within some part of the design.
>
> 3 (Acceptable): Considerations of inherently safer process design have been applied, with an explanation of the rationale. Considerations of process safety are apparent in the design.
>
> 2 (Marginal): MSDS safety sheets have been reviewed. Plans within the design include basic considerations of personal and process safety.
>
> 0-1 (Unacceptable): No meaningful discussion of safety (either personal or process).

Health & Environmental Considerations

> 4 (Exceptional): Green chemistry and engineering are clearly considered in the preliminary evaluation. Regulations and/or appropriate engineering standards have been specifically included. The life of by-products and energy flows are part of the design.

3 (Acceptable): The impact of health and environmental effects were considered in the design. Specific health and environmental considerations are apparent.

2 (Marginal): MSDS safety sheets have been reviewed. General considerations of health and environmental issues are indicated.

0-1 (Unacceptable): No meaningful discussion of safety (either personal or process).

The Cargill safety coaches assist the students only on the last design project. This allowed a comparison of students' ability to perform process hazards analysis with and without receiving input from industrial practitioners. Figure 1 compares the results for these three design reports.

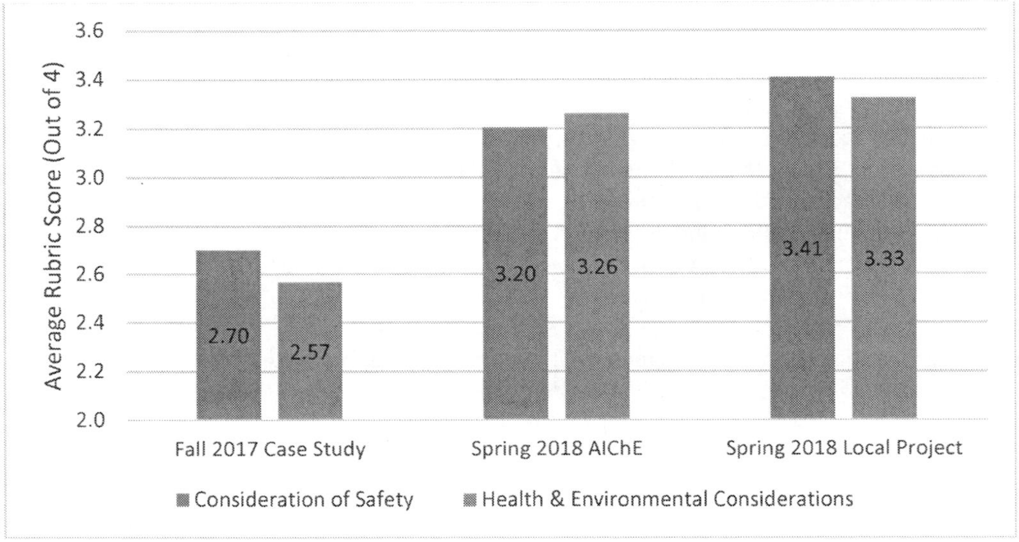

Figure 1. Student performance on design reports on "Consideration of Safety" and "Health and Environmental Considerations", as judged by instructor scoring on a four-point rubric

As seen in Figure 1, student scores markedly improved from the fall, 2017 project to the design reports in the spring for both "Consideration of Safety" and "Health and Environmental Considerations". A further, smaller increase in both areas is noted from the AICHE design report (turned in on March 16th, 2018) and the local design project (turned in on May 4th, 2018).

There can be several explanations for the increases in student performance relative to plant safety and environmental considerations. Students received faculty feedback on each design report to improve performance on subsequent reports. The students in this study took a one credit-hour course (ChE 565, Healthy and Safety in ChE) related to process safety in the 2018 spring semester prior to the submission of their spring design reports. This should have significantly improved their understanding of process safety. Finally, the expectation is that interaction with the Cargill safety coaches with the local project in the spring improved student performance as well. It is difficult to separate out the individual impact of faculty feedback, the undergraduate safety course, and the Cargill mentoring when comparing Fall 2017 data with the Spring 2018 AICHE project data. Comparing the data for the two projects in Spring 2018 may be more instructive, although the change noted is smaller, in part because the student performance had already improved significantly. We hypothesize that the Cargill coaching is responsible for most of the improvement noted between these two projects. Students had already completed the safety course by the time the AICHE design project was submitted (the last day of class for CHE 565 was March 8, 2018), so it is doubtful that it is responsible for any increase. While faculty feedback on the design reports may have played a role, the students would have already received feedback on one report, so the impact of more feedback would probably not have been as significant. We hypothesize that the opportunity to be coached in HAZOP analysis by the Cargill safety coaches improved students understanding and capability to address process hazards analysis.

4.1 *Student perspective*

Students have commented positively on the Cargill mentoring experience. In particular, they have noted how exhaustive the HAZOP process can be in examining every possible cause that could lead to an unsafe situation, as well as how many control systems are involved in the safe operation of different unit operations equipment.
The student perspective was further probed by surveying the students about their motivation, anxiety, and ability to complete various tasks related to process design and process safety. In all, there were 33 questions on student motivation, anxiety and ability. The survey also asked for their opinion on the importance of safety at the industrial level. In 2017, 38 students took the pre-semester survey and 10 took the post-semester survey, while in 2018 39 students took the pre-semester survey and 31 took the post-semester survey. Table 1 shows the questions for which there was a statistical difference between the pre- and post-surveys (based on a t-test for two independent samples). While differences were noted in the pre- and post-surveys for the other questions, the differences were not found to be statistically significant

Table 2. Survey questions that were found to give a statistically significant change from the pre-semester to the post-semester surveys

Statistically significant change between pre- and post-semester survey results	2017	2018
90% significance level	Success in - Identifying a design need	
	Success in – Selecting the best possible design	
	Success in – Conduct a process hazards analysis (PHA)	
	Success in – Identify safety risks in a process design	
	Success in – Implement the principles of inherently safer design	
95% significance level	Success in - Conduct engineering design	Success in - Use the HAZOP approach to consider safety hazards for a unit operation
	Success in - Construct, evaluate and test a design	
	Success in - Construct a prototype by using a process simulator	
	Success in - Use the HAZOP approach to consider safety hazards for a unit operation	

Notably, the student's response on the success at conducting a process hazards analysis and implementing the principles of inherently safer design was statistically higher at the end for the semester at the 90% significance level in 2017, while in both 2017 and 2018 the student ability to successfully use the HAZOP approach to consider safety hazards for a unit operation was statistically higher after the semester had ended. Since the key component of the Cargill mentoring program was the phone conference where the Cargill coaches guided the students through a HAZOP analysis, it is not surprising that the students felt more confident in their ability to conduct a HAZOP analysis.

4.2 Industry Perspective

The unique partnership between Cargill and the Chemical Engineering department at Kansas State University demonstrates the potential for industry to assist in process safety education within chemical engineering education. With such efforts industry can employ chemical engineers with better process safety understanding and build their specific company and industry specific programs on this stronger foundation. These student designers will also enter the workforce better prepared to incorporate process safety considerations into their daily work, designs, and operations.

Cargill has also noted there is significant benefit for the coaches that participate in the program. While all the coaches are familiar and manage process safety regularly, many coaches are not process safety professionals but have other roles within Cargill. These coaches have roles in project management, production management, reliability excellence, and manufacturing technology. By participating in the coaching program, these professionals have the chance to review and refresh the basic principles of process safety and deepen their process safety knowledge. This improves and revitalizes Cargill's process safety focus across its organization and facilities.

Cargill has also noted there is significant benefit for the coaches that participate in the program. During the spring semester of 2019 Cargill is providing approximately 25 coaches who are supporting 48 student design teams. The coaches will invest about 8 hours per design team over the five-month period. We found that this experience creates an awareness within the coaches of how much knowledge and practical experience they have gained since graduation and Cargill's role in their development. Although there are just three sessions with a time investment of approximately eight hours per design team over a five-month period the company finds that the coaches' engagement and enthusiasm for process safety is re-energized. The coaches' feedback taken at the end of the coaching sessions overwhelmingly supported this initiative finding the experience rewarding and a great break from their daily routine.

5.0 Conclusion

Cargill, Inc. and the chemical engineering faculty at Kansas State University believe the education of future engineers is a critical success factor in the pursuit of zero harm to people, communities and our environment. Partnership and collaboration between academia and industry has the untapped potential of enhancing the student's learning experience so that the next generation of engineers are better prepared to design and manage the safety risks within the chemical processing industry which is becoming increasingly complex and more demanding.

5 References

[1] Doing a World of Good, AIChE Foundation

[2] U.S. Chemical Safety and Hazard Investigation Board Investiation Report No. 2008-3-I-FL September 2009

Competence Assurance System (CAS) Implementation in RIL Manufacturing – Refining & Petro-chemicals

Dr. Mat Malladi
Reliance Industries Limited
Mumbai 400701 (India)
Mat.malladi@ril.com

Prepared for Presentation at
American Institute of Chemical Engineers
2019 Spring Meeting and 15th Global Congress on Process Safety
New Orleans, LA
March 31 – April 3, 2019

Competence Assurance System (CAS) Implementation in RIL Manufacturing – Refining & Petro-chemicals

Dr. Mat Malladi
Reliance Industries Limited,
7B, 2nd Floor, RCP, Thane-Belapur Road, Ghansoli, Navi Mumbai 400701 (India)
mmalladi@mac.com

Keywords: Competency, Assessment, Assurance, Job Role, HSEF, Operations, Maintenance,

Abstract

Reliance Industries Limited (RIL), operates multiple assets spread across many geographies. Hydrocarbons and Chemicals handled at these facilities are hazardous and hence require a competent workforce to ensure safe & reliable Operation.

Towards this, RIL designed, developed and implemented a Competency Assurance system (CAS to de-risk and assure Safety & Reliability of operations, people, environment, business, assets and reputation.

The emphasis is on human reliability dimension to drive organization to have the right people with the right skills in the right job. It covered

- Frontline operations & maintenance jobs and activities that pose high HSEF (in other words Safety &Operational Risks-S&OR) risks to people and Assets.
- Supervisory roles that can impact assets and operations significantly from HSEF risks point of view. This involves positions that supervise contractors that perform HSE critical activities.
- Contractors and subcontractors staff performing HSEF critical tasks

The design & developmental activities involved

- Formulating & developing standardized Competence Catalogues for 8 academies Technology, Operations, Engineering & Maintenance and HSEF.
- Developing Corresponding Competence/Content links
- Creating consistent Job Competence Profiles
- Automating process up to Field Assurance –

The document covers design, development and execution of CAS as implemented in RIL.

1 Introduction

Competency development of employees is done through various stages.

The prerequisite for the process is to ensure employee meets the basic academic qualification prescribed for the job before entry into the industry.

The Competency development process begins by first identifying the tasks to be carried out by the employee. The tasks are detailed through JSA where the duration, nature of the job routine non-routine, physical conditions in the field are all clearly identified. A written procedure is then prepared tailored to the level of information needed by the performer. The procedure highlights the risk, KPIs and the Barriers of Protection (Controls), Do's and Don'ts. The content is enhanced and elaborated to higher levels as needed.

The process is time tested and authentic and experienced leaders have developed, reviewed and approved it. The process has three steps- Assessment of Learning gaps: Learning interventions to fill identified gaps followed by field Assurance of HSE critical tasks. The candidate is first assessed for the competencies required as per job description. If any competency gaps are identified, the candidate goes through tailored leaning interventions. The person is trained using a blended learning approach and re-assessed & assured in the field by actual demonstration of procedure (SOP, SMP) driven tasks & certified fit for on the job training.

To prepare the employee for assurance he or she is intensely trained on the job (OJT). In downstream manufacturing, about 80% of employee training occurs on the job). On the job, the employee is assigned a competent mentor who demonstrates how the job is performed in a right manner enough number of times till the mentee gets confidence and then mentor supervises mentee demonstrating the job adequate number of times to his satisfaction with no errors and underpinning knowledge.

Following which the employee is assured by independent assessor for certification prior to assigning independent responsibility (Licensed to operate).

The end-to-end process described above is summarized in the table below:

Table 1. Competence development & assurance (CAS) using blended learning

Blend of Learning interventions ⇓	Competence Development Process				Competence Assurance Process			
	Self-assessment ⇒ L1 review ⇒ learning interventions based on competence gaps ⇒ *move to competence assurance steps*				Online quizzes based on SOPs ⇒ learning interventions based on gaps ⇒ Perform field tasks & interview ⇒ learning interventions based on gaps ⇒ **certified**			
Proficiency scale ⇒	**A**	**K**	**S**	**M**	**A**	**K**	**S**	**M**
Academic qualifications	✓	✓						
1. **eLearning (web based)**	✓	✓						
2. **Classroom**	✓	✓						
3 **Plant Training**		✓	✓				✓	
A. Field Orientation/Training		✓	✓				✓	
B. Simulation (OTS)		✓	✓				✓	
4. **On the Job Training (OJT)**			✓				✓	
C. Assessment Under supervision			✓				✓	
D. Independent assessor certification							✓	
Independent charge (Licensed to Operate)								

A – awareness, K – knowledge, S-skill, M-master

2 Objectives of CAS

The objective of the Manufacturing competency assessment and assurance is to have a uniform practice across all sites, which will:

- Minimize the risks arising out of lack of competencies of employees carrying out S&OR and other critical tasks
- Safeguard people, environment, business, assets and reputation
- Help Manufacturing to ensure and build critical inventory of technical & functional skills
- Effectively address current & future business challenges through competent workforce

- Assure all the stakeholders viz. employees, management, the board, statutory authorities, the shareholders and the society at large
- Workforce Development & Assurance: Safe, Reliable and Productive work force Tailored training and development
- Support Human capital decision making by assessing gaps between competency requirements and available competencies. Hence facilitate recruiting/ Career development/ Succession planning
- Provides employees, organizations, and functional communities with the capability to find employees, positions, or projects that utilize competencies

Best practice on Talent Management (as shown below) recommends competence management as its foundational element. Success of CAS will ensure fulfilling one of Talent Management's end state goals of institutionalizing an Integrated Framework.

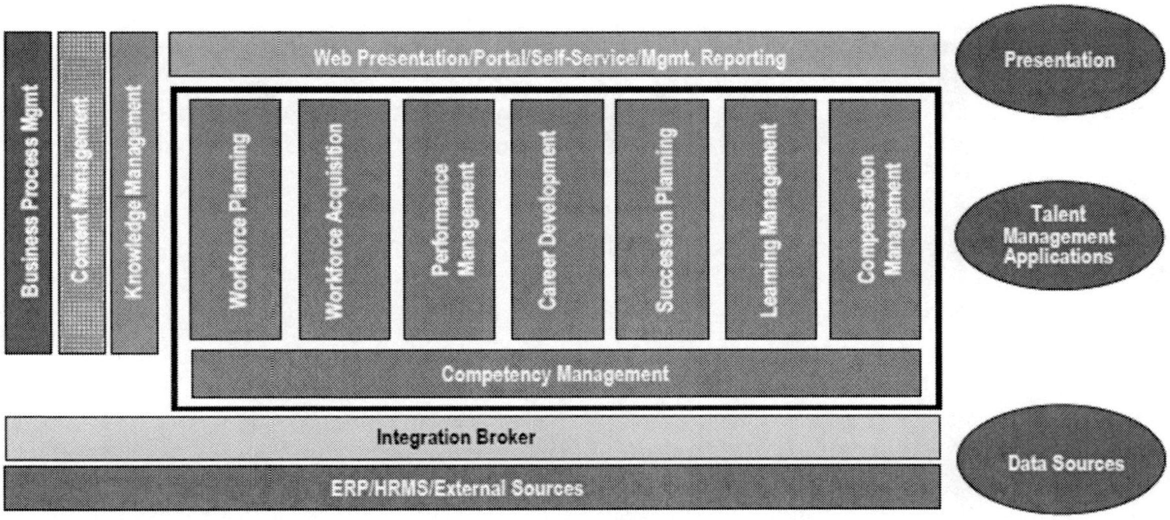

Figure 1. Integrated Talent Management Framework

3 S&OR (Safety& Operational Risk) Critical Competencies and Job Roles

3.1 S&OR Critical competencies

- S&OR competencies involve those that are required to carry out safety critical tasks and significantly impact safe, compliant and reliable operations.
- These competencies may result directly in C+ (purple or blue) risk events on Assessment, Prioritization and Management of Risk.

- S&OR critical competencies are subjected to a more rigorous assurance process to guarantee performance to a proficiency standard defined by RIL and grants an employee - License to Operate; Maintain; & Service Manufacturing Assets.

3.1 S&OR Critical Job Roles

- Tier 2 – S&OR Critical Roles that are directly asset facing, i.e. those who undertake direct Operation & Maintenance of plant equipment & piping and their first line supervisors are categorized under Tier2. Field Operators, Panel Operators, Shift Engineers, Shift Superintendent, Maintenance Technicians and Maintenance Engineers under Tier 2.

- Tier 1 – All S&OR Critical Roles that are not Tier 2 Roles are categorized under Tier 1 Roles. Senior roles in Operations, Maintenance, Engineering, Reliability, Technology, Projects, HSSE and other functional authorities are covered under this category which include Supervisors of shift staff up to Head of Operations & Maintenance and above

4 Outcomes of CAS – what success would like

4.1 Operational Assurance of all S&OR critical tasks

- Level 1 Tasks: Shift Staff includes Frontline Operational/Maintenance/QAQC staff and their leaders including Lab Technicians , Inspectors & contractor employees, if any in defined job roles

- Involves tasks, activities, or events where there are HSEF RAM Red/Yellow 5 A/B risks

- Viewed as mandatory for RIL License to operate in the current job

- Requires individual mentoring, training & tracking, reporting and recertification.

4.2 Technical Assurance of all S&OR critical tasks

- Technical Authority role holders in all Disciplines with HSEF (especially Process safety management) impact.

- Level 2 Tasks: Supervisory / Professional planning /Review / Audit tasks

    - Identify hazards and effects and consequences
    - Assess their significance
    - Provide controls & Mitigation for hazards and effects
    - Provide recovery preparedness & response in the event the controls are lost

- Requires individual tracking, reporting and recertification

4.2 Professional certification

- A 'record 'of a professional certification is essential part of competence development

- Recognized by external body/society or through an internally defined RIL standard
- Structured means to meet or exceed country/Industry standards and expectations

5 Scope and S&OR roles in Manufacturing

Table 2. Scope and S&OR roles in Manufacturing

Unit	Job Roles for Assessment & Assurance (per GDP 2.2-0001)	Functional Competence Catalogs
1) JMD **2)** HMD **3)** VMD **4)** DMD **5)** PMD **6)** NMD **7)** Polyester Sites **8)** Manufacturing Corporate functions	**Roles to be Assured** **1)** Tier 2 roles Assurance: • Field Executive • Panel Officer • Shift Field Engineer • Shift Superintendent • Plant Mechanical Engineer • Plant Electrical Engineer • Plant Instrumentation Engineer • Maintenance Executive – Mechanical • Maintenance Executive – Electrical • Maintenance Executive – Instrument • Officer Maintenance – Mechanical • Officer Maintenance – Electrical • Officer Maintenance - Instrumentation **2)** Contractors that support O&M **Roles to be 'only' Assessed** **3)** Tier 1 roles • Production Manager • Plant Manager • Site Shift Manager • Maintenance Lead/EIC: CES Engineer & Senior Engineer • Sector Chief (Maintenance & Operation) • Plant Safety Manager • Plant Process Engineer (CTS Engineer)	1) Operations 2) S&OR • Health • Safety • Environment • Fire Prevention& Emergency response 3) E&M • Mechanical • Electrical • Instrumentation • Civil 4) Technology • Technical Services • QA/QC 5) Behavioral& Supervisory 6) Leadership

6 CAS Business Process

6.1 Business Process description

It is a 3 step process which is described in the table below:

Table 2. CAS Business Process steps, description, input and output

Process step	Process description	Process Inputs	Process Output
1. Develop Job Competency Profile	• Competencies mapped to JOB ROLE. Job Competency profiles developed • Training and Courses are mapped to job roles/ positions • Identify Critical & S&OR critical positions & Tasks and identify competences required to perform them • Individuals are mapped to JCPs using people-to-position mapping (HR database)	• Competency catalogues with proficiencies • Unique Job Roles • Learning content identified at position level in ELS	• Unique job positions/ individuals mapped to competencies and learning content
2. Competency Assessment to identify learning interventions	• Enable Self-Assessment by employee • Enable Supervisors (L1) Review and Validation • Discuss & Identify all competency gaps with focus on S&OR critical competences • Identify learning needs and prioritize at: Generic, Site, Plant & Job Position • Enable Learning Intervention to address competence gaps identified	• Job Competency Profiles [JCPs] mapped to individuals	• Training Needs Identified • Developed Annual IDP
3. Competency Rigorous Assurance for S&OR Critical Competencies Using two-step process	• When ready (determined by Individual and Supervisor) schedule assurance of Critical competencies with special attention to S&OR critical competences. • Assign certified Assessors for Planning & executing assurance	• Completion of Assessment Process – Self & L1 • Completion of learning as identified in IDP •	• Updated ICP • Certify candidates meeting competency requirement

Process step	Process description	Process Inputs	Process Output
	• Shall follow a 2 step process to build the rigor required in ASSURANCE PROCESS: • Knowledge Test (on-line) • Task Test + Structured Interview		

6.2 Business Process Work flow

The business process workflow is summarized in figure 2 (as given below):

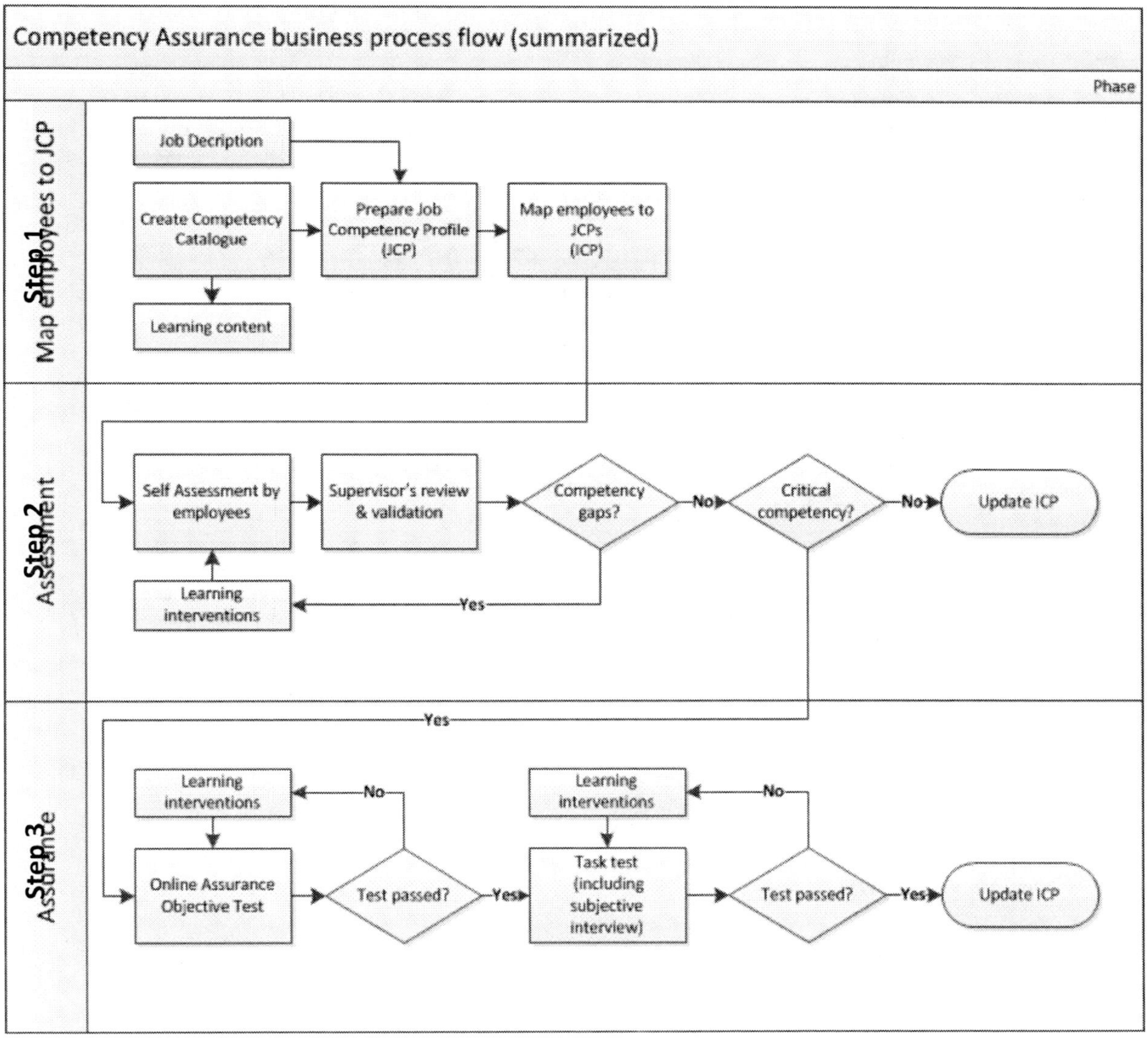

Figure 2. Business Process Work Flow

6.3 Enhanced rigor in assurance process for S&OR critical positions

Since Manufacturing jobs in Hydrocarbon processing carry higher operational & safety risks, a more rigorous with higher levels of assurance is proposed for all S&OR critical competencies. The visual in figure 3 further illustrates the higher degree of assurance introduced for S&OR critical competencies:

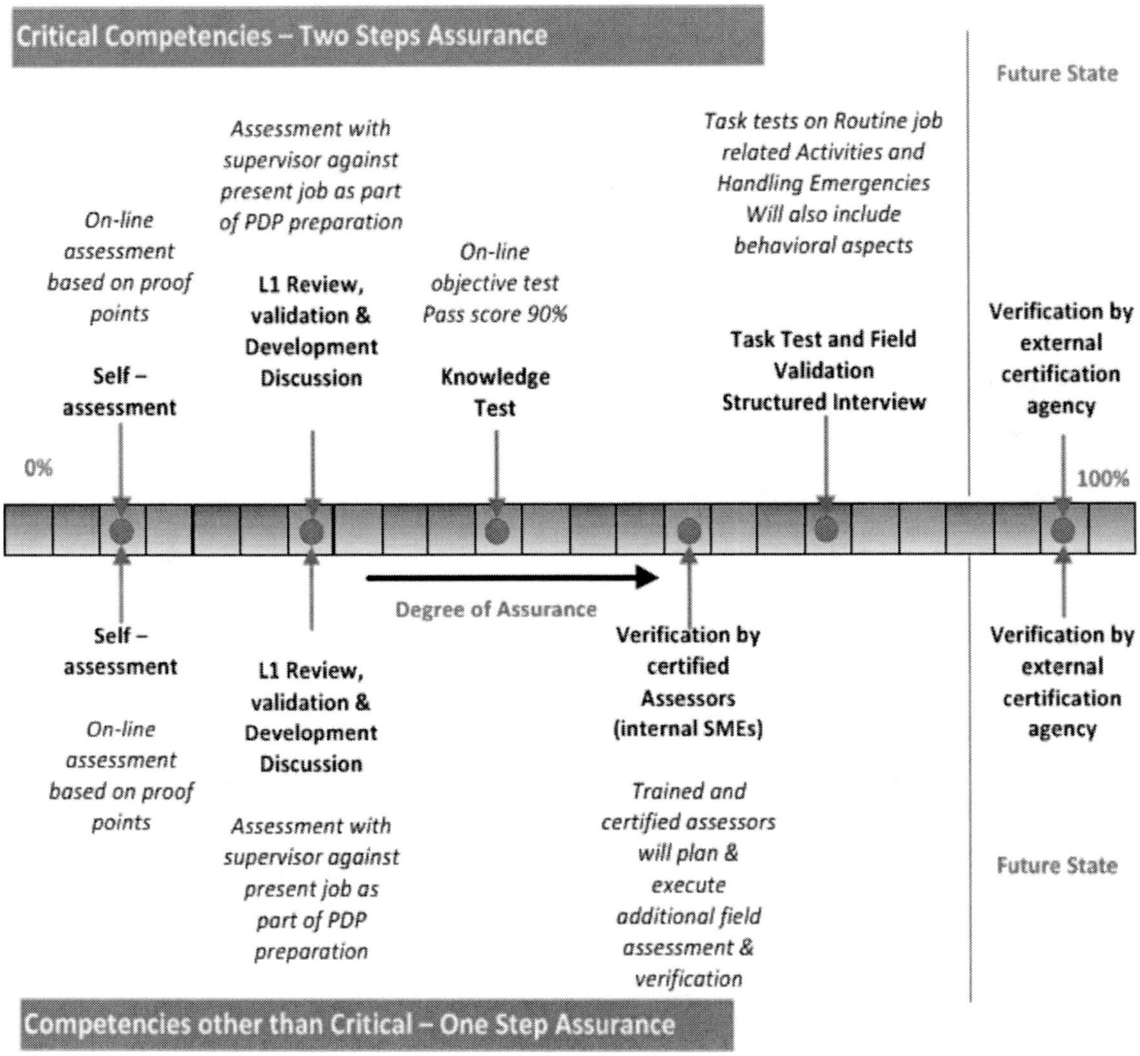

Figure 3. Degree of Assurance

6.4 Frequency of Assurance

- Annual: Emergency response related tasks-
- Three year: All S&OR critical tasks
- Five year: All other competences

6.5 Assessment process & Assessor's role

Table 3. Assessment process & Assessor's role

Phase	Steps	Work break down
PLAN	Assessor Notification	• Assessor is notified that assesse needs to be assessed.
	Assessment Planning	• Assessor collects global competence checklists and site-specific procedures, checklists, and performance criteria. • If applicable, Assessor modifies global competence checklists to site. • Assessor reviews assessee's previous qualifications and assessments (if available). • Assessor plans most effective and efficient evidence collection methods. • Assessor plans location and condition of assessment sessions. • Assessor plans communication and coaching style for each interaction with assesse.
	Assesse Notification and Assessment Scheduling	• Assessor notifies assesse that assessment(s) is due. • Assessor provides assesse with relevant competence statement and performance criteria (for example, the competence checklist). • Assessor contracts with assesse about dates, locations and methods for assessment sessions. • Assessor asks assesse to prepare ahead of time and to notify assessor of any issues (for example, any procedures or trainings that cannot be easily located).
	Assesse Preparation	• Assesse gathers any existing, relevant evidence to demonstrate the criteria has been met, and provides evidence to Assessor. • Assesse reviews procedures, verifies his/her own competence, and requests training if necessary.
ASSESS	Introduction	• Assessor greets the assesse, reviews competence statement, and discusses assessee's current level of performance (awareness, knowledge, skill). For example, they might review past trainings and procedures completed.
	Evidence Gathering and Organizing	• Assessor conducts assessment(s) with Assesse to collect evidence. • Assessor reviews other sources of evidence and meets with others who can provide such sources.

Phase	Steps	Work break down
		• Assessor assures evidence demonstrates the level and breadth of competence being assessed. • Assessor intervenes if necessary
RECORD &FEED BACK	Evidence Evaluation and Documentation	• Assessor compiles and evaluates the evidence to ensure evidence is sufficient. • Assessor determines the extent to which performance criteria has been met. • Assessor determines an assessment result – competent OR not-yet-competent – based on site policy and CD&A standard. • Assessor refers to site policies regarding consequences for not yet competent and notifies Line Management. • Assessor documents the result and all supporting evidence. • Assessor determines and plans feedback approach, e.g. meet with Assesse directly or involve Supervisor.
	Feedback	• Assessor and Assesse (and possibly Supervisor) discuss areas of skill that were competence was not demonstrated. • Assessor and Assesse (and possibly Supervisor) discuss next steps, sign-off as competent, or agree on training plan to close gaps to meet performance criteria. • Assessor and assesse agree and sign the form. • Assessor meets with Supervisor, Learning advisor and/or trainer to discuss results of assessment and plan next steps to improve competence if necessary. (create training plan) • Remediate, Reassess and close gaps.
	Final Signoff	• Line Management approves and signs off on assessment result and next steps.

7 Project development approach

7.1 Project work streams

The project design, development and implementation was divided in 7 project work streams. The project achieved 90% completion of all project related activities in 24 months.

The workstreams and key activities are summarized below:

1. Develop Competency Catalogues
   - ➢ Prepared function wise competency catalogues

- Reviewed & validated by CoEs and Shell Global Solution SMEs
- Uploaded in SAP Production environment

2. Develop Job Competency Profiles
   - Prepared function wise competency catalogues
   - Reviewed & validated by CoEs and Shell Global Solution SMEs
   - Uploaded in SAP Production environment
3. Develop Learning content and question banks
   - Prepared function wise competency catalogues
   - Reviewed & validated by CoEs and Shell Experts
   - Uploaded in SAP Production environment
4. Configure CAS in SAP Environment
   - System developed & UAT completed
   - Production system configuration completed
   - Resolve bugs & configure design changes (based on Pilot recommendations)
   - Set up tech support desk and setup site PCs' with SAP compatible software
5. Assessors Training
   - 120 assessors trained by Shell experts (from different sites)
   - Workshops conducted at Manufacturing Sites for balance assessors
6. Pilot Project
   - Pilot implemented
   - Recommendations being incorporated in design
7. Scale up and Roll out
   - Select and appoint Senior mentor trainers for all plans at a site
   - Implemented across Manufacturing Sites

7.2 Project Governance

- A three-tier approach towards project governance was institutionalized during project phase – plant level, site level and manufacturing level
- Project teams were constituted at each level and project charters and accountabilities defined.
- A rigorous managing system was installed to ensure project objectives & plans were met

8. *CAS Team*:

Dinesh Sud; HK Moorthy; N.Hariharan; S.Kakati; Ashish Joglekar;Shraddha Masare; Maria Paul; Uma Visvanathan; Mahesh Gambhir; Jagrut Vaidya; Kalpesh Pistolwala; Kaushik Chaniyara , Varadendra Koti, Sangawar,Andhare,Mohan Hosur

9. References

[1] Design, Development & Implementation of Competence Assurance system in Manufacturing: Company Internal

[2] Best Practices from Shell Global solutions: Company Internal

Return to Operation-The Final Check

Commissioning of the green field project, integrated refining and petrochemical complex or even single production unit in Hydrocarbon industry poses challenges before the leadership of the organisation. Challenges arsing due to novice or unskilled workforce and unforeseen events become more serious during delayed mechanical completion of turnaround activities. These challenges demand improvements in both organisational and management effectiveness for assuring operational readiness and preventing human error before the hazardous chemicals are introduced on initial commissioning or recommissioning.

The initiative of flawless start up drives to achieve world-class commissioning, start-up and operational excellence for mega projects including the first cycle of operation. Major objective of this program is to ensure functional integrity of equipment, instruments, components so that it results in No Harm to people, environment, asset or reputation of the company. The program is designed based on risk based approach. This paper addresses the essence of flawless commissioning of the plant, i.e. introduction of Hydrocarbon after successful pre-commissioning phase.

We at Reliance Industries Limited operate 200 odd process plants at different geographical locations. All these plants have been commissioned in last 40 years and nearly half of these plants have been commissioned in last 15 years. Numerous issues like connection of utility lines to Hydrocarbon lines, commissioning the lines with no end blinds, Commissioning of steam headers without draining condensate, plugging of control valves, configuration of wrong logics, equipment box up without internals, workplace incidents due to poor access to the equipment etc. has been experienced by our teams and has been documented. This paper would focus on major problems faced and improvements done in processes and procedures to eliminate human errors and ensure safe introduction of Hydrocarbon in the equipment. The paper would address how Risk based approach helped organisation in recent commissioning of World's largest Refinery Offgas Gas Cracker (ROGC) plant at Jamanagar, India.

- **Flaws during start up**

  Major accidents that happened during ethylene plant commissioning are related to temporary modifications, taking short cuts, unhealthy safety barriers etc. Effect of these accident is either Process safety incident involving loss of containment, explosion, equipment failure (such as brittle failure etc), off-specification product generation etc ultimately affecting the start-up schedule. The flaws are caused by factors such as human error during schedule pressure, human tendency to take short cuts, poorly trained workforce etc. Trevor Kletz said that history of these process incidents shows that many incidents are repeated after few years. The people involved in older safety incidents either retire or change the employment and thus reasons for some operating instructions and design preferences are being lost.

FLAWLESS commissioning in ROGC

<u>Design of ROGC is influenced by three key categories</u>

1. Feedstock affecting design requirements
2. Design requirements based on RIL(Reliance Ind Ltd.) desirable features
3. Latest technological trends – with best Energy, operability and safety standards

<u>Implemented/Executed during Construction</u>

1. <u>Align Construction team with plant Operation & Maintenance team</u>

Systematic construction approach by segregating the entire plants into sections. Due to large plot size, hundreds of equipment, thousands of km of pipelines, variety of pipe specifications, thousands of km of cables, the engineering work was divided into smaller sections. This concept helped not only in getting early deliverables from DEC(Detail Engg Contractor)) to client but it helped client to align construction workforce right from engineering phase.

Dividing entire plant into smaller sections and dividing total manpower into smaller teams lead by area managers was key step to ensure critical scrutiny of all safety elements. Forming operations team section wise also helped to collaborate with construction teams to ensure operations requirement of pre-commissioning, commissioning and even steady state are properly addressed during construction phase.

Close monitoring of construction progress to ensure early pre-commissioning initiated by a core team comprising of Operation & Maintenance. Daily progress review meetings by each sections helped to stick to plan and also helped to align the plan based on recent developments. Close aligning with construction team helped to direct construction progress from the point of plant start-up sequence.

2. CLEAN Built efforts and preservation of Equipment:

Majority of efforts in pre-commissioning is to remove debris, water, foreign materials, oil and grease etc from pipelines and equipment. The debris typically gets deposited in opened equipment and pipes either at store or construction site. Since construction team and operating teams are typically different, history of equipment and pipeline cleaning not maintained till pre-commissioning. Large unskilled / poorly skilled workforce involved in construction typically has very little knowledge about process plant operation in which equipment or pipeline being construction are going to be used. The efforts of operation and plant maintenance people for ensuring cleanliness prior to start up can greatly reduced if apt care is taken during installation. This "Clean Build" concept was started with explaining the process sections to construction team and explaining them significance of equipment and pipelines cleanliness in operation. This was win-win for both construction and operation/maintenance team to know areas of concern from each other and work hand in hand.

3. Systematic review of cleaning work
   a. Cleaning of large cooling water network

   Manual cleaning of main cooling water underground heaters and Chemical treatment/Flushing. Extensive use of CW network PIPENET software simulation model to predict the required velocities in different networks during cleaning. Based on the model, systematic approach was taken to carry out CW network header cleaning in steps.
   b. Chemical cleaning of Compressor Lube Oil circuit

   Compressor circuit pressure drop and velocities were estimated at different steps of high velocity blowing and chemical cleaning. The entire progress of cleaning was monitored based on pre-ascertained conditions. This gave confidence of effective cleaning of complex system.

c. Furnace Steam and Hydrocarbon Convection banks physical checking for cleanliness and blowing critical components with air.

Based on experience from earlier commissioning it was decided to do air blowing of furnace convection section to ensure no blockage is observed in flanged and welded sections.

4. P&ID walk through in field

A multidisciplinary team did the section wise P&ID walkthrough which surfaced out many small but equally important factors

a. Direction of critical isolation valves :

Directions of critical valves such as Triple eccentric butterfly valves and Cryogenic service cavity valves are key for effective isolation. The effectiveness of these valves to hold the pressure is only ensured if they are planed in correct direction of flow. During initial installation of these valves in different sections, different methodology was used by different sub-contractor of constructions. However with detailed discussions with valve manufactures, the corrective action was initiated timely.

5. When different layers of contractors are involved

Jobs involved in multiple contractors and sub-contractors poses a chance of potential miss which could lead to process incidents. These factors typically comes out as surprises and could even impact the schedule.

a. Incorrect fabrication of distillation column internals
b. Improper support strength of de-mister pads in vessels and columns
c. Oil soaked de-mister pads in cryogenic services
d. Lube oil overhead tank incorrect elevation
e. Improper alignment in vendor skids with outside pipelines.
f. Modifying 8 leg vortex breaker to 4 leg vortex breaker to overcome flow restrictions in vessels bottom.
g. Initiative taken to check the quality of Alumina balls (by *In situ* impact test and detailed laboratory analysis). This lead to selection of high purity alumna balls from different vendor.

Despite best efforts to align O&M team with construction to ensure Clean Built, the efforts on cross –confirmation of cleaning were not reduced as the schedule impact not only delays the start-up but it also affect the morale of team after tiring efforts.

- **Pre-commissioning good practices which were adopted in ROGC and giving an excellent result during start up and steady state:**

1. Cleaning of large diameter / long length pipelines :
   High velocity / momentum blowing using Air Separation Unit (ASU) air compressor done. To reduce the total air blowing duration a logical approach was used to maximize the use of available air from ASU by creating parallel air blowing paths. Need of air quantity required for each of the header was estimated based on momentum which helped in effective cleaning of pipelines and also helped in drying of cryogenic sections in plant.

2. Cleaning of long pipelines connected to cryogenic columns : Approach of using column as large inventory vessels for blowing connected pipelines was used. Required cleaning velocity was achieved by pressurizing the columns and using a controlled blowing. Welding tapes in cryogenic pipelines could be effectively removed using this approach.

3. Small bore pipelines: Based on pre-identified criteria based on service, application of cleaning method for small bore pipeline were decided. Additionally control valves were provided with inlet strainers to capture commissioning mucks during start up.

4. High Pressure Leak check. No Flange leaks (in HC service)

5. Dry out loops mark-up & execution. Not a single instance of moisture ingress in cold sections. High volumetric flow to achieve (more than 30m/sec velocity) cryogenic dew points in fastest possible time.

6. Used external fuel oil burner for refractory dry out of all furnaces.

- **Best Start up and steady state approaches**

A. Known Challenges

How to get the people ready for start up :

1. With limited manpower with cracker experience and no Panel officer and field executive with cracker experience a mammoth challenge was to train operators before pre-comm and commissioning. OTS (Online Training Simulator) was developed and the young and fresh workforce was trained for all start ups and critical trouble shooting exercises in a structured manner.
OTS was utilised extensively for training for all possible emergencies. This has helped in updating SOPs (e.g. ERC, Ethylene compressor side out control scheme was modified after start up review or PSV popping of ERC side out was first experienced in ERC start up in OTS. Side out flow control is outcome of further deliberation with licensor.)

Close coordination with CCC (Compressor Control Corp.) during main machine start up Ensuring training by CCC expert and sharing of experience of other crackers with CCC.

B. Unforeseen Surprises

1. MIC (Microbes induced corrosion) related pipelines identification & replacement upfront to air blowing.

2. Chemical cleaning of Finned tube carbon steel exchangers

   Finned tubes of carbon steel exchangers were prone for corrosion despite preservation attempts from the time of manufacturing till it's taken in service. Based on experience these fixed tube sheet exchangers were cleaning by chemical cleaning method.

3. Column internals incorrect fabrication :
   Suite-at-site modifications in column internals to address incorrect fabrication
4. Compressor reverse rotation :
   Efforts were made to ensure causes of reverse rotation are eliminated and thus suitable modifications are done.
5. High vibration / noise in pipeline downstream of restriction orifices. – Provision of multi hole orifices in place of single hole helped to fulfil site requirement.
6. First start-up of front end hydrogenation reactor… Provision of CO soaking in reactor

7. Frequent cleaning requirement of strainers at the inlet of Aluminium cold box : Rectification and strengthening of all cold box strainers & blowing of upstream lines to remove foreign particles executed during opportunity.

What can we do to address human error?

Let's take a look at the environment. We cannot change human condition but we can change the conditions where humans work, so let's review our systems and operation control. Some areas to evaluate are:

1. Management systems: documentation control, investigation management, risk management and project management are important to set the bases for the rest of the operation.
2. Procedures: these need to be accurate, human-engineered, available and enforceable.
3. Human factors engineering: Work areas need to be designed with human factors and human capabilities in mind. Excessive monitoring, mental calculations, housekeeping, and work layout, among others, become the main reasons for error in this category.
4. Training: training needs to include the whys as well as the whats and hows. Also, on the job training and qualification (OJT) is necessary, especially for critical tasks and activities.
5. Immediate supervision: pre-job briefs, walkthroughs, presence and instructions to workers are necessary. We need to have supervisors on the floor, not in the office.
6. Communication: between groups, shifts, radio communication rules and training. Employees should know what needs to be achieved daily and the proper way to do it.
7. Individual Performance through CAS (Competency Assessment System): Evaluate conditions that could potentially create cognitive overload that creates attention and memory failures. Some of these conditions include available time for the job, fitness for duty or fatigue management, and complexity and task design.

Prepared by Dipankar Bandyopadhyay
Sr.Vice President
Head Operation C2 Complex.
Reliance Industries Ltd.
Jamnagar . Gujrat
India.361140
Mail Id: Dipankar.bandyopadhyay@ril.com
Mobile: +91 9970050031
Landline:+912884035017

19 AIChE Spring Meeting
15TH GLOBAL CONGRESS ON PROCESS SAFETY
NEW ORLEANS • 2019

Study of the Initiator Mischarging Effect on Runaway Reaction of Styrene Polymerization

Lin Zhao
Artie McFerrin Department of Chemical Engineering, Texas A&M University, 3122 TAMU, College Station, Texas 77843-3122, USA
Mary Kay O'Connor Process Safety Center, Texas A&M University, 3122 TAMU, College Station, Texas 77843-3122, USA
linzhao@tamu.edu

Maria I. Papadaki
Mary Kay O'Connor Process Safety Center, Texas A&M University, 3122 TAMU, College Station, Texas 77843-3122, USA
Department of Environmental and Natural Resources Management, School of Engineering, University of Patras, Seferi 2, Agrinio 30100, Greece
marpapadaki@upatras.gr

M. Sam Mannan
Artie McFerrin Department of Chemical Engineering, Texas A&M University, 3122 TAMU, College Station, Texas 77843-3122, USA
Mary Kay O'Connor Process Safety Center, Texas A&M University, 3122 TAMU, College Station, Texas 77843-3122, USA

Mustafa Akbulut
Artie McFerrin Department of Chemical Engineering, Texas A&M University, 3122 TAMU, College Station, Texas 77843-3122, USA
makbulut@tamu.edu

Prepared for Presentation at
American Institute of Chemical Engineers
2019 Spring Meeting and 15th Global Congress on Process Safety
New Orleans, LA
March 31 – April 3, 2019

Study of the Initiator Mischarging Effect on Runaway Reaction of Styrene Polymerization

Lin Zhao
Artie McFerrin Department of Chemical Engineering, Texas A&M University, 3122 TAMU, College Station, Texas 77843-3122, USA
Mary Kay O'Connor Process Safety Center, Texas A&M University, 3122 TAMU, College Station, Texas 77843-3122, USA
linzhao@tamu.edu

Maria I. Papadaki
Mary Kay O'Connor Process Safety Center, Texas A&M University
Department of Environmental and Natural Resources Management, School of Engineering, University of Patras

M. Sam Mannan
Artie McFerrin Department of Chemical Engineering, Texas A&M University
Mary Kay O'Connor Process Safety Center, Texas A&M University

Mustafa Akbulut
Artie McFerrin Department of Chemical Engineering, Texas A&M University

Keywords: Reactive chemical, Calorimeter, Polymerization, Thermal runaway

Abstract

Polymerization reactions are prone to runaway risks due to the unstable nature of monomers and complex exothermic interactions between reactants. Deviation from a normal recipe or designed operation condition may lead to auto-accelerated temperature and pressure rise, followed by rupture of reaction vessels, fire, and explosion. Mischarging of initiators (radical generators) is one of the most common credible mal-operation cases, which has only been studied to a limited extend. In this study, benzoyl peroxide (BPO) and azobisisobutyronitrile (AIBN) were used as initiators for styrene polymerization as a function of high concentrations to mimic the mischarging scenarios. Both screening and adiabatic tests were performed to obtain the most critical safety parameters such as onset temperature, adiabatic temperature rise, and time to maximum rate. It was found that there exists a significant difference in reaction pathways and safety indicators with respect to initiator type and concentration. Uncontrolled initiation reactions not only lead to subsequent faster polymerization but also pose threats to relief valve design due to the generation of non-condensable gas. We anticipate that this study will provide beneficial guidelines for polymer manufacturers on the initiator selection and also thermal runaway prevention.

1 Introduction

Styrene monomer is a thermally unstable vinyl type monomer with the potential of self-polymerization and triggering uncontrolled thermal runaway reaction during handling, storage, and polymer production. The polymerization reaction is highly exothermic and auto-accelerates due to the drop of termination rate hampered by the long polymer chains, which may result in unexpected temperature increase beyond design cooling capacity, higher pressure building-up in vessels, loss of containment, fire, and explosion. According to the statistics, 48% of the runaway incidents in the UK during 1962-1987 were polymerization reactions.[1] A more recent study from the Chemical Safety Board (CSB)[2] revealed that polymerization thermal runaways contributed to 15% of total runaway incidents in the US between 1980-2001. Uncontrolled polymerization of styrene kept reoccurring all over the world, resulting in fatalities and asset loss. Thus, understanding the thermal runaway characteristic of the styrene polymerization reaction is vitally important regarding estimating the risks of runaway reactions and designing proper protection measures.

Thermal runaway of reactions can be caused by various types of deviations from the normal operation condition, and the mischarging of reactants has been identified to be one of the most common scenarios.[3-4] For a typical polymerization process of styrene, thermally unstable initiators are used to generate free radicals and trigger the chain-addition polymerization of styrene in the presence of a small portion of diluent (solvent). The type and the initial concentration of initiators strongly affect the initiation process and also the overall polymerization behavior. Maschio[5] investigated the free-radical polymerization of methyl methacrylate (MMA) using an adiabatic calorimeter and identified that the celling temperature, and initial polymerization rate varied with initiator type. The study of Theis[6] on copolymerization of MMA and styrene with a pseudo-adiabatic calorimeter showed an increase of over 100 °C/min on self-heating rate when the initiator concentration was doubled. A more recent study on the polymerization of styrene with different benzoyl peroxide (BPO) concentration[7] identified that higher BPO dosage affected adiabatic temperature rise of the runaway reaction.

In this work, a comprehensive study on the runaway behavior of styrene polymerization in the solvent was conducted with two different types of initiators. Benzoyl peroxide (BPO) and azobisisobutyronitrile (AIBN) were used at a series of elevated concentrations to mimic the initiator mischarging scenarios. Both screening and adiabatic calorimeter test were performed to understand the thermal activity at small and bulk scale. Important safety indicators such as "onset" temperature of polymerization, the time to maximum rate, the heat of reaction, as well as pressure increase during the exothermic were reported.

2 Experimental

2.1 Materials

In this paper, the two initiators, benzoyl peroxide (Sigma-Aldrich, 98%) and 2,2'-Azobis(2-methylpropionitrile) (Sigma-Aldrich, 98%) were used without any further purification. Ethylbenzene (Sigma-Aldrich, ReagentPlus, 99%) was used as a diluent to reduce the viscosity of the system and mitigate the gel effect. Styrene (Sigma-Aldrich, ReagentPlus, 99%) monomer stabilized with 15 ppm of 4-tert-butylcatechol (TBC) was purified with a packed column filled with alumina inhibitor remover (Sigma-Aldrich) powders. The inhibitor-free styrene monomer was mixed with ethylbenzene and initiators to prepare a serious of reaction solutions and was used immediately. The mass fraction of ethylbenzene was 15% for all polymerization samples, and the initiator mass fraction varied from 500 ppm to 10000 ppm based on the total styrene mass. Initiators were also dissolved into ethylbenzene at 5% to investigate their thermal decomposition in solvent.

2.2 Differential Scanning Calorimeter (DSC)

The test samples were first measured in a Q20 differential scanning calorimeter (TA Instruments, New Castle, DE) to obtain preliminary thermal activity results during the reaction. Approximately 5 mg of test samples were sealed in a high-pressure (15 MPa), stainless-steel sample capsule and tested between 30 °C and 300 °C in a modulated DSC (MDSC®) mode. The MDSC operates differently from the traditional DSC that a sinusoidal temperature change which oscillates around the average heating rate is applied, and the total modulated heat flow signal obtained can be separated into two components based on whether it responds to the modulated heating rate.[8] The reversible heat flow represents the heat capacity related thermal activity, i.e., heat capacity change, glass transition, and melting. The non-reversible heat flow stands for kinetic involved thermal behaviors such as reaction, decomposition, curing, etc. The monomer-contained samples and pure initiators were heated up at a fixed average heating rate of 4 °C/min oscillating at 0.6 °C per 60 s. In addition, the 5% initiator in ethylbenzene samples were tested at four different average heating rates of 2, 4, 8, 10 °C /min to obtain the kinetic information.

2.3 Automatic Pressure Tracking Adiabatic Calorimeter (APTAC™)

Adiabatic calorimeter tests using APTAC™ were performed on the styrene-ethylbenzene-initiator system for different initiator dosages to investigate the runaway behavior of bulk samples. Purified styrene was mixed with ethylbenzene at a fixed monomer mass fraction of 85%. The initiators were then dissolved in the monomer-solvent mixture and used immediately. The initiator ratios used in adiabatic tests were 500 ppm, 2500 ppm, and 10000 ppm based on the mass of the monomer for both BPO and AIBN. In this work, *ca.* 10 g of mixture sample was loaded into a glass test-cell, and the corresponding thermal inertia factor (φ-factor) was around 1.26. An initial nitrogen pad pressure of 1000 kPa (147 psi) was used to suppress the boiling of the reactant. During the heat-wait-search operating, a temperature increment of 5 °C, a heating rate of 5 °C/min, and a 30 min waiting period between each temperature step were used. The "onset" of exothermic activity was defined as when the self-heating rate exceeded 0.1 ℃ /min. The temperature and pressure during the exothermic were recorded, and the rate of self-heating and pressure-rise were calculated.

3 Results and Discussion

3.1 DSC results of initiators

The thermal decomposition of the two solid initiators was first examined by MDSC, and the results are summarized in Figure 1. The non-reversible heat flow curve denoted one single exothermic peak, which was the thermal decomposition process. The reversible heat flow curve showed a single endothermic peak, overlapped with the non-reversible heat flow, which was identified as the melting process of the initiator from the solid state to the liquid state. This coupled phenomena of heat absorption and generation has been observed by many other researchers, and efforts have been made to decouple the endothermic peak from the decomposition peak.[9] As it is clearly shown in Figure 1, MDSC provided a convenient way to separate overlapping thermal activities and rule out the phase transition from the hazardous exothermic behavior. As a result, the determination of the "onset" temperature of the thermal decomposition became straightforward. The non-reversible heat flow denoted the "onset" temperature of solid BPO and AIBN was 103 °C and 100 °C, respectively. The heat of decomposition integrated from the non-reversible heat flow was 1352 J/g for BPO, and 1140 J/g for AIBN, which were slightly higher than the heat of decomposition obtained from the traditional method,[10-11] implying the MDSC decoupling method provided a more conservative estimate of the reaction heat and thermal hazard.

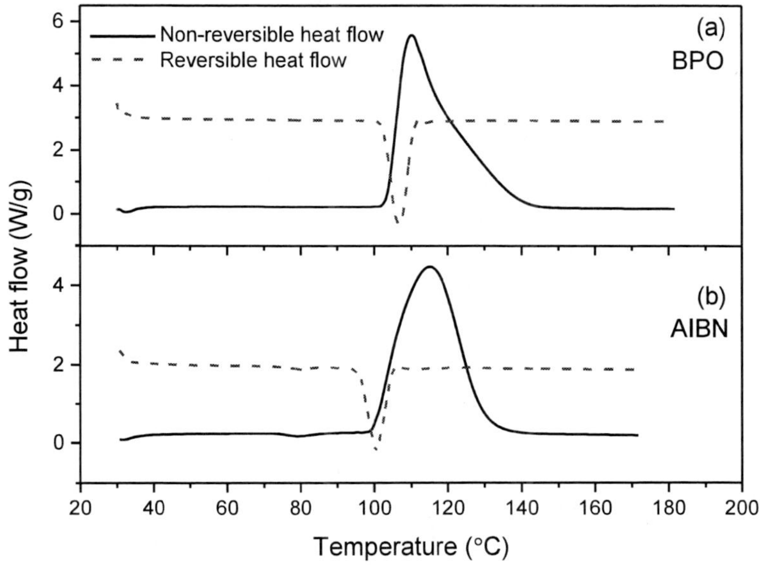

Figure 1. Non-reversible heat flow (—) and reversible heat flow (- -) of the two solid initiators during thermal decomposition from the MDSC experiments. (a) BPO. (b) AIBN.

In general, industrial practices utilizes initiators which dissolved in the solvent to trigger the polymerization reaction, and the thermal decomposition of the two initiators was examined in ethylbenzene following the ASTM E698 standard (Ozawa method).[12] Four different scanning rates were applied to the 5% initiator-ethylbenzene mixture as described in the *Experimental* section. The corresponding results were summarized in Table 2.

Table 2. The peak temperature and the heat of reaction of the 5% initiator-ethylbenzene solution at different heating rates in MDSC

β (℃ /min)	*BPO*		*AIBN*	
	T_{max} (℃)	ΔH (J/g)	T_{max} (℃)	ΔH (J/g)
2	114.9	1341	111.5	1394
4	125.9	1320	118.5	1478
8	138.1	1319	127.1	1477
10	140.5	1257	129.5	1304

In Table 2, the peak temperature (T_{max}) corresponding to the maximum heat generation was recorded for each heating ramp, and the activation energy (E_a) of thermal decomposition in the solvent was obtained by using Equation (1), where E_a is the activation energy of thermal decomposition, R is the gas constant, β is the heating rate in ℃ /min, and T is the peak temperature in Kelvin.

$$E_a \cong -2.19R[\mathrm{d}\ \mathrm{Log}_{10}\beta/\mathrm{d}(1/T)] \qquad (1)$$

When initiators were dissolved in ethylbenzene, the phase transition was eliminated, and the reversible heat flow became a trivial part. The overall heat flow obtained was identical to the non-reversible heat flow through the process of thermal decomposition. As shown in Figure 2, when initiators were dissolved in ethylbenzene, the thermal decomposition became detectable at a significantly lower temperature when compared to the concentrated solid form. For the heating rate of 4 ℃ /min, 5% BPO in ethylbenzene "onset" at around 84.8 °C, and the 5% AIBN in ethylbenzene "onset" at around 70.7 ℃ . The maximum temperature of the exothermic peak increased with a higher heating rate. The activation energy E_a obtained of 5% initiators in ethylbenzene was 79.7 kJ/mol for BPO and 108.1 kJ/mol for AIBN. The experimental results in the solvent denoted that AIBN was able to thermally decompose and trigger polymerization at a lower temperature when compared to BPO which usually had a higher activation energy and was more sensitive to temperature.

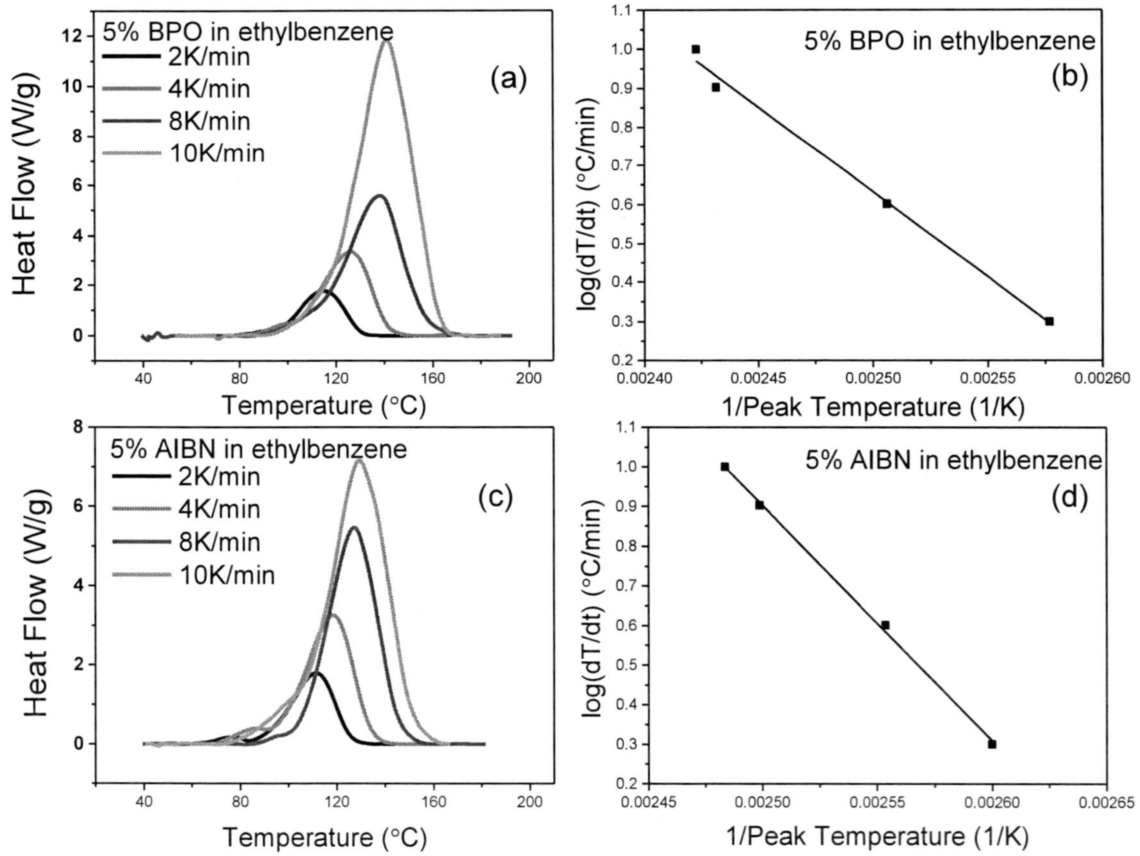

Figure 2. Heat flow of the 5% initiator-ethylbenzene solution at different heating rates in the MDSC and the kinetic fitting. (a), (b) BPO. (c), (d) AIBN.

3.2 DSC results of polymerization systems

After the thermal hazard of the initiators was assessed in the solvent, the initiators were then mixed with 15% styrene-ethylbenzene at a series of initiator concentrations and tested with MDSC. The heat flow results are depicted in Figure 3 and Figure 4 for BPO and AIBN, respectively. For free-radical polymerization triggered by both initiators, the total heat flow was equal to the non-reversible flow, and the kinetic related thermal activity dominated the whole reaction.

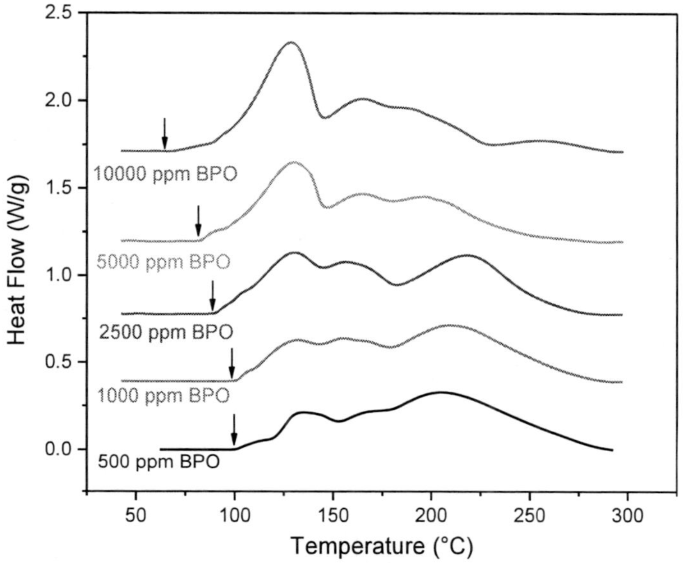

Figure 3. Heat flow of the polymerization of 85% styrene-ethylbenzene with different BPO concentrations in MDSC at 4 °C/min

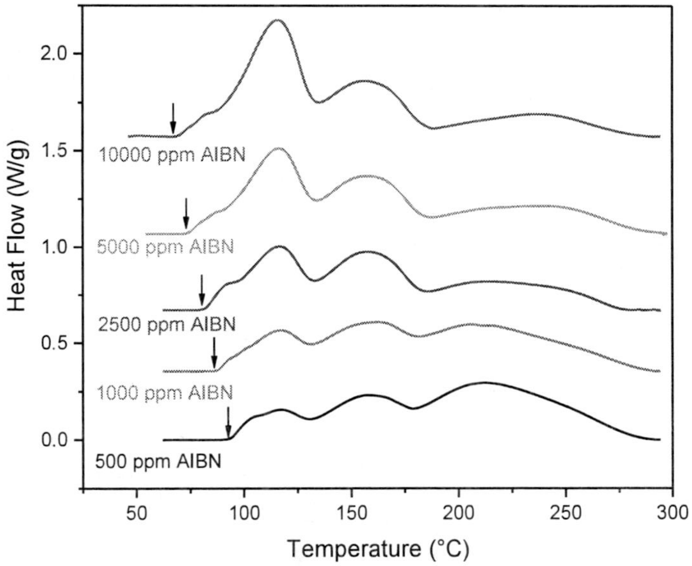

Figure 4. Heat flow of the polymerization of 85% styrene-ethylbenzene with different AIBN concentrations in MDSC at 4 °C/min

The polymerization with the two initiators expressed some common trends. First, more than one exothermic peak emerged upon heating, denoting a complex reaction pathway including initiation, chain propagation, chain-transfer, and termination. Compared to the thermal polymerization of styrene without an initiator, Liao[13] identified two major peaks to represent two different types of reaction patterns. However, when the initiators were present, peak numbers increased, and no clear conclusion can be drawn whether the peaks were associated with a specific reaction step as they highly overlapped with each other. A significant difference regarding the "onset" temperature and the relative peak area was noticed under different initiator dosages. For both BPO and AIBN, the free-radical polymerization reaction started "onset" at a lower temperature when the initiator concentration was increased. The lowest "onset" temperature for AIBN initiated polymerization was 71.3 ℃ , which was identical to the starting temperature of the thermal decomposition of AIBN in ethylbenzene at 5%. For BPO initiated system, the lowest "onset" temperature obtained at 10000 ppm BPO in styrene-ethylbenzene was 70.1 °C, which was even lower than the thermal decomposition of BPO in ethylbenzene (84.8 °C). This finding implied that the presence of monomer further promoted the decomposition of BPO.

Another difference noticed from the DSC results was that the main exothermic peak shifted to a lower temperature upon the addition of the initiators. The area under the heat flow curve represents the heat generated during the reaction, and thus the relative size and position of the peak can be used to estimate the heat release and at different temperatures, in other words, assess the severity of the thermal hazards. When BPO and AIBN dosage was low (500 ppm), the major exotherm took place at the late stage of the reaction at around 210 °C. With the addition of the initiators, the main heat release was observed at a much-lowered temperature. For systems with 10000 ppm of BPO or AIBN, the maximum heat generation took place at around 120 °C; and the sharper, narrower peak denoted a sudden heat release during a short period.

The overall heat of reaction was calculated for all test samples and summarized in Table 3 together with the "onset" temperature. The change of these important thermal hazard indicators obtained from the MDSC study with respect to initiator dosage was depicted in Figure 5.

Table 3. Thermal kinetic data for the polymerization of 85% styrene-ethylbenzene with different initiator concentrations in MDSC at 4 °C/min

Initiator concentration (ppm)	*BPO*		*AIBN*	
	T_{on} (℃)	ΔH (J/g)	T_{on} (℃)	ΔH (J/g)
500	102.0	478.0	94.1	503.7
1000	100.0	531.8	87.5	498.9
2500	91.1	558.0	80.9	499.9
5000	83.5	584.7	75.1	565.3
10000	70.1	596.1	71.3	637.2

As shown in Figure 5(a), when BPO was used to trigger the polymerization of styrene in ethylbenzene, the reaction heat increased monotonically upon BPO addition, and this increase was most significant when the BPO concentration was below 5000 ppm. At the same time, adding BPO to the monomer-solvent system also lowered "onset" temperature, and the lowest "onset" temperature obtained was around 15 °C lower than the initial thermal decomposition temperature of BPO itself when dissolved in ethylbenzene at 5%. When compared to the case of AIBN in figure 5 (c) and (d), it was noticed that the reaction heat roughly remained constant when AIBN concentration was below 2500 ppm, and the further charging of AIBN resulted in a sharp increase of reaction heat. The "onset" temperature of the polymerization system with different AIBN dosage showed a similar declining trend as what was observed for the case of BPO. However, the lowest "onset" temperature was not affected by the monomer.

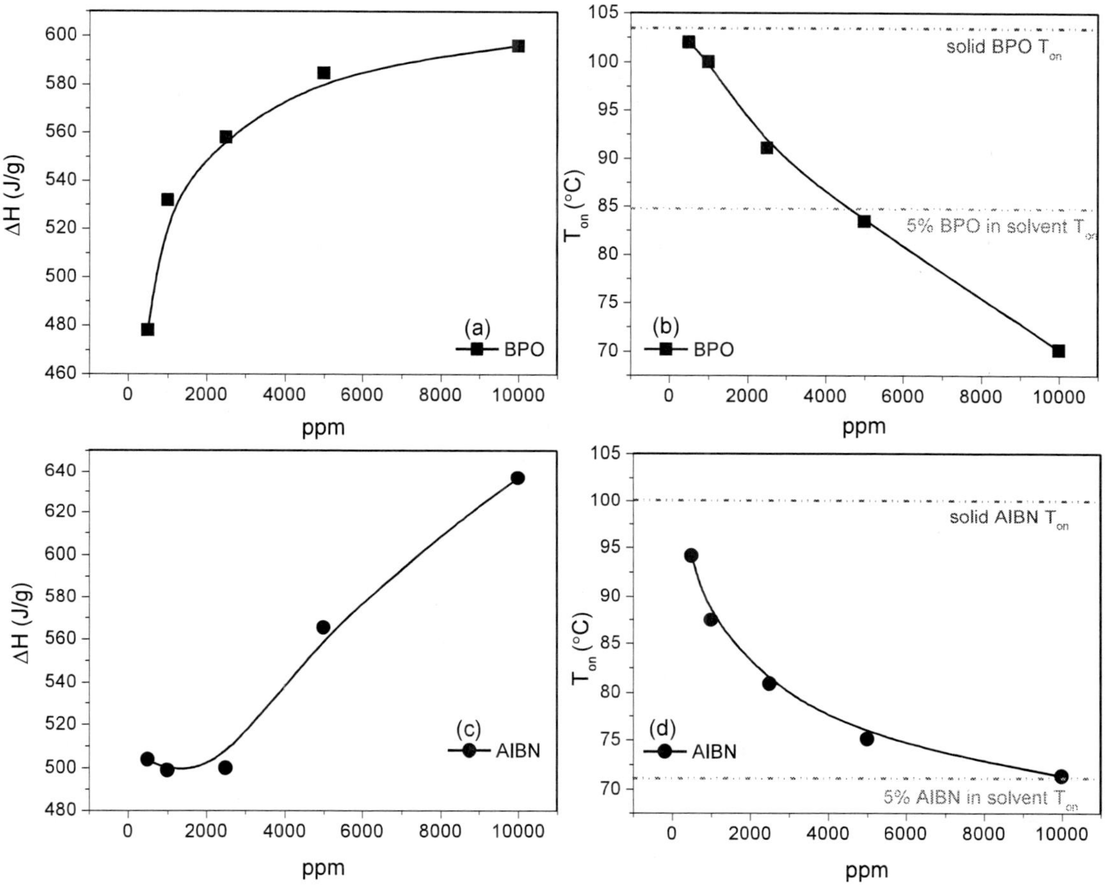

Figure 5. The "onset" temperature and heat of reaction for the polymerization of 85% styrene-ethylbenzene with different initiator concentrations in MDSC at 4 °C/min. (a), (b) BPO in styrene-ethylbenzene. (c), (d) AIBN in styrene-ethylbenzene

3.2 Adiabatic calorimeter results of polymerization systems

The DSC provided some preliminary heat generation information of the samples at a small scale, where the heat and mass transfer did not play an important role in the overall reaction. The thermal runaway behavior of the polymerization process with various initiators dosages was further investigated in APTAC to obtain bulk temperature and pressure data under adiabatic condition. BPO and AIBN were added into the 85% styrene-ethylbenzene system at 500 ppm, 2500 ppm, and 10000 ppm to represent the normal operation and two initiator mischarging scenarios. The adiabatic temperature rise with respect to time and the self-heating rate with temperature was shown in Figure 6. The detected "onset" temperature, time to maximum rate under adiabatic conditions (TMR), the maximum self-heating rate (dT/dt_{max}) and the maximum pressure-rising rate (dP/dt_{max}) were calculated and summarized in Table 4.

Table 4. The adiabatic thermal kinetic data for the polymerization of 85% styrene-ethylbenzene with different initiator concentrations

Initiator (ppm)	*BPO*				*AIBN*			
	T_{on} (°C)	*TMR* (min)	dT/dt_{max} (°C/min)	dP/dt_{max} (kPa/min)	T_{on} (°C)	*TMR* (min)	dT/dt_{max} (°C/min)	dP/dt_{max} (kPa/min)
500	83.8	206.7	3.8	22.8	71.4	236.8	2.8	17.7
2500	73.4	78.3	3.6	17.5	62.8	68.6	3.9	18.4
10000	62.7	70.2	37	105.5	55.1	73.4	25.0	93.4

The adiabatic "onset" temperature monotonically decreased with initiator dosage for both BPO and AIBN, which was consistent with the observations in MDSC tests. It is also noteworthy that the APTAC identified the thermal activity at a lower starting temperature when compared to DSC, highlighting that the "onset" threshold value strongly depends on the sensitivity of the instrument, and can only be used as a rough estimate of the thermal hazard. At the same time, the results emphasized the necessity of employing adiabatic calorimeter in thermal hazard assessment, as more conservative and realistic value can be obtained. The lowest "onset" temperature obtained in this study was 55.1 °C when 10000 ppm of AIBN was charged into the 85% styrene-ethylbenzene system. This value was around 50 °C below the regular recognized thermal decomposition temperature of pure AIBN,[10, 14-15] and also much lower than the "onset" of thermal polymerization temperature of styrene of around 100 ℃ .[16] Mischarging of a small quantity of AIBN into the system may result in an unexpected thermal activity at a very low temperature.

At the same time, a significant decrease in TMR was noticed when the initiator concentration was higher than 2500 ppm for both BPO and AIBN. This can be observed in Figure 6 (a) and Figure 6 (c) that as BPO and AIBN was 500 ppm in 85% styrene-ethylbenzene, the polymerization reaction activity became notable between 70-80 °C, and the temperature building-up was relatively slow as only around 20 °C increase in temperature was observed in the first 2 hrs. The main temperature increase was noticed after 200 min of the "onset", where a temperature increment of over 100 °C was achieved within less than an hour. The time used for the system to reach its maximum self-heating rate (TMR) was 206.7 min for BPO and 236.8 min for AIBN, respectively. When initiators were mischarged into the polymerization system at 2500 ppm, the TMR was shortened to around 80 min for both initiators; a steep temperature increment was observed, denoting that the runaway behavior changed to almost explosive type upon initiator mischarging. The corresponding massive heat release in such a short time may result in a huge increase in cooling demand which might not be able to meet by equipment design.

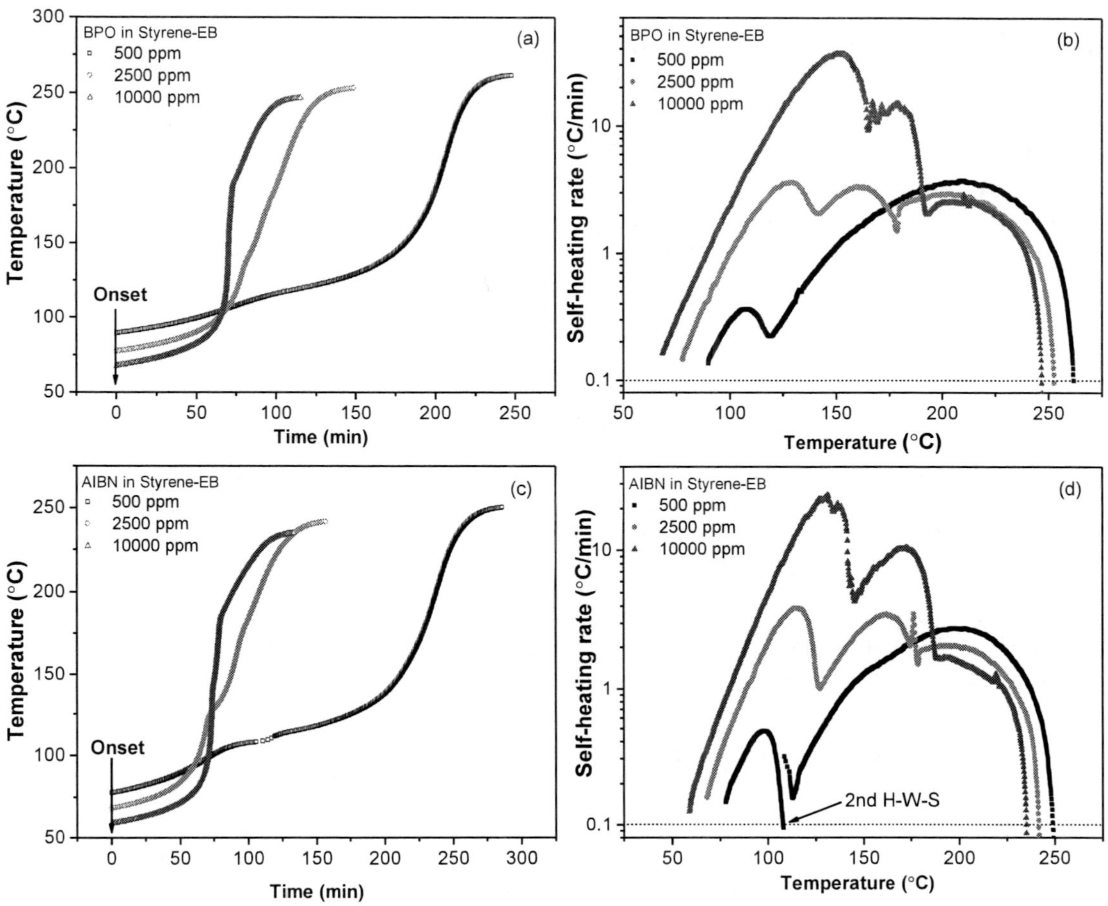

Figure 6. The adiabatic runaway behavior of 85% styrene-ethylbenzene system with various initiator charges. (a), (b) BPO. (c), (d) AIBN.

The energy release, or the heat generation, that needs to be taken care of by proper cooling is expressed as the self-heating rate versus temperature in Figure 6 (b) and Figure 6 (d). The polymerization of 85% styrene in ethylbenzene initiated by BPO and AIBN went through multiple exothermic steps, which resulted in several peaks on the curve. When BPO dosage was 500 ppm, the self-heating rate exceeded 0.1 °C/min at 83.8 °C and resulted in a mild exothermic peak with a maximum self-heating rate of 0.2 °C/min before 125 °C. This peak was followed by a major exotherm with a peak temperature at over 200 °C. The latter reaction step also possessed a characteristic of auto-acceleration as the self-heating rate was first linearly related to the reciprocal of temperature (pseudo-first-order) and then had a negative deviation (higher than first order) after around 145 °C. This unique heat generation pattern represented a typical free-radical polymerization process with a gel-effect[17]. The gel-effect poses a great threat to the polymer production as the polymerization rate, and the heat production rate suddenly increases due to the reduced mobility of the long polymer chains after the conversion achieved a certain critical point[18]. For the case of 500 ppm in AIBN, similar temperature pattern was observed only that the initial exothermic step was not able to generate enough heat to sustain the following reaction at around 110 °C. The APTAC entered heat-wait-search (H-W-S) mode and then heated the system up again to 120 °C; the following reaction developed to complete after that. The temperature rate curve confirmed that when BPO and AIBN dosage was at 500 ppm, the thermal risk associated with the polymerization reaction was due to the sudden heat release affected by the gel-effect at high temperature, and the maximum self-heating rate was below 4 °C/min. When initiator dosage was increased to 2500 ppm, three exothermic peaks were observed for both BPO and AIBN. The maximum self-heating rate of the reaction was achieved in the first reaction step at a lowered temperature of around 125 °C for BPO and 114 °C for AIBN, and the value of this peak self-heating rate had no obvious change when compared to the case of 500 ppm initiator. Further increasing of initiator concentration to 10000 ppm resulted in a remarkable change of maximum self-heating rate which was almost ten times higher than that of the 500-ppm case. The heat release rate of the first reaction step overwhelmed the following reactions.

The change of the maximum pressure-rising rate with respect to initiator concentration also followed the same trend. As summarized in Table 4, there is no significant change in the peak pressure-building up rate until the initiator concentration increased from 2500 ppm to 10000 ppm. The polymerization of the styrene-ethylbenzene system was a *vapor system* according to the Design Institute for Emergency Relief Systems (DIERS) classification, as such the pressure increase was solely related to the vaporization of the volatile species. When initiators were present, the decomposition of the BPO resulted in the release of oxygen, while the AIBN generated nitrogen upon heating, which also contributed to the total pressure of non-condensable gases. However, in this study, the initiator dosage was remained at a very low concentration (less than 1%), and the contribution of oxygen/nitrogen generation was trivial. The pressure building-up was related to the vapor pressure change caused by the temperature rise.

4 Conclusion

In this study, the runaway behavior and thermal hazard of mischarging of two initiators (BPO and AIBN) into the styrene-ethylbenzene polymerization system were investigated using screening and adiabatic calorimeters. The thermal decomposition of solid initiators was first characterized by a novel MDSC method to decouple the endothermic phase change and the exothermic decomposition reaction. The "onset" temperature and the activation energy of the thermal decomposition of initiators in ethylbenzene were investigated by MDSC, revealing that when dissolved in ethylbenzene, the thermal decomposition of initiators became notable at a much-lowered temperature when compared to their solid state.

MDSC studies on the runaway behavior of styrene-ethylbenzene with different initiator charges discovered that the thermal hazard of free-radical polymerization was strongly related to the initiator dosage regardless of the initiator type. When the BPO and AIBN were mischarged into the system at an elevated concentration, the exothermic onset temperature dropped 30 °C when compared to the thermal polymerization temperature of styrene and also the thermal decomposition temperature of the initiators themselves. Furthermore, the total heat generated by the system with over-dosed initiators increased monotonically with the increase of BPO concentration and had an approximate linear increment after AIBN concentration was over 2500 ppm. The overcharging of both initiators also resulted in a shift of main exothermic activity temperature from around 200 °C to around 120 °C.

The adiabatic runaway test of bulk samples also demonstrated a similar trend that the free-radical polymerization started "onset" at lower temperature upon initiator dosage, and the time required for the reaction to develop to its maximum self-heating was significantly reduced. With an elevated initiator dosage of 10,000 ppm, increased peak heat-release was observed at a lowered temperature associated with higher pressure building-up. The adiabatic data provided extremely useful information regarding thermal behavior and pressure profile for large-scale samples regarding scenarios of mischarging initiators into the polymerization system.

5 Reference

1) Barton, J.; Nolan, P., Incidents in the chemical industry due to thermal runaway chemical reactions. *Hazards X: Process Safety in Fine and Speciality Chemical Plants* **1989,** (115), 3-18.
2) CSB, Improving Reactive Hazard Management. *US Chemical Safety and Hazard Investigation Board (CSB)* **2002**.
3) Saada, R.; Patel, D.; Saha, B., Causes and consequences of thermal runaway incidents—will they ever be avoided? *Process Saf. Environ. Prot.* **2015,** *97*, 109-115.
4) Sales, J.; Mushtaq, F.; Christou, M.; Nomen, R., Study of major accidents involving chemical reactive substances: analysis and lessons learned. *Process Saf. Environ. Prot.* **2007,** *85* (2), 117-124.
5) Maschio, G.; Feliu, J.; Ligthart, J.; Ferrara, I.; Bassani, C., The use of adiabatic calorimetry for the process analysis and safety evaluation in free radical polymerization. *J. Therm. Anal. Calorim.* **1999,** *58* (1), 201-214.
6) Theis, A. E.; Burelbach, J. P.; Askonas, C. F., Safely scale-up processes and accommodate recipe changes. *Process Safety Progress* **2009,** *28* (2), 135-140.
7) Wang, W.; Fang, J.; Pan, X.; Hua, M.; Jiang, J.; Ni, L.; Jiang, J., Thermal research on the uncontrolled behavior of styrene bulk polymerization. *J. Loss. Prevent. Proc.* **2019,** *57*, 239-244.
8) Thomas, L. C., Modulated DSC® Paper# 1 Why Modulated DSC®?; An Overview and Summary of Advantages and Disadvantages Relative to Traditional DSC. *TA Instruments Technol. Pap* **2005**.
9) Zhang, C.-X.; Lu, G.-B.; Chen, L.-P.; Chen, W.-H.; Peng, M.-J.; Lv, J.-Y., Two decoupling methods for non-isothermal DSC results of AIBN decomposition. *J. Hazard. Mater.* **2015,** *285*, 61-68.
10) Liu, S.-H.; Lin, W.-C.; Hou, H.-Y.; Shu, C.-M., Comprehensive runaway kinetic analysis and validation of three azo compounds using calorimetric approach and simulation *J. Loss. Prevent. Proc.* **2017**.
11) Tsai, L.-C.; Chen, J.-W.; Hou, H.-Y.; Liu, S.-H.; Shu, C.-M., Exothermic behaviors in decomposition of three solid organic peroxides by DSC and VSP2. *J. Therm. Anal. Calorim.* **2012,** *109* (3), 1303-1309.
12) ASTM., Standard test method for Arrhenius kinetic constants for thermally unstable materials using differential scanning calorimetry and the Flynn/Wall/Ozawa method. West Conshohocken New York (NY): 2005.
13) Liao, C.; Wu, S.; Su, T.; Shyu, M.; Shu, C., Thermokinetics evaluation and simulations for the polymerization of styrene in the presence of various inhibitor concentrations. *J. Therm. Anal. Calorim.* **2006,** *85* (1), 65-71.
14) Liu, S.-H.; Shu, C.-M.; Hou, H.-Y., Applications of thermal hazard analyses on process safety assessments. *J. Loss. Prevent. Proc.* **2015,** *33*, 59-69.

15) Guo, S.; Wan, W.; Chen, C.; Chen, W., Thermal decomposition kinetic evaluation and its thermal hazards prediction of AIBN. *J. Therm. Anal. Calorim.* **2013,** *113* (3), 1169-1176.
16) Chen, C.-C.; Duh, Y.-S.; Shu, C.-M., Thermal polymerization of uninhibited styrene investigated by using microcalorimetry. *J. Hazard. Mater.* **2009,** *163* (2), 1385-1390.
17) Leung, J. C.; Fauske, H.; Fisher, H., Thermal runaway reactions in a low thermal inertia apparatus. *Thermochim. Acta* **1986,** *104*, 13-29.
18) Achilias, D. S., A review of modeling of diffusion controlled polymerization reactions. *Macromol. Theory Simul.* **2007,** *16* (4), 319-347.

19 AIChE Spring Meeting — 15th Global Congress on Process Safety — NEW ORLEANS • 2019

A semi-quantitative tool to estimate risks of combustible dust fires and explosions from process equipment

Emmanuel Kwasi Addai
DEKRA Process Safety
113 Campus Dr
Princeton, NJ, USA
emmanuel.addai@dekra.com

Jeff Foisel
DEKRA Process Safety
113 Campus Dr
Princeton, NJ, USA
jeff.foisel@dekra.com

John Wincek
DEKRA Process Safety
113 Campus Dr
Princeton, NJ, USA
john.wincek@dekra.com

Christopher Headen
Cargill, Inc.
15407 McGinty Road West
Wayzata, MN, USA
Christopher_Headen@cargill.com

Prepared for Presentation at
American Institute of Chemical Engineers
2019 Spring Meeting and 15th Global Congress on Process Safety
New Orleans
March 31 - April 4, 2019

UNPUBLISHED

A semi-quantitative tool to estimate risks of dust fires and explosions from process equipment

Emmanuel Kwasi Addai
DEKRA Process Safety
113 Campus Dr
Princeton, NJ, USA
emmanuel.addai@dekra.com

Jeff Foisel
DEKRA Process Safety
113 Campus Dr
Princeton, NJ, USA
jeff.foisel@dekra.com

John Wincek
DEKRA Process Safety
113 Campus Dr
Princeton, NJ, USA
john.wincek@dekra.com

Christopher Headen
Cargill, Inc.
15407 McGinty Road West
Wayzata, MN, USA
Christopher_Headen@cargill.com

Abstract

Consistently assessing risks associated with a particular process or piece of equipment can be difficult, even for very low risk operations. Furthermore, when utilizing industry experts to complete risk assessments, inconsistencies remain in results due to the still highly qualitative nature of combustible dust risk assessments. In response, the industry is advancing efforts to bring a higher level of consistency in evaluating risks for combustible dust processes [1]. In line with the industry focus on improving combustible dust risk assessments, this work presents a semi-quantitative tool to consistently and effectively estimate risks of combustible dust fires and explosions in non-complex unit operations. This tool has been piloted for bucket elevators in the grain industry. Additional tool modules are in development and validation phases that are designed to analyze unit operations such as: grain dryers, dust collection systems, hammer mills and belt conveyors. The function of this tool is to identify and evaluate the most common preventive, protective and mitigative measures used in industry to prevent dust fires and explosions. When this measures evaluation is combined with combustible dust test data (such as explosibility properties of the material; minimum ignition energy, minimum ignition temperature, deflagration index, maximum explosion pressure, self-ignition temperature, limiting oxygen concentration, minimum explosible concentration etc.), a more complete picture of the risk is obtained. In addition to providing an estimated risk level, the tool identifies potential safeguarding opportunities and thus providing guidance for improvement.

Keywords: dust explosion, risk assessment, semi-quantitative risk tool, bucket elevator

1. Introduction

Dust explosions present serious and widespread hazards in processing industries such as the grain industry [2, 3]. Incidents involving dust explosions often lead to equipment failures, injuries to people, negative impacts to the surrounding environment, plant shut-downs and sometimes destruction of buildings resulting in fatalities and enormous financial losses. For many years in the processing industry, efforts have been in place to develop and improve measures intended to prevent or mitigate combustible dust explosions [4, 5, 6]. Despite extensive research, increasing regulations, standardization and advancements in technologies surrounding combustible dust hazard prevention and mitigation, these types of incidents still occur. In 2006, the US Chemical Safety Board (CSB) [7] conducted a study that reported 281 dust explosions occurred between 1980 and 2005 in the USA. As a result of these dust explosions, 119 workers were fatally injured while 718 experienced life-altering injuries. Since publication of the 2006 study, the CSB has confirmed an additional 105 combustible dust incidents [8]. The CSB conducted in-depth investigations of 5 of these high-profile incidents, including most recently the Didion Milling dust explosion in Cambria, Wisconsin. 27 lives were lost and dozens more injured as a result of these 5 incidents investigated by the CSB. In the processing industry, approximately 70% of dusts handled are classified as combustible. Most reported dust explosions involve organic materials from industries such as agricultural, food, fuel and pharmaceuticals.

Preventing these catastrophic dust explosions requires a comprehensive assessment of materials and processes to identify operations posing risks. One of the many challenges for organizations is determining how to consistently assess the risks for even common low risk operations. A qualitative dust hazard analysis relies heavily on expertise of the assessment team and facilitator. Experience shows a high level of variation in qualitative risk assessment results due to bias and a lack of industry accepted combustible dust data. In order to obtain an overview of the risk of fires and explosions in a process, this work presents a semi-quantitative tool to consistently estimate the risks of combustible dust fires and explosions in a unit operation.

This tool has been piloted for bucket elevators in the grain industry. Tool modules are in development and validation phases and are designed to analyze unit operations, such as grain dryers, dust collection systems, hammer mills and belt conveyors. The tool is designed for use by on-site supervisors and local Subject Matter Experts (SMEs) trained in the hazards of combustible dust and use of the tool. It uses multiple-choice and yes/no questions to assess both likelihood and severity of risk. Values are assigned to each response, providing the opportunity to weigh the relative importance of each question. The values for Severity and Likelihood are used to display the mitigated risk in real-time, as questions are answered.

The function of this tool is to establish a process baseline using material properties of the dust (such as explosibility properties of the material; minimum ignition energy, minimum ignition temperature, deflagration index, maximum explosion pressure, self-ignition temperature, limiting oxygen concentration, minimum explosible concentration etc.), the location or siting of equipment and management system elements (i.e. Asset Integrity) in place. The tool allows users to identify and evaluate the most common preventive, protective and mitigative measures taken to address combustible dust fire and explosion risks. As a result, a more complete picture of the risk is obtained. In addition to providing an estimated risk level, the tool identifies potential safeguarding opportunities and the impacts safeguards on level of risk.

2. Tool Creation

The key deliverables for this tool were to provide:

1. Consistent and effective risk identification
2. Consistent assessment of risk based on data
3. Customized risk reduction recommendations based on facility data
4. Data driven logic
5. Ease of use

Microsoft Excel was chosen as a development platform for the tool because it provides A) a familiar platform, B) programming capability, and C) a straightforward method to perform calculations associated with determination of mitigated risk.

The first step in tool creation involved applying core concepts of risk management, operating knowledge and experience, industry data, expert opinion and incident case studies in designing a background worksheet (see Table 1) that includes factors affecting frequency and consequence. These sheets are integral to tool functionality and allow for:

A. Questions compilation – All questions are compiled allowing experts to review progress as the tool is developed. Capturing the right questions in ways that are clear to the user is critical. The questions are limited to true / false, or a selection from a short list of options to further aid the user in responding.
B. Maximum scoring – This is the percentage of overall risk factor assigned to a specific question/response. This value is used to determine risk increase based on a given response. For example, though both are important and contribute to the overall risk factor, locating the bucket elevator outside and limiting people near it has a much greater impact on the consequence than having the explosion suppression device on a preventive maintenance program.
C. User Value Entry Sheet – Values returned from a form control in an Excel worksheet are connected to a cell. By placing the return value into the Background Sheet, it keeps the response directly connected to the other information associated with that particular question.
A. Scoring – This is the value from the form control that is converted to the score for a given question on the basis of the maximum score. This could also be called the weighted score for the question. The weighted score is foundational to tool functionality and, as aforementioned, is built upon operating knowledge and experience, industry data, expert opinion and incident case studies.
B. Logic – Several logic steps are used to evaluate relative sections, conditional scoring limits, and even tool use boundaries communicated directly to the user. For example, if the parameters selected in response to certain questions place results outside the intended tool scope, the user is notified by a prompt box (i.e. low MIE powders, self-heating materials, etc.)
C. Additional information, such as considerations to decrease the risk.

Table 1: Typical Layout of Background Sheet

Type of Text	Text / Question	Yes/Option Buttion Return	No	Index for Option Button Lookup	Score	Max Response Score	Max Question Score	Section Score	Section Max Score	Calculation Logic	Stop Logic	Considerations
H	1.0 Minimum Ignition Energy							40	45			
TF	Is the Minimum Ignition Energy <30 mJ?	FALSE	TRUE		0						Stop if True	*** Tool Not Valid for
SH	1.1. Electrical and Electrostatic Discharge	FALSE	TRUE									
TF	Is the belt constructed with conductive or static di	TRUE	FALSE		5	5	5					Consider using a belt
TF	Are the buckets made of non metal or other non-c	TRUE	FALSE		5	5	5					Consider using bucket
SH	1.2. Hazard Area Classification											
TF	Has the location (bucket elevator) completed a su	TRUE	FALSE		5	5	5					Consider completing
TF	If an area was found to be at electrical hazard loca	TRUE	FALSE		5	5	5					If an area was found t
TF	Is there routine inspection of electrical equipmen	TRUE	FALSE		5	5	5					Ensure routine inspec

The second step in tool creation involved developing worksheets for user interface (see Figure 1 – Example User Entry **Sheet**). The user interface is built with a series of standard Excel controls such as check boxes and multi-button or option button controls. The result is a clean user entry sheet that enables the user to focus on answering questions presented via the form tools. As responses are entered, resulting values are linked to cells on the Background Sheet. The responses are then converted from Excel form tool responses into the scored values for the questions.

Questions from the User Entry Sheets are designed for the appropriate level of technical information to be entered. Most questions are simple yes/no type and describe physical properties of materials being handled, bucket elevator configuration, existing protections in place, and operational controls used to prevent incidents. The chosen format is helpful to the user by allowing a simple click rather than typing lengthy responses, and further helps reduce errors by displaying the responses selected.

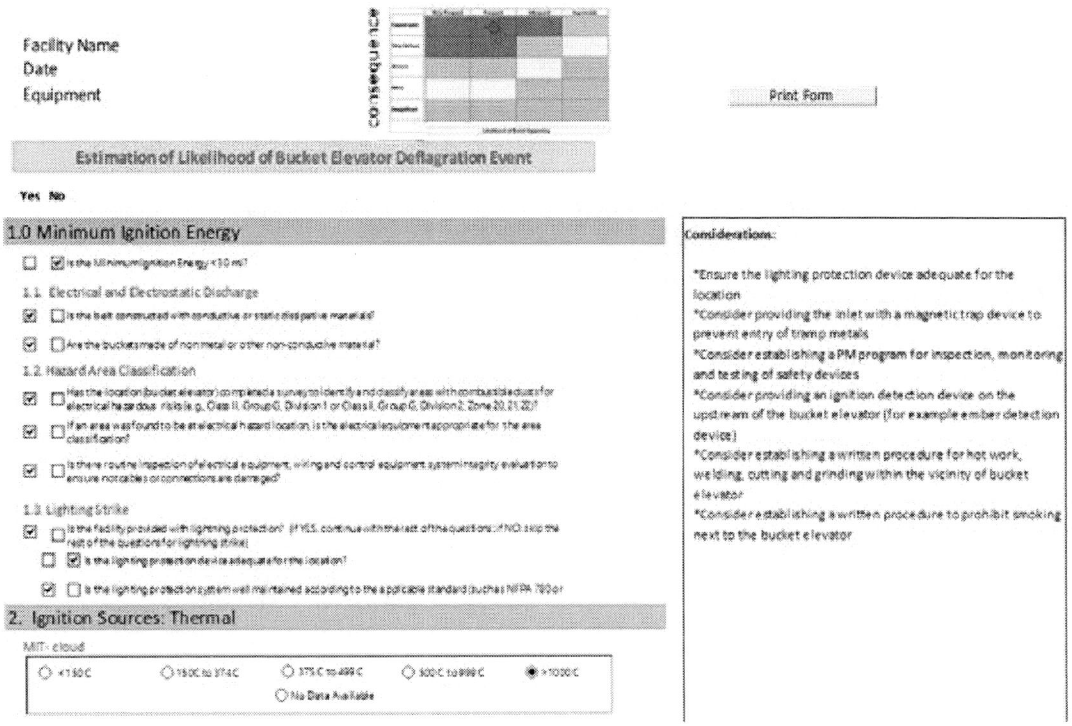

Figure 1 – Example User Entry Sheet

The user interface page is enhanced with a mitigated risk graph depicting risk relative to an organization's risk tolerance levels. A digital readout of the values shows the overall change as data is entered. Lastly, there is a place where the user can view suggestions for reducing overall risk.

3. Risk Concepts

According to Risk Based Process Safety concepts, there are three main questions associated with risk. They include [9]:

Hazard - What can go wrong?

Consequences - How bad could it be?

Likelihood - How often might it happen?

These questions are fundamental to adequately understand the risk at hand. When we ask these questions, the objective is to perform an analysis sufficient to understand the level of risk and to determine if existing levels of management system elements and safeguards employed align with the organization's risk tolerance [1].

Each person and organization have different risk mitigation approaches and level of risk tolerance. The organization's risk acceptance criterion is defined in the tool as a matrix (see Figure 2) displaying the range from broadly acceptable risks through risks that are completely intolerable. Though this is not an extremely precise value, it provides a visual representation of risk level based on a more quantitative approach. A more quantitative approach reduces personal biases in interpreting risk and aligns results to an organization's expectations, consistently.

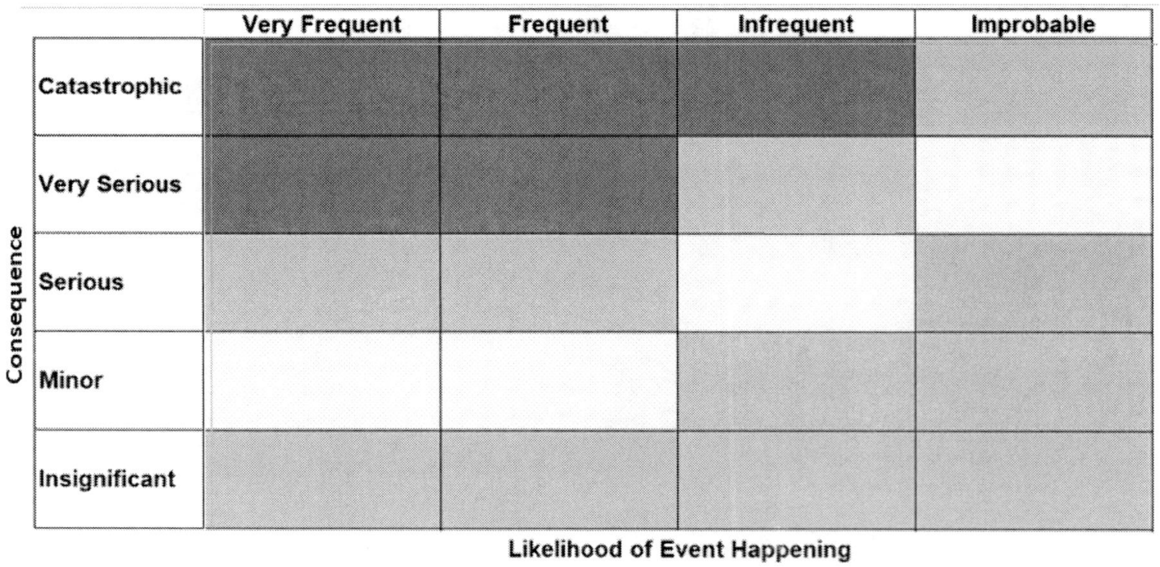

Figure 2: Example Risk Map

Green	–	**Low risk:** risks here are insignificant with no further actions required
	–	**Medium:** risks ranked here require reasonable steps to reduce; not the most urgent priority; action plan
Orange	–	**High risk:** risks here are high priority and call for immediate action; action plan
Red	–	**Extreme:** risks here are critical and the highest priority; immediate action; action plan

Likelihood and Consequence factors are used to distill conditions and controls associated with a bucket elevator into values used to prioritize implementation of potential solutions. For the tool, these factors are built into a summary form on a sheet and into the headers of the User Entry sheets for both the likelihood and consequence. This approach to risk, when combined with the quantified questions used on the User Entry Sheets, provides users with a mitigated risk aligned with an organization's risk tolerance.

4. Results and Discussions

As a result of this tool and associated work, an end user is able to bring a higher level of consistency to the analysis. The user is also able to be more quantitative when identifying dust hazards. The tool takes into consideration the various factors which could cause combustible dust explosions. As generally known, three elements are required for a fire (1) a fuel; (2) an oxidant, typically the oxygen in air; and (3) a sufficiently energetic ignition source. If any one of these three elements are removed, a fire cannot be initiated. The first two of these elements (fuel and oxidant) when in an appropriate ratio, are referred to as the flammable atmosphere. For an explosion involving a combustible dust to occur, two additional requirements are necessary: (4) suspension or mixing of the combustible dust in air at or above the lower explosible concentration, and (5) confinement. When a combustible dust cloud is ignited, a flash fire (flame propagation without significant pressure effects) will ensue and the developing fireball can cause serious injuries to people and damage to process equipment. If the dust-air mixture is confined in a vessel or building, the expansion of the hot combustion materials will be impeded while the pressure increases, leading to an explosion. Due to the increase in pressure, equipment and building damages are likely.

This tool also takes into consideration the possibility of fuel generation, ignition sources consequences. In the case of event consequences, various parameters such as construction and design, equipment location, personnel and public risk, and mitigating measures such as explosion protection (relief venting or suppression and isolation) are evaluated. Moreover, housekeeping in the process area being assessed is considered. Detailed questionnaires based on operation experience, industrial data, operational principles of the equipment are developed for each parameter mentioned above and values are assigned for likelihood and consequence. The values generated from each questionnaire are summed and plotted in the risk matrix as a mitigated risk. In order to test the tool, five grain elevator scenarios were used as presented in the Figures 3 to 7.

Figure 3 displays the mitigated risk map for scenario A. In this example, the bucket elevator is well designed and located outdoors with personnel egress restrictions. The bucket elevator is provided with explosion relief venting according to NFPA 68 and equipped with upstream and downstream isolation. It has all recommended safety devices and a monthly preventive maintenance program for inspection that includes monitoring and testing. As detailed above, personnel restrictions are in place and operators work around the bucket elevator once a week. The mitigated risk of an explosion is at the intersection of "infrequent" and "minor". No additional requirements are required for this bucket elevator.

Facility:	Facility Name: Example 1
Date:	Date: 01/11/2019
Equipment Evaluated:	Equipment: Bucket elevator

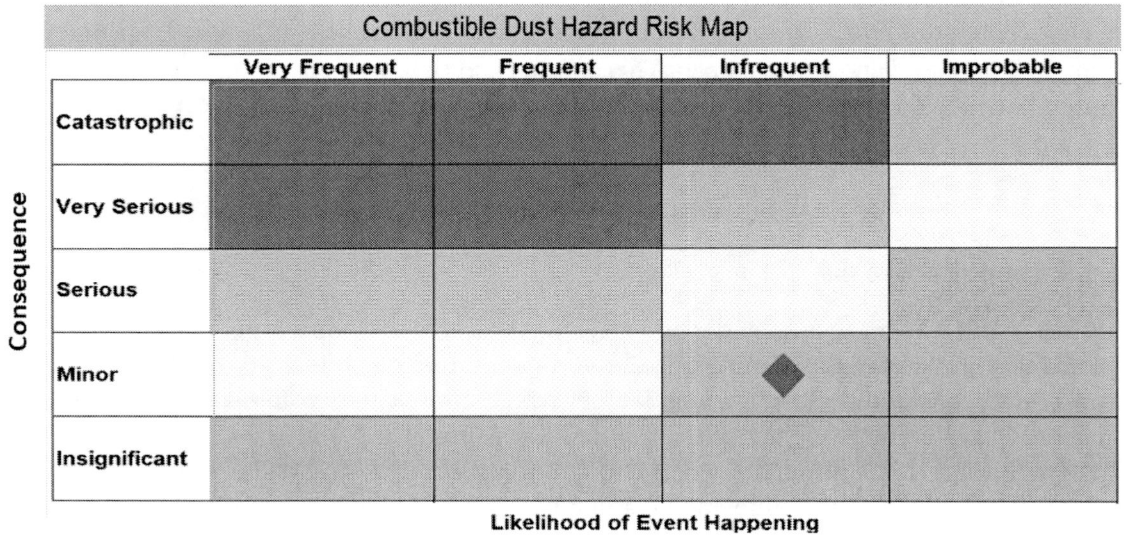

Figure 3: Example Bucket Elevator Scenario A

Figure 4 presents the mitigated risk map for scenario B. In this example, the bucket elevator is designed with safeguards and located indoors with no personnel egress restrictions (that is personnel work 25 feet around it while the bucket elevator is running). The elevator leg is provided with explosion relief venting ducted outdoors and equipped with explosion isolation devices. It has all required safety devices and a monthly preventive maintenance program for inspection, monitoring and testing. In this case, the bucket elevator is located indoors with evidence of housekeeping issues. There is an established housekeeping program, but it is not strictly followed. The likelihood of experiencing an explosion inside the bucket elevator is "infrequent" and the consequence of explosion is defined as "serious".

With regards to scenario C represented in Figure 5, the bucket elevator is located outdoors. It is equipped with explosion relief venting, but no isolation. The bucket elevator has all the required safety devices including a monthly preventive maintenance program for inspections, monitoring and testing. The likelihood of experiencing an explosion inside the bucket elevator is "infrequent" and the consequence of an explosion in the bucket elevator is categorized as "very serious". Any deflagration event inside the bucket elevator could propagate to upstream and downstream operations, thereby potentially harming the operators in these locations.

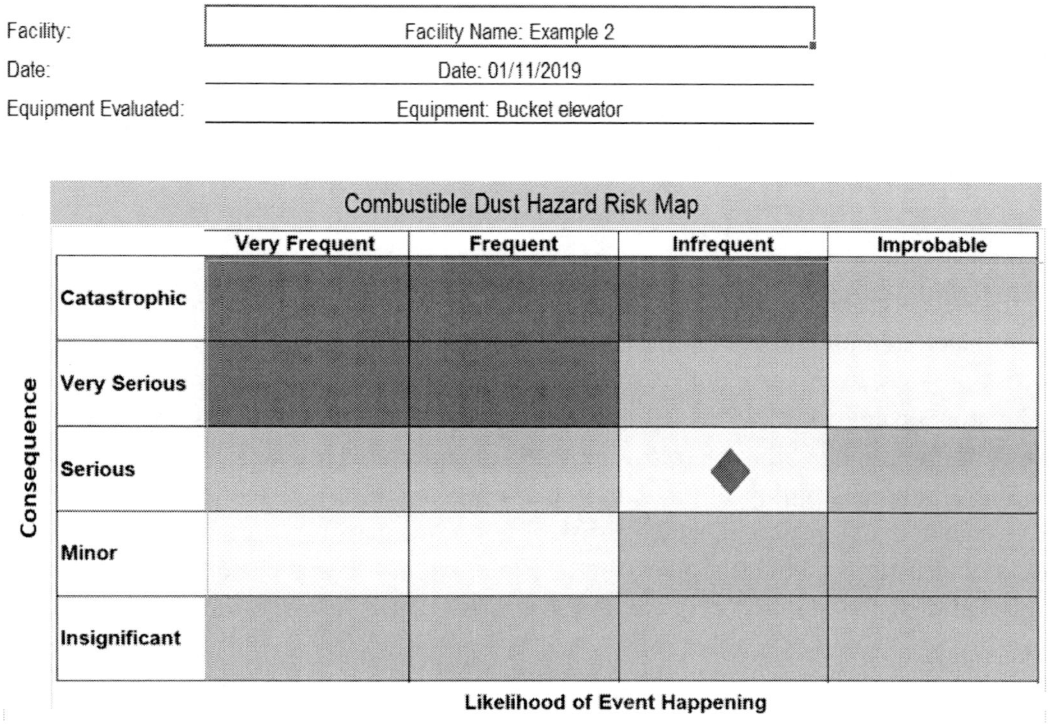

Figure 4: Example Bucket Elevator Scenario B

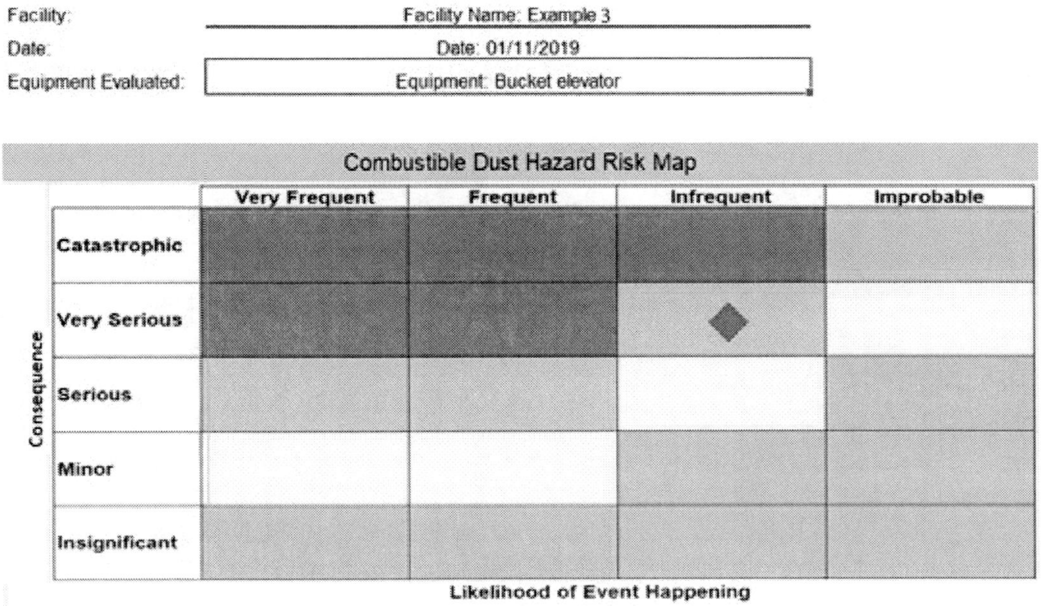

Figure 5: Example Bucket Elevator Scenario C

Figure 6 presents the mitigated risk map for scenario D. In this example, the bucket elevator is located indoors where personnel work within 25 feet of the elevator while it is running. The elevator is provided with explosion relief vents routed outdoors without explosion isolation. The consequence of an explosion in the bucket elevator is categorized as "very serious". Any deflagration event inside the bucket elevator has potential to propagate upstream and downstream to adjacent operations, with the potential to harm operators in these locations. This example simulates a facility that has limited explosion prevention measures. For example, there are no belt and pulley alignment sensors to detect misalignment and there is no preventive maintenance program to inspect, monitor and test safety devices. The likelihood of fire or explosion inside this bucket elevator is "frequent".

Finally, Figure 7 depicts the mitigated risk map for example scenario E. The bucket elevator is located indoors with personnel working within 25 feet while the bucket elevator is running. It is neither provided with explosion relief venting nor explosion isolation. There is no housekeeping program in place to maintain accumulation levels below acceptable limits. Dust accumulations in this process area create a secondary explosion hazard and the consequence of explosion in the bucket elevator is categorized as "catastrophic". Moreover, this example was populated to simulate the impacts of operations without explosion protection measures in place. Electrical equipment are not classified, belt and pulley alignment sensors are not installed to detect misalignment, no preventive maintenance program, no established hot work permit program, no upstream ignition detection devices, etc. The likelihood of fire or explosion inside this bucket elevator is considered "very frequent"

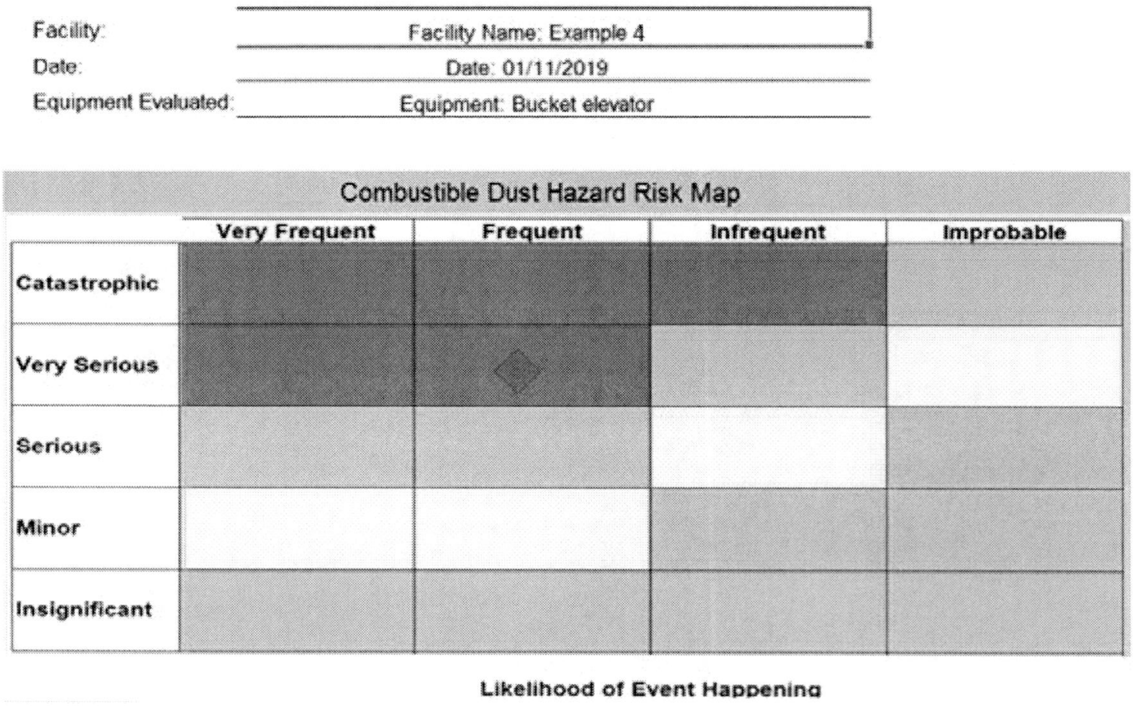

Figure 6: Example Bucket Elevator Scenario D

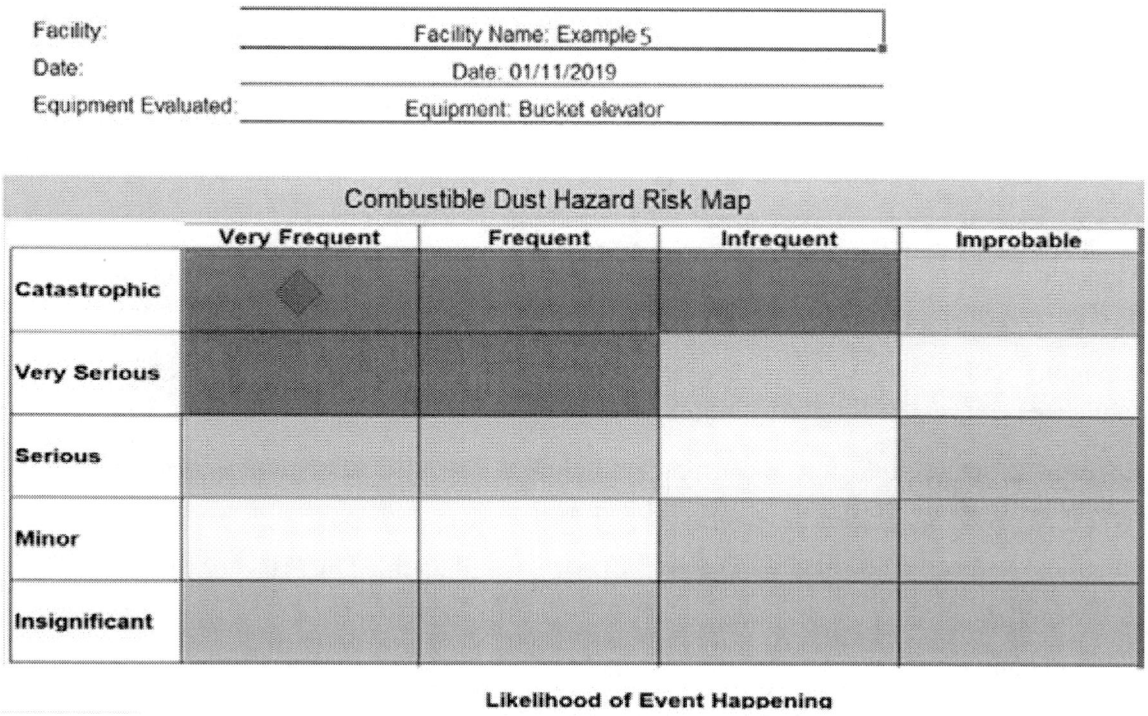

Figure 7: Example Bucket Elevator Scenario E

5. Future actions

Future actions include applying similar methods in developing a suite of customized tools designed to address common solids handing unit operations such as hammer mills, belt conveyors, silos, pneumatic conveying, dust collection system, etc. Future work plans include continued validation of tool modules by applying actual facility data and benchmarking with incident histories. These iterative steps allow for further calibration of the tool modules.

6. Conclusion

A semi-quantitative risk-based screening tool has been developed to consistently and effectively evaluate dust explosion risks for bucket elevators with future tool modules in development. This tool is designed for use by on-site supervisors and local subject matter experts (SMEs) trained in combustible dust hazards and the tool. The tool uses multiple-choice and yes/no questions to assess both likelihood and severity of risk. Values are assigned to each response, providing the opportunity to weigh relative importance of each question, in real-time. Values for Severity and Likelihood are used to automatically show the mitigated risk plotted on a Risk Matrix. The function of this tool is to identify and evaluate the most common preventive, protective and mitigative measures an organization has taken to prevent combustible dust related fires and explosions. When this measures evaluation is combined with combustible dust test data (such as explosibility properties of the material; minimum ignition energy, minimum ignition temperature, deflagration index, maximum explosion pressure, self-ignition temperature, limiting oxygen concentration, minimum explosible concentration etc.), a visual representation of risk is obtained. In addition to providing an estimated risk level, the tool identifies potential safeguarding opportunities and thus providing guidance for improvement.

References

[1] CCPS Guidelines for Combustible Dust Hazard Analysis, 2017

[2] NFPA 61 Standard for the Prevention of Fires and Dust Explosions in Agricultural and Food Processing Facilities, 2019.

[3] Mannan, S., Lees' Process Safety Essentials: Hazard Identification, Assessment and Control. 2013: Butterworth-Heinemann.

[4] Khan, F.I. and S. Abbasi, Major accidents in process industries and an analysis of causes and consequences. Journal of Loss Prevention in the process Industries, 1999. 12(5): p. 361-378.

[5] Eckhoff, R., Dust explosions in the process industries: identification, assessment and control of dust hazards. 2003: Gulf professional publishing.

[6] Chemical Safety and Hazard Investigation Board, Combustible Dust Hazard Study, 2006-H-1, November 2006,

[7] CSB incident database, as of 4/23/2018.

[8] There were 14 fatalities and 36 injured at Imperial Sugar (2009), three fatalities and one injured at AL Solutions (2014), five fatalities and three injured at Hoeganaes (2012), seven injured at US Ink (2012), and five fatalities and 14 injured at Didion Milling (2017). The CSB also investigated West Pharmaceutical Services (2004), CTA Acoustics (2005), and Hayes Lemmerz (2005). In total, these eight incidents cost the lives of 41 individuals and injured 142 others.

[9] CCPS Guidelines for Risk Based Process Safety, 2007

548425 Building Hazard Recognition Competency

Understanding hazards and risk is essential for the success of process safety as well as occupational safety. Process safety involves managing energy transfers throughout a process life cycle in-order-to minimize the risk of releasing highly hazardous chemicals that can result in large-scale harm. The goal of the process industry is to manage transfers and transformations of energy that result in the safe manufacture of products. Occupational safety involves managing energy as workers perform jobs

involving routine operations, maintenance, or repair in-order-to prevent personal injuries and illnesses, damage to facilities, harm to the environment as well as prevention of process related incidents. A consistent and systematic approach to hazard recognition can help workers understand risks more clearly and improve the efficiency and effectiveness of both process safety and occupational safety.

Establishing an effective safety culture requires solid hazard recognition and evaluation principles coupled with effective communication of process and occupational hazards. A hazard can be defined as the potential for an uncontrolled release of, or an unwanted contact with, energy that can cause harm to people, the environment, and/or equipment and facilities. In some circumstances, a loss or lack of energy can initiate events that result in harm from the interaction of other forms of energy. Both process and occupational hazards stem from the potential for energy transfer or transformation in an undesirable manner that can result in harm. The type and magnitude of energy are general characteristics that differentiate process hazards from occupational hazards.

Process hazards are potential catastrophic events such as loss of containment (uncontrolled release) of highly hazardous chemicals resulting in spills, toxic vapor clouds, fires, explosions, and other largescale events. These events can harm many people, the process unit, the processing facility, local communities, public areas, and the environment well beyond the facility boundaries. Process safety incidents often involve the shut-down of part or all of a process.

Occupational hazards are often viewed as localized circumstances that can injure or cause illness to one or more workers and may result in damage to the environment, equipment, or parts of a facility. The process itself may or may not be involved in these types of events. Energy contained within a process can certainly present occupational hazards to operators, maintenance personnel, and others who are required to work with the process.

A harm-producing incident cannot take place without an energy transfer or transformation. Understanding the energy associated with a process or job and preventing an uncontrolled release of energy or unwanted contact with energy is necessary to ensure that incidents are prevented – both process and occupational. An excellent model for understanding this relationship of energy to both process and occupational hazards is the The Management Oversight and Risk Tree (MORT).

MORT is a logic diagram that maps incident structure and includes energy as a prominent factor in all incidents. In Gate SA1 of the MORT diagram there are four conditions that are necessary for an incident to take place. An incident is not possible if any one of the four conditions is missing. Three of these conditions are: (1) the presence of potentially harmful energy, (2) the lack of adequate barriers and controls to manage the potentially harmful energy, and (3) the presence of vulnerable people or objects that can be exposed to the potentially harmful energy. The fourth condition is a sequence of events that leads to an uncontrolled release of energy, unwanted contact with energy, or incompatible interactions among the energies present. An incident occurs when people or objects are harmed because of this sequence of events.

Perceptions of risk vary depending upon a worker's knowledge of energy and whether the worker is focused on process or occupational hazards. This creates a need for common terminology about hazards so that the risk associated with a process or job can be effectively communicated, especially when process and occupational hazards are present simultaneously. Knowledge of the energy involved in a job or process is paramount to determining possible harmful outcomes.

A pillar of process safety is to help workers effectively answer risk-based process safety questions. "What can go wrong?", "How bad could it be?", and "How often might it occur?" are rooted in identifying and evaluating the energy involved in a job or process. Teaching workers how to use a simple and effective energy-based hazard

recognition tool for both process and occupational safety provides an efficient and effective way to improve process and job knowledge. This will result in improved hazard identification competency, risk reduction plans, and communication among work groups with respect to hazards, risk evaluations, and subsequent risk reduction actions.

548376 Ignorance Is Not Bliss with Reactive Chemical Hazards

Many highly hazardous chemicals have been identified by the Occupational Safety & Health Administration (OSHA) Process Safety Management (PSM) regulation. Exclusion from the regulatory lists, however, must not lull you into the mistaken belief that your chemistry does not have reactive hazards that must be identified, understood and controlled.

A recommended approach for reactive chemical characterization includes reactivity data, thermal and chemical stability data as well as hazardous effects of inadvertent mixing of materials or other credible upset conditions that could potentially occur. This presentation will provide practical tips for reactive chemical hazard identification and characterization that need to be reviewed at both the design stage as well as full production scale up. Several industry accidents will be used to illustrate key points for characterizing and recognizing chemical reactivity hazards as well as identifying potential safeguards or alternate designs that could be used to mitigate the identified risks.

550289 Process Safety Improvements – Things That Help Me Sleep at Night

Process safety focuses on preventing fires, explosions and accidental chemical releases in chemical process facilities or other facilities dealing with hazardous materials such as refineries, and oil and gas (onshore and offshore) production installations. There has been a continuous increase with improvement in focus and attitude toward process safety within the oil and gas industry and particularly within operating companies over the history. Within the last 15-20 years, the improvements have taken exponential shift in the positive direction. Companies have adopted programs, some in reaction to catastrophic events, but just as many if not more programs were initiated in response to employees' desire to be safe, coupled with management's desire for the employees to be safe. Large industrial programs – High Temperature Hydrogen Attack studies, LOPA, Emergency Isolation Valves etc. are often born from catastrophic industrial accidents. These programs are directed at eliminating or safeguarding against a known potential problem. However, equally important are the company or plant-level initiated programs—behavioral based safety programs, joint job site visits including operating personnel and personnel performing hot work, safety observations, focus on near misses, categorized incident investigations, etc. At the end of the day these site programs have a cost for implementation that is absolutely dwarfed by the comparison to the potential negative consequence the programs are aimed to eliminate. All of the programs are driven toward ensuring that our fellow workers go home in a condition at least as good as when they reported to work.

552927 The Day I Got a Second Chance

This presentation recounts a major process safety accident that occurred in 1989. The accident released a large volume of hot flammable liquid hydrocarbons from an 8" diameter pipe that experienced a guillotine type failure. Miraculously the flammable gas cloud generated by the release failed to ignite. The events leading up to the accident are examined and the lessons learned on both a professional and personal level are discussed.

You've Received Your PHA Report: What Now?

Rene' Murata (presenter),
Sean Classen, M. Jacob Alvarez, Janet Benaquisto
Risk Integrity Safety Knowledge, Inc. (RISK, Inc.)
Rene.Murata@psmrisk.com

Merrill Childs
Cargill Incorporated
Merrill_Childs@cargill.com

Prepared for Presentation at
American Institute of Chemical Engineers
2019 Spring Meeting and 15th Global Congress on Process Safety
New Orleans, LA
March 31 – April 3, 2019

You've Received Your PHA Report: What Now?

Rene' Murata (presenter), Sean Classen, M. Jacob Alvarez, Janet Benaquisto
Risk Integrity Safety Knowledge, Inc. (RISK, Inc.)
Rene.Murata@psmrisk.com

Merrill Childs
Cargill Incorporated
Merrill_Childs@cargill.com

Keywords: PHA, Draft Report Review, Quality Assurance

Abstract

This paper provides guidance for the review of a Process Hazard Analysis (PHA) draft report. The draft phase of a PHA is the final chance for site personnel to make updates and corrections before the PHA becomes a permanent part of a facility's Process Safety Management (PSM) records. There are several things that can and should be checked before the final report is issued.

Guidelines are provided that can be used by those reviewing a PHA draft. Key quality control points are defined to ensure that minimum requirements are met in various areas including meeting auditable requirements, appropriate documentation depth, consistency for risk ranking, and robust PHA recommendations. With this quality assurance in place, the resulting integration of the PHA into a site's PSM program has positive ripple effects, improving the quality and effectiveness of other PSM elements such as Procedure Development, Process Safety Information, Training and Management of Change.

Introduction – Importance of PHA Report Review

Following the conclusion of a PHA, the results are documented in a PHA report. The first submitted draft of the PHA report is not final and should be reviewed before it is updated to a final version. Even though this draft is not final, it should be as complete and correct as possible. A quality assurance process, such as peer review should be used to identify and correct as many errors or deviations from best practice and procedure requirements as possible before delivering it to the reviewer.

The final stage of verifying a PHA is conducted at the site level, where site management is responsible for ensuring the report is reviewed. If the responsible manager is not familiar with the unit, they may assign this review to qualified personnel. During this final review several aspects of a PHA can and should be checked before the final report is issued. Because the PHA report serves a key function in feeding information to multiple areas, including the recommendation tracking system, this is an important opportunity to ensure information is accurate and clearly documented to provide a complete summary of the hazards and resulting actions necessary by site staff to prevent a future incident.

This review has the potential to be a daunting task since a multi-week PHA can stretch across hundreds of pages. Experienced process safety professionals at a site are generally the preferred personnel to conduct this review, since they are aware of the site's technology, corporate requirements, involved personnel, and are aware of known pitfalls and key areas to check before a PHA report is accepted.

Industry Perspective

Cargill has been developing a PHA Draft Review process and shared what they found important and what they found challenging in putting a process in place. From their perspective, skipping the review step or not doing it well can result in:

- More time and effort to make corrections later
- Scenarios coming up again in revalidations, and
- Risk of an incident where the intended safeguard was not implemented because it was not clear.

They found it important to define who will perform the review, recognizing that it could be more than one person. Specific recommendations and responsibilities for each reviewer need to be defined. Also key is communicating and gaining acceptance of updates with the PHA team for improved wording or substantive changes.

A Manufacturing Technology Lead or similar technology representative is a key reviewer, specifically to confirm:

- Consistent risk assessment with similar sister plants
- Scenario development and documentation are complete such that it will make sense a few months later
- Consequence probability assignment is appropriate
- Safeguards are credited correctly
- Rules of independence are followed, and
- Solutions are right sized to solve the risk gap

The biggest challenges have been time to do a thorough review and getting key team members back together to review specific scenarios if needed. When there is a new facilitator, a new Manufacturing Technology Lead or a new process, the review is taken to a business unit or central PSM management level for added scrutiny. Deep reviews by business unit or central management review also contributes to identification of leading practices and opportunities for improvement in the PHA process.

PHA Draft Review Process

A draft review process can identify issues within a specific PHA. It can also identify systematic problems that occurred during a specific PHA so that they can be corrected in future PHAs at that site. A review of OSHA citations shows that the PHA element is cited routinely during audits and post incident investigations. Industry examples of inadequate hazard analysis range from subtle deficiencies that lead to cost inefficiency to major incidents that grab headlines. Some of the key deficiencies commonly cited include[1-7]:

- Prior incidents not identified during the PHA
- Not considering all modes of operation
- Failure to consider early warning/detection systems
- Failure to ensure that the process hazard analysis addressed possible safety and health effects from failure of engineering and administrative controls.
- Failure to address the all hazards of the process and operating procedures
- PHA did not address the hazards appropriate to the complexity of the process
- Failure to address process hazard analysis deficiencies
- Failure to address the hazards associated with inadvertent manual valve operation.
- Inadequate or missing documentation
- Failing to address the findings and recommendations of the process hazard analysis team
- Failing to ensure that the process hazard analysis is accurate

The following sections provide key areas to review in your draft report to improve the robustness of PHAs.

Guidelines for PHA Draft Report Review

1. First Steps

Administrative Details

Upon receiving the PHA draft report the reviewer should ensure that it contains all the required elements. This is comprised of the name of the unit studied, the dates that the study was conducted, as well as team members names, expertise and attendance. The report must include a list of all Process Safety Information (PSI) that was referenced during the PHA including, but not limited to, PFDs, P&IDs, MSDS and procedures. The reviewer should spot check to verify that the most up to date and accurate version of the PSI information was used.

Next, the reviewer should evaluate the Node descriptions. Are they clearly identified both in the description and on the P&IDs? They should also analyze the node descriptions to ascertain if there is a clear process description for the node, one that includes the design conditions and safe operating limits. This is important to later determine if the appropriate scenarios were reviewed within the node. For example; if temperature is a concern, then the safe operating limit should be identified. This helps the team understand and identify what the actual hazard is, or can be, when the temperature goes beyond the safe operating limits, such as exceeding metallurgical limits of equipment leading to failure of the equipment.

In addition to the node description, the reviewer should evaluate if it is clear what P&ID/PDF drawings were used within the node (normally with the nodes of study delineated by color), as well as the equipment covered within each node.

Missing PSM Data

The PHA worksheets should be spot checked to reveal inadequacies in documentation. In a situation where such documentation is unavailable, a recommendation that deficient PSM information will be complied and brought up to date should be made and the analysis should be continued as well as can be managed with the available data.

All referenced documents must be up to date and accurate. P&IDs should have a specific name or reference number and a date when was last revised. Equipment and instruments referenced should be called out by their unique ID numbers. PSVs used as safeguards should reference the relief device's set point, the MAWP of the vessel protected, and if the PSV has been sized for the scenario it is being used for.

2. Auditable Requirements

In addition to the above items; reviewing administrative information, appropriateness of depth, consistent scenario development, severity and likelihood ranking, as well as recommendations, there are several other auditable requirements that one must consider.

Causes

The reviewer should spot check to verify that Causes (or what-if questions) address initiating causes/events rather than consequences. Additionally, they should verify that the causes address equipment failure, failure of administrative controls, and human error.

Safeguards

One must review safeguards used to determine if they are appropriate for the scenario as written. Again, this can be done, initially, by spot checking. For example, if the scenario includes a valve malfunctioning closed ultimately leading to overpressure of a vessel, but the safeguard is a high-level alarm, is it clear how a high-level alarm will help prevent the scenario? If not, then the safeguard should not be considered as a valid safeguard and should be flagged for review. The reviewer should also do a spot check of safety instrumented functions, if appropriate.

Additional concerns around safeguards would include whether there is enough information to appropriately identify it for inclusion in maintenance programs or procedures (i.e., is there a tag number or other way to clearly identify it). Also, if there are corporate (or site) requirements about how a safeguard is written for inclusion in a maintenance or 'IPL management like' program afterward, have the safeguards been written to meet those requirements?

Industry and Corporate Standards

Other areas for evaluation would include compliance with the company's written policy on conducting a PHA. There are often differences in what items may be considered safeguards (i.e., preventative vs. mitigative), specific language preferred in a scenario, (i.e. personnel injury vs. fatality), and guidance on how to risk rank particular types of scenarios (i.e. overpressure). This is particularly important when the facilitator is a 3rd party facilitator that works with multiple clients. The reviewer should verify that the report contains a description of the risk matrix that was utilized during the PHA.

The reviewer should also consider if there are industry standards for the reviewed process. They can verify that those standards referenced and utilized during the PHA along with appropriate (or recommended) safeguards (i.e. Chlorine).

Additional Study Information

Reviews of the Human Factors and Facility Siting Checklists, Hazardous Incidents/Near Misses and MOCs since last PHA must be reviewed as part of the PHA.

Once the reviewer has confirmed that these checklists and reviews have been performed, those sections may also be spot checked for appropriate depth. Do the checklist responses include sufficient detail such that the full impact is understood or is the entire checklist a string of "yes" answers? If there has been a serious incident, near miss or MOC, is it referenced in a specific PHA scenario? Software "search/find" functions can help with this type of checking.

3. Appropriate Scenario Development

There are several things the reviewer(s) should consider when reviewing the scenarios throughout the PHA. As previously discussed, this can be done initially through spot-checking, then, if determined to be necessary, a more thorough review can be conducted.

The reviewer must consider if the consequences clearly illustrate the hazard and how it impacts employees and the public.

The reviewer(s) should consider if the consequences are appropriate for the process being reviewed and if the worst credible case is considered. For example, if the process is light hydrocarbons, highly flammable, but the consequences do not include fire, then a deeper dive into the PHA may be necessary.

Additionally, the reviewer should look to see if safeguards were considered in the development of the consequence. In other words, did the scenario stop at a PSV lifting? Or did the team take it all the way to vessel failure then call out the PSV as a safeguard.

Next thing to review is whether or not upstream and downstream consequences were both considered. Too often, a team will be concerned about a scenario in one direction and miss potential hazards in the other.

4. Consistency for PHA Scenario Development

PHA consistency challenges and techniques is an on-going challenge for sites and inconsistency between PHA's can occur for a variety of reasons. In some instances, as PHA sessions progress, the understanding of the PHA process improves leading to changes in how scenarios are identified and developed. This then can lead to inconsistencies between the beginning and the end of the PHA. Occasionally, changes in PHA team personnel can lead to different conclusions. In instances such as these, different conclusions can lead to inconsistent risk ranking if not managed and corrected for consistency. A reviewer of a draft PHA should look for the following key areas.

Consistency for Overall Scenario Development

Achieving overall PHA scenario consistency within a company and even between PHAs conducted at a single site is an ongoing challenge. There exists various methods to achieve consistency and the PHA team and the draft PHA reviewer are key players in this process. Consistency should be applied in all aspects of a PHA, including scenario selection, consequence development, risk ranking, safeguard selection, and even gap closure/recommendation development. Various techniques used to assure consistency include:

- Application of generic/template PHA scenarios
- Alignment with industry codes/standards for applied safeguards
- Inclusion of corporate subject matter experts to assure alignment.
- Use of existing site PHAs to cross-check appropriate scenario development

The use of templates is a popular option for companies that have similar technology installations at multiple sites. Alignment with industry codes is useful for well-understood hazards in industries that have applicable institutions (i.e. The Chlorine Institute).

A PHA draft review may also have access to previous PHAs performed on similar units at other sites, corporate subject matter experts and other PHAs performed on the same site which include similar scenarios. While it is possible for potential hazardous scenarios may deviate from each other in consequence and frequency, even for similar units, these resources may be used to drive general consistency between PHAs.

Consistent Ranking for Severity

Scenarios with a similar outcome should have consistent results. For example, a PHA consequence that ends with "…loss of containment of hexane leading to possible fire and personnel injury" should be treated consistently throughout a PHA. If a subsequent scenario includes a consequence with "…loss of containment of hexane leading to possible fire, explosion and fatality" and different severity ranking without sufficient context, the PHA study may be seen as inconsistent. Variations may exist due to potential release quantities and locations. In these cases, the PHA should provide sufficient details so that differences in severity ranking are clear.

Consistent Ranking for Likelihood

In a HAZOP, as opposed to a LOPA, where likelihood is qualitative in nature, the likelihood of a scenario is subjective and dependent on the experiences of the PHA team. Due to this, likelihood rankings in a HAZOP are inherently prone to inconsistency if care is not applied. This risk is especially relevant in cases where safeguards and surrounding conditions are situationally dependent. Valid discrepancy may occur when items such as occupancy, release orientation, or complexity associated with safeguard variable. For this reason, PHA Draft reviewers should spot check the likelihood of similar scenarios to ensure that the likelihood is either reasonably constant or there is sufficient explanation to justify the difference.

5. Appropriate Level of Depth

It is important to determine that an appropriate level of detail was used is to ensure that the PHA can be understood by someone who was not a member of the PHA. Without an appropriate level of depth, scenarios that were completely understood when the PHA was being performed can be baffling to those who helped develop them only a few years later.

Some errors may be readily apparent and other will be subtle errors. Spot checking is the most practical method to find subtle errors that may be systemic to how the study was facilitated. It is normally impractical to completely review an entire multi-week PHA line-by-line but if a more rigorous method is required to reveal systematic subtle errors, a single node or section of a node made be reviewed line-by-line.

Risk Gaps Not Addressed

One of the most glaring depth of documentation errors which may be discovered when reviewing a PHA is a risk gap not addressed by a recommendation. Spot checking is sufficient for most other level of depth concerns but not unclosed risk gaps. Most PHA software offers functions to quickly check for unclosed gaps but even if the PHA was performed using a spreadsheet program, it is necessary to completely review that all gaps have been adequately addressed. In this case the reviewer must scroll through all PHA worksheets looking for unclosed gaps. A similar error is a cause/consequence pair where the risk ranking has not been completed (i.e., a scenario where either the likelihood, severity or both is not recorded).

It is also possible to have "orphan recommendations" not associated with any cause/consequence pair or checklist item. While there may be cases where the team wanted to make a general recommendation, orphan recommendations are normally artifacts of adding a recommendation to a scenario and later removing it without deleting it from the "master" recommendation list. A reviewer should confirm with the report author whether any orphan recommendations found are intentional or should be deleted.

<u>Missing Hazards</u>

The reviewer of a PHA draft report should check that all obvious major hazardous scenarios which exist in the unit have been caught by the PHA team. This can be more difficult than catching other errors in documentation because there is no existing scenario to check for deficiencies. The reviewer should be familiar with all the known major hazards of the unit and check that each has been covered by the PHA. Ideally, the reviewer for this part is an "expert" in the process. However, a list of these scenarios, either compiled by subject matter experts or from industry experience of previous incidents in similar units, could be compiled and checked off as they are located in the PHA worksheets. This is more critical if the PHA team was composed of less qualified personnel. Example: A unit has a batch reactor with a potential to run away resulting in overpressure and loss of containment – the reviewer should check the node containing the reactor to make sure that this scenario has been covered.

6. Robust PHA Recommendations

Upon completion of a PHA, recommendations are vetted for acceptance by the site, assigned to responsible parties, and tracked to completion. Most sites now use electronic tracking systems which means they no longer have the context of the full PHA worksheet and by necessity must provide complete information to communicate the full nature of the hazard and the action(s) to be completed.

A robust PHA recommendation is SMART:

- ***S***pecific
  - Recommendations should be clear and complete. It simplifies the review and closure of recommendations immensely if their wording is such that each recommendation can stand alone. A Recommendation that "stands alone" allows for anyone reading it to have the complete story on why there is a concern, where the concern is and what the intention of the team was. The recommendation should include equipment numbers, lines numbers and/or drawings where relevant.
- ***M***easurable, therefore able to close
  - Recommendations must be written so that they are accomplishable and have a clear point of closure.
- ***A***ccountable
  - During the closeout meeting all recommendations must be assigned to a responsible party. This may occur after the PHA meeting and so, not present in the draft report. –
- ***R***elevant
  - Recommendations should have a direct impact on the risk ranking of the scenario for which they are made.
- ***T***ime Limited

- Recommendations are expected to be completed in a timely manner. The language used in the recommendation should reflect this expectation. Individual companies often have specific rules around when recommendations should be closed based on ease of closure and risk ranking. If a recommendation requires a shutdown to be implemented or a significant engineering study to be performed, a recommendation may be split into multiple recommendations – some of which may be performed more immediately.

The following recommendations were cited in an OSHA report[7] and provide and provide a good example of weak recommendations which are not explicit in how they are meant to close a specific risk gap. If the reviewer were to encounter similar recommendations, they should consider how those recommendations could be revised to be ***SMART***.

"Consider installing a level gauge on the pentane tank (vessel E1) to monitor tank level"

"Consider locking valves in pentane piping system in desired positions"

"Consider providing a safeguard to alert operators when the ventilation system above the laminator fails"

"Consider implementing a site mechanical integrity program to include tanks, piping, valves, and components"

Context for why these recommendations were not completed in time is not available, initial inspection reveals the following:

- They don't provide context or basis for why a recommendation is required. Rather, the reader is left to assume the surrounding PHA context
- The actions required are ambiguous as it relates to closing an identified risk gap

To avoid weak recommendations, the PHA draft reviewer should inspect that a recommendation contains sufficient information to stand alone and is clear as to how its closure will address an identified risk. Recommendations associated with cause/consequence pairs should effectively prevent or mitigate the consequence of every scenario they are referenced in. Just because a recommendation addresses the consequence of a cause doesn't mean it is effective at preventing the same consequence from a different cause.

Is the Recommendation stand-alone?

PHA Draft reviewers should ensure that PHA recommendations are written with sufficient detail such that they can be completed accurately and confidently with the information provided in the recommendation.

Generally, a PHA recommendation is ultimately exported into an action tracking database that contains three key pieces of information: The PHA Recommendation, the person responsible for closing it, and a due date. The full PHA report for surrounding scenario context is generally not attached, so it's critical that a recommendation include sufficient detail for its audience: the personnel assigned to close the recommendation. Critical information to include is:

- The action that needs to be achieved (such as "provide effective overpressure protection on Vessel XXX for the liquid overfill scenario…")
- The risk that exists, and how the desired recommendation would mitigate the risk
- Any flexibility that may exist to the recommendation closure task.

Would completion of the recommendation close the identified risk gap?

PHA Draft reviewers should ensure that closure of a given PHA recommendation would unambiguously lead to closure of an identified risk tolerance deficiency. Earlier, it was discussed how every unclosed risk gap should be addressed by a PHA recommendation. All recommendations must also be reviewed to ensure that they unambiguously close the risk gap of all scenarios they are utilized in.

Example: A recommendation for upsizing a PRV currently in service is used in multiple blocked vessel outlet scenarios. Some of these scenarios result in vapor overpressure and some result in liquid overpressure. The reviewer must check that the recommendation specifies a PRV appropriate for both liquid and vapor scenarios and that a liquid valve is feasible for this application in order to take credit for the recommendation.

Recommendations that do not address a gap between an identified risk exposure compared to a risk tolerance threshold represent pitfalls to a PHA program and erode confidence in the PHA process. This can occur in instances where well-intentioned operability improvement opportunities are not separated from risk-driven requirements. In other cases, use of the word "consider" as a prefix to a recommendation sometimes can be interpreted as if the recommendation merely needs to be subjectively considered as a useful improvement for sufficient closure. This is rarely the desired intention of the PHA evaluation conclusion.

To re-iterate a point made above, any flexibility that may exist to the recommendation closure task should be clearly delineated. For example, the example recommendation above of "Consider installing a level gauge on the pentane tank (vessel E1) to monitor tank level" could possibly be improved to the following:

"Provide a preventative safeguard to protect against overfill of the E1 pentane tank. Possible measures to consider:

- *Install a level gauge, independent of the existing E1 level control loop, with a high alarm*
- *Install a high level trip on E1 to close inlet flow in the event of a high high level*

Basis: In the event that the existing level control gauge in E1 malfunctions, there is the potential to overfill E1 pentane tank, leading to a spill of pentane to the E1 diked area with possible flammable vapor cloud formation, fire/explosion, and personnel injury/fatality."

The above recommendation example is sufficiently detailed so as to not require reference to the original PHA documentation, and gives the personnel assigned an action a clear closure plan. Conversely, this recommendation is not excessively rigid and is not overly prescriptive on a specific design solution.

Conclusions

Upon completion of the review(s), there may be cause to update the PHA with corrections and/or changes. Depending on the level of change required, there may a number of ways that this can be completed. If the changes are minor in nature, in other words, they do not effect a change in how the scenario is described, this may be done without team involvement. However, if there are questions about the legitimacy of a scenario, the effectiveness of a safeguard or the validity of a ranking, the team, or its equivalent, must be convened to revisit the PHA.

The appendix of this paper includes a simple Quality Assurance checklist to assist reviewers in completing site reviews of PHAs. It could be augmented with specific corporate standards to personalize it to your company.

Having an effective Quality Assurance process in place to review a PHA draft report is particularly beneficial to a site. There are ripple effects to the PHA process and to other aspects of PSM. It is one way to improve overall PHA robustness which in turn strengthens the site personnel's understanding of hazards and communication of the hazards to better manage them. Better management of hazards leads to fewer incidents which thereby reduces risk for both site personnel and the surrounding community.

References

[1] Center for Chemical Process Safety, AIChE. Guidelines for Risk Based Process Safety. Wiley, 2007

[2] Citation and Notification of Penalty, Inspection Number 315298307, Occupational Safety and Health Administration, https://www.osha.gov/ooc/citations/Bostik_Inc_315298307.pdf (accessed Dec 14, 2018)

[3] Citation and Notification of Penalty, Inspection Number 1312169, Occupational Safety and Health Administration, https://www.dol.gov/sites/dolgov/files/OPA/newsreleases/OSHA20181543.pdf (accessed Dec 14, 2018)

[4] U.S. Chemical Safety and Hazard Investigation Board, "Factual Investigation Update, April 26, 2018 Husky Superior Refinery Explosion and Fire", https://www.csb.gov/assets/1/6/husky_factual_update_-_2.pdf (accessed Dec 14, 2018)

[5] Citation and Notification of Penalty, Inspection Number 844143, Occupational Safety and Health Administration, https://www.osha.gov/ooc/citations/PilgrimsPride_844143_0718_13.pdf

[6] Citation and Notification of Penalty, Inspection Number 844143, Occupational Safety and Health Administration, http://www.osha.gov/ooc/citations/ThreeRiversEnergyLLC_1117233.pdf

[7] Citation and Notification of Penalty, Inspection Number 312927312, Occupational Safety and Health Administration, https://www.osha.gov/ooc/citations/DunaUSAInc_312927312_0927_11Citations.pdf (accessed Jan 10, 2018)

Appendix A:

PHA Draft Review Checklist

Quality Assurance

PHA Questions	
Question	**Comment/Finding**
1. Are the PHA team members identified?	
2. PHA Facilitator name and company	
3. Is complete team member information provided (title, department, experience, expertise)?	
4. Are sessions listed?	
5. Is attendance tracked?	
6. Is the study scope, purpose and methodology described?	
7. Are drawings listed, including rev number and rev date?	
8. Are study nodes clearly defined, both in tabular form and on the P&IDs?	
9. Is the design intent of each node clearly defined and understandable? [Note: Are design & operating limits defined (i.e. MAWP at xF)]	
10. Do Causes (or what-if questions) generally address initiating causes/events rather than consequences?	
11. Do the what-if questions or HAZOP causes generally address both equipment failure and human error?	
12. Spot check P&IDs to ensure that obvious what-if questions or HAZOP causes are being identified and documented. (For example, check to make sure that all control valves fail open and close, pumps stop or do not start, etc.).	
13. Spot check to ensure that consequences are being developed without safeguards.	
14. Spot check to ensure that consequences are being developed in what appears to be chronological order.	
15. Have upstream and downstream consequences been considered and documented?	
16. Are the Consequences fully and comprehensively developed to "worst plausible" consequence?	
17. Are consequences documented consistently and appropriately for the process under study?	
18. Do the consequences described in the worksheets illustrate the hazard(s) and how employees and the public may be affected by the scenario?	
19. Do the recommendations appear to address the identified cause/consequence scenarios?	
20. Are Recommendations fully developed, clear and concise such that they remain comprehensible when extracted from the worksheets?	
21. Is there a description of the risk-ranking matrix, and how it is used as a tool to qualitatively evaluate a range of safety and health effects on employees and the public?	
22. Compare Risk Priority Ratings of similar scenarios to determine if risk factors are applied consistently.	
23. Are Severity and Likelihood assignments consistent with the Consequence and process under study?	
24. Spot check safeguards versus initiating cause to ensure that common-mode failure is considered.	
25. Are recommendations SMART (specific, measurable, accountable, relevant, time limited)?	

Safety Integrity Level Determination and Categorization

Anca Elena Costea, Andreea Iordache, Andreea Elena Catana
Siemens SRL
Preciziei Street, No. 24
Bucharest, Romania
anca.costea@siemens.com, andreea.iordache@siemens.com, andreea.catana@siemens.com

Prepared for Presentation at
American Institute of Chemical Engineers
2019 Spring Meeting and 15th Global Congress on Process Safety
New Orleans, LA
March 31 – April 3, 2019

Safety Integrity Level Determination and Categorization

Anca Elena Costea, Andreea Iordache, Andreea Elena Catana
Siemens SRL
Preciziei Street, No. 24
Bucharest, Romania
anca.costea@siemens.com, andreea.iordache@siemens.com, andreea.catana@siemens.com

Keywords: safety, integrity, risk, hazard, functions

Abstract

The standards IEC 61508[1] (umbrella standard, the first standard from which the others are derived) and IEC 61511[2] (standard for process industry) define a measure for the reliability of the safety systems, known as the Safety Integrity, which comprises four levels. These four levels of Safety Integrity are named SIL 1, SIL 2, SIL 3 and SIL 4, SIL 1 being the lowest and SIL 4 – the highest level.

The article provides examples for a SIL review assessment including its reporting and flags required input data. It refers to the relevant norms and standards and explain how SIL categorization is involved in certain studies (e.g, HAZOP).

The four levels of Safety Integrity are used in a HAZOP study for instrumentation analysis. The whole system is assessed to determine the level of protection during any type of scenario. Each type of instrumentation (e.g.: level indicators, temperature alarms, pressure alarms) is compared to other instrumentation. The SIL level is then determined, and as long as the overpressure protection comes mostly from the safety functions and not from the pressure relief devices, the safety integrity level increases and reduces the occurrence of catastrophic events.

1 Introduction

Absolute safety without risk cannot be achieved; however, the risk can be reduced to an acceptable level by implementing the ALARP principle. This means that the risk should be reduced to a level that is as low as reasonably practicable. This reduction is achieved by implementing various safety-related systems. These different systems provide safety barriers, also called protection layers, which are independent of each other, meaning that failure in one layer does not lead to failure in others.

During the basic design (or Front-End Engineering & Design) of an industrial plant, all safeguards and Layers of Protection need to be properly identified and defined. The Safety Instrumented Function (SIF) represents the third protection layer. The SIF is required to interfere in case the basic process control system as well as the process alarms could not bring back the process values under normal control. [5]

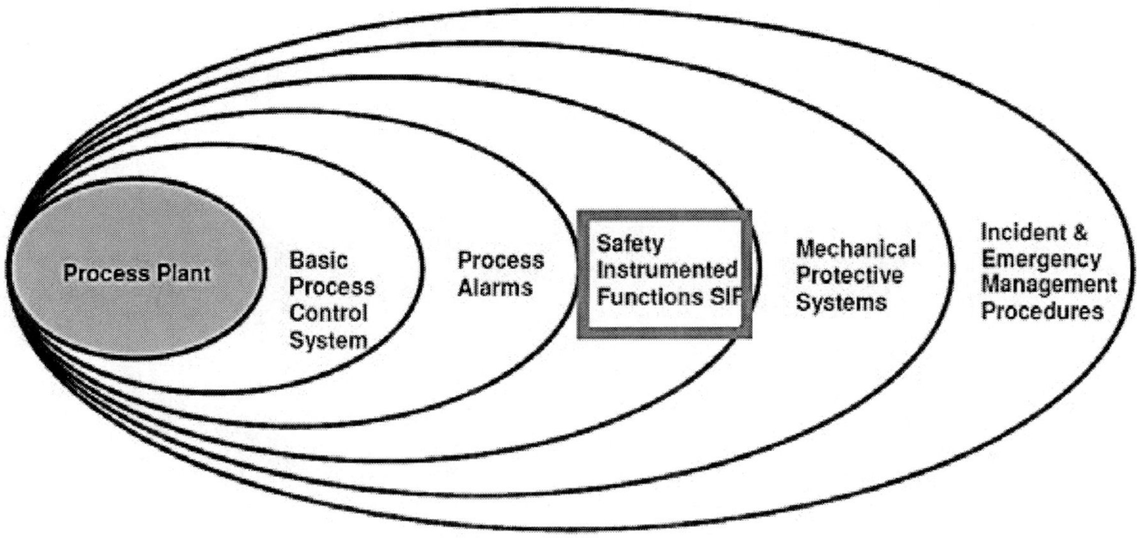

Figure 1. Process Protection Layers [5]

The intention of a safety function is to prevent or mitigate the consequences of a hazardous event. Therefore, it needs to function properly, and it needs to be reliable.

Reliability is defined as the ability of a system to function as planned and can be expressed by probabilities and expected values. The level of this ability may be measured in different ways in different situations, such as:

- Average lifetime
- Average number of failures per unit of time (frequency of failures)
- Probability that the component/system will function at a specific point in time
- Average portion of time that the component/system will function [7]

Reliability can also be measured using the average probability of failure on demand (PFD) and it is often used for systems that take actions when a dangerous condition is detected.[8]

Achieved risk reduction by other safety related measures and systems are compared against acceptable risk and residual risk is found. If the residual risk is seen as unacceptably high, a risk reduction factor (RRF) is determined. This factor expresses by how much the risk should be reduced. The risk is then allocated to the SIS and the associated safety instrumented functions (SIFs), where the reliability target of the functions is expressed as SIL. Higher RRF yields higher SIL. Figure 2 illustrates how different barriers influence the risk reduction.[4][5]

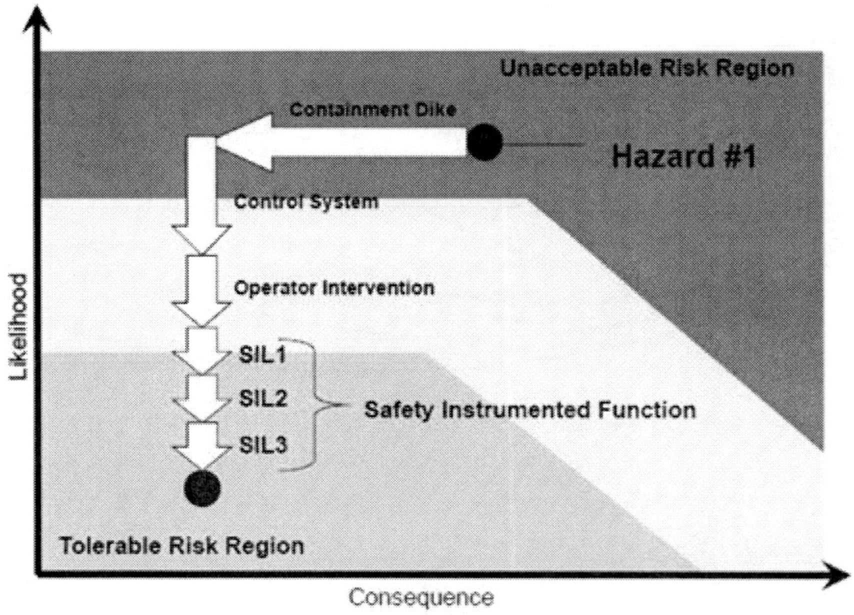

Figure 2. Risk reduction achieved by SIS[4]

2 Safety Instrumented Function

The IEC 61511[1] standard defines a SIF as a "safety function with a specified safety integrity level, which is necessary to achieve functional safety". Moreover, a safety integrity function is defined as a "function to be implemented by a Safety Integrity System, other technology safety-related system, or external risk reduction facilities, which is intended to achieve or maintain a safe state for the process, with respect to a specific hazardous event."[2]

The two terms, SIF and SIS are often used interchangeably, but it should be mentioned that a SIS represents a combination of two or more SIFs. Also, the SIF can be considered as a barrier function, while the SIS is considered as a barrier system.

A Safety Integrity Function consists of input elements, logic solvers and final elements for the purpose of taking the process back to a safe state when predetermined conditions are violated. The safe state is achieved when the SIS performs the intended SIFs.[6]

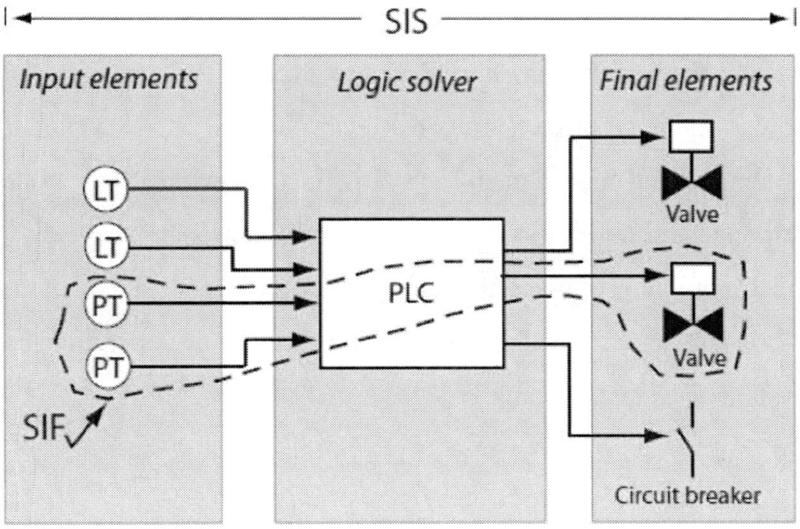

Figure 3. The distinction between SIS and SIF[4]

SIFs are divided into two groups : SIFs with a protective function that can be activated on demand and SIFs provided with instrumented control functions that are operated in continuous mode. If the system fails to perform the intended functions, the dangerous event might escalate and result in a much serious accident. For that reason, each SIF is required to be highly reliable. The reliability is measured as a safety integrity level (SIL).[4]

3 Safety Integrity Level (SIL)

Safety Integrity is defined as "Probability of a safety-related system satisfactorily performing the required safety function under all the stated conditions within a stated period of time" (IE-61508, 2004). In general, the IEC 61508 and OLF 070 state that three main types of requirements need to be fulfilled in order to achieve a given SIL. These three types are: quantitative, semi-quantitative and qualitative requirements. If one of the three categories fails to meet a specific SIL, say 3, the SIF cannot be classified as a SIL 3 function, even if SIL 3 is supported by the calculated PFD. [1]

The IEC 61508 and IEC 61511 standards distinguish between four discrete safety integrity levels, which are used as a measure of reliability for SIFs, SIL 4 has the highest level of safety integrity, while SIL 1 has the lowest. The higher the SIL value, the higher is the associated level of safety and the lower is the probability of failure. This basically means that a high SIL value gives lower tolerance of dangerous features. The IEC 61508 standard makes a distinction between low demand and high demand systems. Low demand safety systems are activated on demand to respond to abnormal situations, not more than once per year. Typical examples are Process – and Emergency Shutdown system (PSD and ESD). High demand systems may be systems that are subject to frequent demand, or continuous operation mode (e.g. dynamical positioning system or ballast system). [1][2]

4 SIL usage during HAZOP

There are various safety studies which are carried in the process industries such as HAZID, HAZOP, HAZAN, QRA, etc. SIL study is also one of the safety studies which is carried out to prevent hazard in process industries. Generally, SIL study is carried out after the HAZOP study is completed for a project. There would be a question on the need of SIL study when HAZOP is already conducted. HAZOP study makes the process as a whole robust and reliable by adding redundant instruments, additional alarms, etc. But HAZOP study does not address the problems of required reliability of the system provided. [5]

SIL study provides solution to above mentioned problems. As HAZOP study ensures the process to be safe or reliable, similarly SIL study ensures that each component of the loop to be robust and reliable. In other words, we can say that HAZOP study ensures safety and reliability at process level and SIL study ensures availability or robustness at component level in a process industry. Hence, we can say that HAZOP and SIL study are equally important for a process industry and both the studies play important role to ensure safety and reliability of process industries. Similar to HAZOP, there are well established procedures and codes to carry out SIL study.

Adequate definition of a safety function can only be achieved if a full understanding of its demand scenarios, design intent and consequences of failure on demand are known:

a. The "demand scenario" describes the initiating events (e.g. failure of control instruments or failure of equipment), which will ultimately lead to a demand on the SIF.
b. The "design intent" specifies the released hazard to be averted (e.g. loss of containment).
c. The "consequences of failure on demand" describe the ultimate consequences (of a SIF failure) and the way they are achieved.[5]

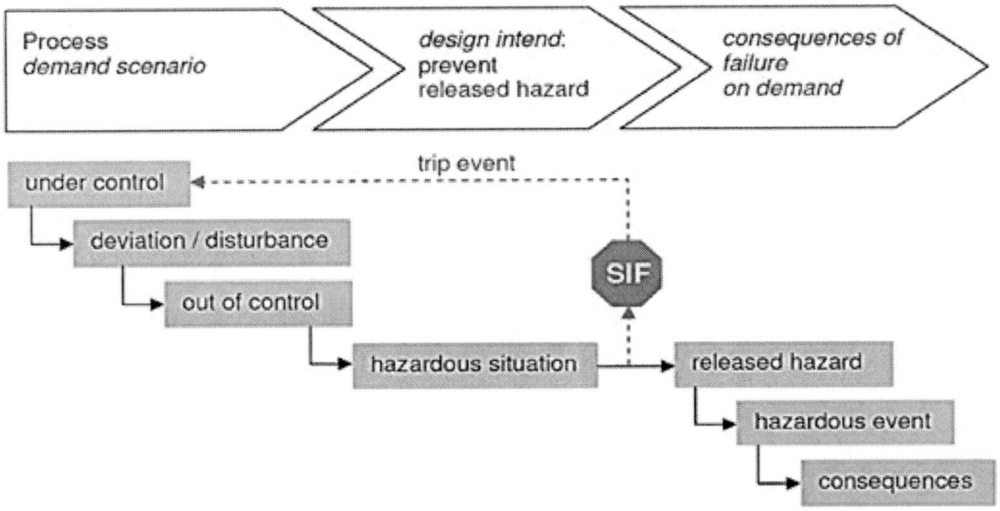

Figure 4. SIF designed to stop / avert a specific hazardous event [5]

During the "Hazard and Operability" (HAZOP) study potential out-of-control process situations are identified, by analyzing variations of process parameter values (e.g. more/ less/ none/ negative pressure). Variations can cause a potential hazardous situation which can lead to a hazardous event (e.g. leakage of toxic products, explosion, fire, etc.). The consequences of the hazardous event can have an impact on human safety, environmental pollution, damage to the installation, production loss and other negative effects. As a result, the HAZOP study identifies the required safeguards for the process. [5]

In a SIL classification the demand scenario frequency and the severity of the consequences of the hazardous event are used to establish the Safety Integrity Level, which determines the required level of risk reduction. The SIL indicates the minimum probability that the equipment will successfully do what it is desired to do when it is called upon to do it. As soon as the required risk reduction is known, it can be converted into a Probability of Failure on Demand (PFD) which is required for the SIF and which results into the SIL. [4]

Each SIL has a specified target range for the PFD, where each level corresponds to a risk reduction factor (RRF). If the RRF has a factor of 1.000 it means that the PFD of the safety function should be lower than 10^{-3}, to fulfill the SIL 3 requirements. To verify that the necessary risk reduction is achieved, all components in the safety function have to be included in the PFD calculation.

Table 1. SIL Definition (for Low Demand Application) as per IEC 61508-1 [8]

Safety Integrity Level (SIL)	**Probability of Failure on Demand (PFD)**	**High Demand or Continuous Mode of Operation (PFD per Hour)**	**Risk Reduction Factor (RRF)**
4	10^{-4} to 10^{-5}	10^{-9} to 10^{-8}	10.000 – 100.000
3	10^{-3} to 10^{-4}	10^{-8} to 10^{-7}	1.000 – 10.000
2	10^{-2} to 10^{-3}	10^{-7} to 10^{-6}	100 – 1.000
1	10^{-1} to 10^{-2}	10^{-6} to 10^{-5}	10 – 100
0	Basic Process Control		

The norms are focused primary on personnel safety and do not provide any quantitative values for tolerable risk. The tolerable risk has to be defined by company standards and/or local authorities. Many companies have developed their own internal standards which provide the necessary quantitative values for the tolerable risk.[4]

5 Conclusions

There are various layers of protection in the process industry; some are hazard prevention-oriented, while others are used for the mitigation of the hazard. The Safety Integrity Systems are part of the prevention, as well as the mitigation category.

The four levels of Safety Integrity are used in a HAZOP study for instrumentation analysis. The whole system is assessed to determine the level of protection during any type of scenario. In general, the SIL study is completed at the end of a HAZOP study for a project.

The consequences of the hazardous event can have an impact on human safety, environmental pollution, damage to the installation, production loss and other negative effects. The four levels of Safety Integrity are used to reduce the risk of a hazardous event and his negative effects. SIL 4 is the highest level and SIL 1 is the lowest, each one of them providing reduction of the risk.

SIL solutions are not always most effective way of decreasing risk. Also, the instrumentation components need to be regularly maintained and tested at well-defined intervals.

References

[1] International Electrotechnical Commission, IEC Standard 61508 "*Functional Safety of Electrical/Electronic/Programmable Electronic Safety-related Systems* ", 2010

[2] International Electrotechnical Commission, IEC Standard 61511 *"Functional Safety – Safety instrumented systems for the process industry sector "*, 2003

[3] J. Capewell and L. Fernie. "Instrumented Protective Systems – SIL Classification and Lifecycle Integrity Management to comply with BS EN 61508 / 61511", AK EHS & Risk

[4] I. Eliassen. "Management of change – with the main focus on Safety Instrumented Systems", Master's Thesis, University of Stavanger, Faculty of Science and Technology, 2013

[5] T. Walk. "Safety Integrity Level (SIL) Assesment as key element within the plant design", 6th Pipeline Technology Conference, 2011.

[6] "Safety Integrity Level (SIL) study", February 2018, https://www.tce.co.in/blog/?p=207 , (accessed Jan. 2, 2019)

[7] T. Aven, "Reliability and Risk Analysis", October 1992, Springer, pg. 4 – 10

[8] K.H Eikeskog, "Reliability as a decision tool against SIL requirements", Master's Thesis, University of Stavanger, 2012

19 AIChE Spring Meeting
15TH GLOBAL CONGRESS ON PROCESS SAFETY
NEW ORLEANS • 2019

Implications of the International Building and Fire Codes on New and Existing Processes

Michael Snyder, PE CSP CFPS
DEKRA Process Safety
113 Campus Drive
Princeton, NJ 08540
Mike.Snyder@Dekra.com

John Wincek, CCPSC CSP
DEKRA Process Safety
113 Campus Drive
Princeton, NJ 08540
John.Wincek@Dekra.com

Prepared for Presentation at
American Institute of Chemical Engineers
2019 Spring Meeting and 15th Global Congress on Process Safety
New Orleans, LA
March 31 – April 3, 2019

Implications of the International Building and Fire Codes on New and Existing Processes

Michael Snyder, PE CSP CFPS
DEKRA Process Safety
113 Campus Drive
Princeton, NJ 08540
Mike.Snyder@Dekra.com

John Wincek, CCPSC CSP
DEKRA Process Safety

Keywords: Occupancy classification, Hazardous materials, Group H occupancy, Building code, Fire code, Hazardous materials categorization, Exempt amounts, Combustible dust,

Abstract

The use and storage of hazardous materials in any structure can introduce process operations to a number of regulatory and permitting requirements. One of the most overlooked and misunderstood requirements is how hazardous material utilization affects the building (and fire) code occupancy classification of your facility, and affects the allowable design and layout features of the structure. Occupancy classification can present a significant operational issue when an existing facility plans to increase or change the type or quantity of hazardous materials being used.

The model building codes in the United States, most commonly the International Code Council Series, establish strict requirements for the use hazardous materials through the occupancy classification process. Since these are generally adopted by local and state-levels of government, the model codes serve as legal requirements. These codes rely on the accurate determination of high-hazard, or Group H, occupancy designation to ensure proper building features are provided for occupant and facility safety.

Recent updates to the code requirements have affected occupancy classification based on the use of combustible dust. With increasing regulatory focus on combustible dust, the methods used for occupancy classification involving this hazard will become more impactful to many manufacturing facilities.

1 Introduction

The use and storage of hazardous materials in any structure can introduce process operations to a number of regulatory and permitting requirements. One of the most overlooked and misunderstood requirements is how hazardous material utilization affects the International Building Code (IBC)[1] and International Fire Code (IFC)[2] occupancy classification of your facility, and affects the allowable design and layout features of the structure. Occupancy classification can present significant operational issues when an existing facility plans to increase or change the type or quantity of hazardous materials being used.

The model building codes in the United States, most commonly the International Code Council Series, establish strict requirements for the use hazardous materials through the occupancy classification process. Since these are generally adopted by local and state-levels of government, the model codes serve as legal requirements. These codes rely on the accurate determination of high-hazard, or Group H, occupancy designation to ensure proper building features are provided for occupant and facility safety.

While appearing complex on the surface, the consistent use of a three-step process will help ensure accurate occupancy classification: (1) Appropriate categorization of hazardous materials used; (2) Determining use category of each hazardous material; and (3) Properly accounting for the quantity of each category of hazardous material within each control area of a facility.

Strategic use of control areas, to manage segregated areas of hazardous materials inventory, can be used to develop strategies for select utilization of hazardous materials without triggering all high hazard occupancy construction methods and restrictions.

Recent updates to the code requirements have affected occupancy classification based on the use of combustible dust. With increasing regulatory focus on combustible dust, the methods used for occupancy classification involving this hazard will become more impactful to many manufacturing facilities.

2 Overview of the Occupancy Classification Process

The Building (and Fire) Code utilizes occupancy classification to provide a rationale criteria that matches a building's use and occupancy with the features required to address fire hazard and life safety considerations. This designation is fundamental in establishing the features of construction, occupant safety requirements, building height and size limitations, means of egress, fire protection systems, and allowable interior finishes.

In general, the occupancy classification is set at the time of construction or initial occupancy of a building. Operations must conform to the requirements and restrictions imposed by this classification for the facility's lifetime, which is particularly important as the type and quantity of hazardous materials use changes over time.

Typical occupancy classifications in the manufacturing & industrial sectors include:

- Group F: Factory Industrial
- Group S: Moderate- and Low-Hazard Storage
- Group H: High Hazard

The IBC further categorizes High Hazard (Group H) occupancies into one of five subgroups. As detailed in Figure 1, the most severe hazardous occupancy group is H-1 and the hazard severity gradually decreases through H-3. H-4 is used for occupancies containing materials presenting a health hazard, and H-5 is a special classification for the unique hazards presented by semiconductor fabrication facilities.

Often, building and structures containing hazardous materials may represent hazards in more than one high hazard subgrouping (H-1, H-2, H-3, H-4). In these cases, the structure must conform to the code requirements for each of the classified occupancies, and use the most restrictive limitations when multiple groups apply.

Table 1. High Hazard Subgroups Used in the IBC & IFC

High Hazard Group	IBC Reference	Description
Group H-1	307.3	Occupancies containing materials with a detonation potential.
Group H-2	307.4	Occupancies containing materials that have a deflagration potential or that create a hazard from accelerated burning.
Group H-3	307.5	Occupancies containing materials that are easily combustible or pose a physical hazard.
Group H-4	307.6	Occupancies containing materials that pose a health hazard, specifically including toxic, highly toxic, and corrosive chemicals.
Group H-5	307.7	Occupancies containing *hazardous production materials* (HPM) used in semiconductor fabrication, and research and development laboratories.

3 A 3-Step Process for Accurate Occupancy Classification

While appearing complex on the surface, the consistent use of a three-step process will help ensure accurate occupancy classification: (1) Appropriate categorization of hazardous materials used; (2) Determining use category of each hazardous material; and (3) Properly accounting for the quantity of each category of hazardous material within the each control area within a facility. This approach should be used during the initial design phase of a facility, or as part of the planning phase for occupying an existing structure. In addition, the limits established by this analysis should be used for ongoing oversight of the operations, particularly when operational changes are contemplated (Management of Change).

This analysis must be documented, and submitted to applicable building and fire officials in accordance with the requirements of IBC Section 414.1.3.

3.1 Categorization of Hazardous Materials

In the IBC (and IFC), hazardous materials are sorted into two major categories; physical hazards and health hazards. The IBC defines each category as follows:

Physical Hazard: A chemical for which there is evidence that it is a combustible liquid, compressed gas, cryogenic, explosive, flammable gas, flammable liquid, flammable solid, organic peroxide, oxidizer, pyrophoric or unstable (reactive), or water-reactive material.

Health Hazard: A classification of a chemical for which there is statistically significant evidence that acute or chronic health effects are capable of occurring in exposed persons

The IBC further subdivides the major categories into a number of subcategories that are used in the occupancy classification process (and for other code requirements). Table 2 lists the sub-categorization of materials for each of the major hazard sections. Some materials may have more than one classification. In these situations (for example, a material that is classified as both a health hazard and a flammable liquid or both a flammable liquid and water reactive material), all classifications should be documented to compare against allowable quantities for the occupancy determination.

Determining a chemical's hazard category is not difficult, but may take some research from common reference sources. One of the most common methods is to review a copy of the Safety Data Sheet (SDS). The recent incorporation of the Globally Harmonized System (GHS) into the Occupational Safety and Health Administration (OSHA) Hazard Communications Standard [3] establishes a mandatory 16-section format for SDSs, which provides more standardized location of critical process safety information in the document.

Section 14, Transport Information, of a GHS-compliant SDS will often provide information required for a hazard category determination. This section may display the hazardous materials classification exactly or provide reference to a hazard class that can be referenced against the building code definitions for each hazard category in Chapter 3 of the IBC. An example of the SDS information available from Section 14 is shown in Figure 1.

Table 2. Hazardous Materials Categories Used in the IBC & IFC

Physical Hazards	Health Hazards
• Combustible Dust • Combustible Liquid • Combustible Fiber • Consumer Fireworks • Cryogenics, Flammable • Cryogenics, Inert • Cryogenics, Oxidizing • Explosives • Flammable Gas • Flammable Liquid, Combination • Flammable Solid • Organic Peroxide • Oxidizer • Oxidizing Gas • Pyrophoric Material • Unstable (Reactive) • Water Reactive	• Corrosive • Highly Toxic • Toxic

Another useful reference is found in Appendix E of the IFC. Titled "Hazard Categories", this non-mandatory appendix provides a collection of useful information for hazard classification. Chemical examples are provided in sections organized by hazard classification. Terminology used in this appendix closely follows that used in the code text, so it can provide useful information that directly applies to the IBC classification. This appendix also contains useful information for evaluating the classification of mixtures.

A final source of information to aid in this evaluation is NFPA 400 (2019), *Hazardous Materials Code* [4]. Definitions aligned with many of the IBC hazard classifications are provided and the appendix material in the document provides additional classification guidance.

Should these foundational resources not provide the details required for appropriate classifications, review of available testing information may be required to compare the material against definitions in the reference standards. In cases dealing with complex mixtures or newly developed molecules, laboratory classification testing may be required.

14. TRANSPORT INFORMATION

US DOT (shipped by land):

DOT Shipping Name:	UN2903, Pesticides, Liquid, Toxic, Flammable, N.O.S. (Propargite, Isopropyl Alcohol), 6.1, 3, PG II
Emergency Response Guide:	151
Marine Pollutant:	Yes
RQ:	10 lbs. (14.4 gallons of product)

IMDG (shipped by vessel):

UN2903, Pesticides, Liquid, Toxic, Flammable, N.O.S. (Propargite, Isopropyl Alcohol), 6.1, 3, PG II, Marine Pollutant

IATA (shipped by air):

UN2903, Pesticides, Liquid, Toxic, Flammable, N.O.S. (Propargite, Isopropyl Alcohol), 6.1, 3, PG II

Figure 1: Section 14 from the SDS of a Pesticide in a Flammable Liquid Carrier. The SDS provides information to aid in classification of this material as both a Flammable Liquid (Physical Hazard) and Toxic Liquid (Health Hazard). [5]

3.2 Use & Storage Classification

Once the categories of hazardous materials are identified, it is necessary to determine the quantities of material that will be "stored" and the quantities considered "in use" in each facility zone, which is called a Control Area in the code. We will first explore the storage and use definitions, and then explore the concept of Control Area in the next section.

The major definitions used for this evaluation are:

Storage: The keeping, retention, or leaving of hazardous materials in closed containers, tanks, cylinders or similar vessels; or vessels supplying operations through closed connections to the vessel.

Use (Closed System): The use of a solid or liquid hazardous material involving a closed vessel or system that remains closed during normal operations where vapors emitted by the product are not liberated outside the vessel or systems and the product is not exposed to the atmosphere during normal operation; and all uses of compressed gases.

Use (Open System): The use of a solid or liquid hazardous material involving a vessel or system that is continuously open to the atmosphere during normal operation and where vapor are liberated, or the produce is exposed to the atmosphere during normal operations.

The code recognizes that the hazards presented by closed systems is less than that of open systems, and generally allows larger quantities of materials in the exempt allowance for closed systems.

3.3 Comparing Quantities with the Exempt Allowances of the Code: Introducing the Control Area

A control area is defined in the IBC as "spaces within a building where quantities of hazardous materials not exceeding the maximum allowable quantities per control area are stored, dispensed, used or handled." IBC Section 414.2 establishes the construction criteria for control areas, which consist of fire barriers per IBC Section 707 and horizontal assemblies per IBC Section 711. The fire-resistance rating for the fire barriers and horizontal assemblies are dependent upon the height of the control area in stories above grade per IBC Table 414.2.2.

The maximum number of control areas is limited to a maximum of four for the ground floor, with reductions taken for floors above and below grade. Note, also, that the amount of hazardous material allowed per control area is reduced for all levels below or above grade. Details are provided in UBC Table 414.2.2. Control areas are not permitted any lower than two stories below grade.

If a building has no identified control areas, or no fire-resistive construction required by other provisions of the IBC, the entire building must be considered a single control area.

IBC Tables 307.1(1) and 307.1(2) provide the maximum allowable quantities of hazardous materials within a control area for physical hazards and health hazards, respectively. The basic approach for this determination will be to compare the amount of hazardous material present (for each hazardous material category in each control area) against the limits from IBC. If the quantity is exceeded, this area will become classified as a Group H occupancy. Obviously, if can maximize the number of control zones on each floor of a building, you can maximize the amount of hazardous materials stored and/or used without triggering the Group H occupancy classification by keeping each control area just below the maximum allowable quantities.

3.4 Comparing Quantities of Hazardous Materials In Each Control Area with the Exempt Amounts: Am I a Group H Occupancy?

With all the basic background information collected, Section 307 of the IBC provides hazardous materials quantity limits for each physical and health hazard family. If the quantity of hazardous materials exceeds this amount, this area must be classified as high hazard or Group H occupancy.

When the category of a chemical is confirmed, locate the category in column (1) of either Table 307.1(1) (See Figure 2) or Table 307.1(2) (See Figure 3). The next step is to determine the quantity allowed based on how the material is used and stored. Column (3) provides maximum quantities for storage, column (4) for closed systems, and column (5) for open systems. If the actual quantities used or stored exceed the quantities listed in the table, then the building must be classified per the occupancy group indicated in column (2). Table 307.1(2) does not have this column, so quantities exceeding those in this table are classified as Group H-4.

There are numerous footnotes to each of these tables, and each applicable footnote provides valuable information for the analysis. The following footnotes include some of the ones most commonly applied to provide additional hazardous materials quantity allowance per control area:

- Both tables have footnotes that allow some quantities in the tables to be increased 100% if a sprinkler system is installed per National Fire Protection Association's standard NFPA 13, *Installation of Sprinkler Systems*. [6]
- Both tables have footnotes that allow some quantities in the tables to be increased 100% if storage is in an approved storage cabinet, enclosure, or safety.
- Both of these footnotes are accumulative, which means that if both footnotes are applicable, then the allowable quantity in the table may be increased four times.

In addition, Table 414.2.2 establishes the maximum number of control areas per floor and reduces the amount of hazardous material allowed for each floor above and below grade.

The "basic math" in this evaluation process is:

- Take the exempt amounts listed in Tables 307.1(1) or Table 307.1(2)
- Adjust the quantities for any applicable footnotes to the Tables (generally, an increase)
- Reduce the quantities for building floors above or below grade (always a decrease)
- Compare the modified quantities with the planned inventory of each hazard class being used in the control area.

Since the basis for the quantity assumptions can be complex, it is recommended that all expansions and reductions of the hazardous materials allowances be documented in the summary report.

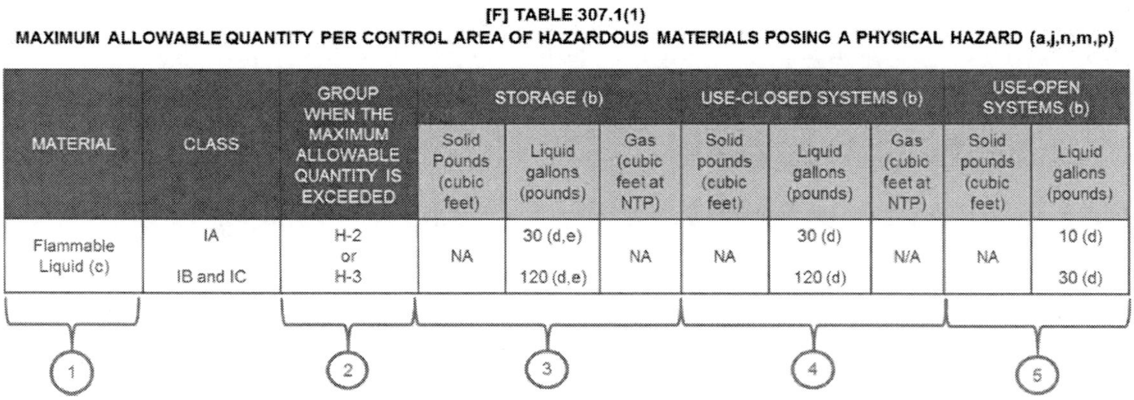
[F] TABLE 307.1(1)
MAXIMUM ALLOWABLE QUANTITY PER CONTROL AREA OF HAZARDOUS MATERIALS POSING A PHYSICAL HAZARD (a,j,n,m,p)

MATERIAL	CLASS	GROUP WHEN THE MAXIMUM ALLOWABLE QUANTITY IS EXCEEDED	STORAGE (b)			USE-CLOSED SYSTEMS (b)			USE-OPEN SYSTEMS (b)	
			Solid Pounds (cubic feet)	Liquid gallons (pounds)	Gas (cubic feet at NTP)	Solid pounds (cubic feet)	Liquid gallons (pounds)	Gas (cubic feet at NTP)	Solid pounds (cubic feet)	Liquid gallons (pounds)
Flammable Liquid (c)	IA	H-2 or H-3	NA	30 (d,e)	NA	NA	30 (d)	N/A	NA	10 (d)
	IB and IC			120 (d,e)			120 (d)			30 (d)

Figure 2 - A portion of IBC Table 307.1(1) showing maximum quantities of a hazardous material in a control area.

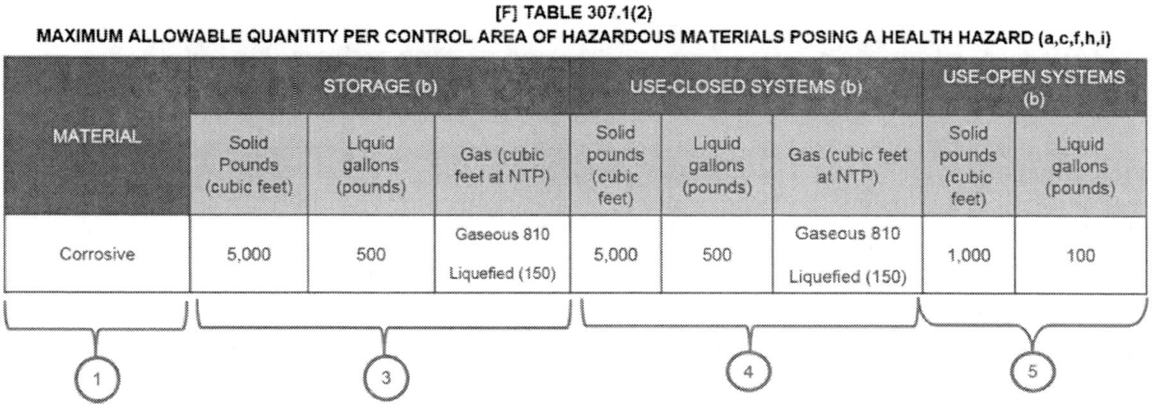

[F] TABLE 307.1(2)
MAXIMUM ALLOWABLE QUANTITY PER CONTROL AREA OF HAZARDOUS MATERIALS POSING A HEALTH HAZARD (a,c,f,h,i)

MATERIAL	STORAGE (b)			USE-CLOSED SYSTEMS (b)			USE-OPEN SYSTEMS (b)	
	Solid Pounds (cubic feet)	Liquid gallons (pounds)	Gas (cubic feet at NTP)	Solid pounds (cubic feet)	Liquid gallons (pounds)	Gas (cubic feet at NTP)	Solid pounds (cubic feet)	Liquid gallons (pounds)
Corrosive	5,000	500	Gaseous 810 Liquefied (150)	5,000	500	Gaseous 810 Liquefied (150)	1,000	100

Figure 3 - A portion of IBC Table 307.1(2) for maximum quantities of a hazardous material in a control area. Since these are for chemicals posing a health hazard, if maximum quantities are exceeded, then the building or portion thereof must be classified as a Group H-4.

4 Code Requirements for Group H Occupancies

Once the requirements for Group H occupancy are met, a number of additional requirements arise. IBC Section 415 details many special provisions for Group H occupancies; some are general in scope and others are applicable to a specific Group H occupancy group.

General requirements include:

- Automatic fire detection system (IBC Section 415.3) and;
- Sprinkler system designed to applicable National Fire Protection Association (NFPA) standards, installed throughout (IBC Section 415.4).
- Specific fire separation distances from other occupancies and facility structures that must be carefully considered (IBC Section 415.6)

Special and unique requirements include:

- Explosion control is required for some chemicals that have explosive characteristics, whether they are classified as explosives or not. IBC Table 414.5.1 lists the material categories that pose an explosive hazard. For each category, the table indicates whether a barricade or explosion venting/prevention system is required.
- Ventilation of vapors, dusts, and other emissions (IBC Section 414.3)
- Spill control and secondary containment is required for indoor and outdoor storage of hazardous materials. The intent of spill control is to prevent the flow of hazardous liquids into adjacent areas (IBC Section 414.5.3, making heavy reference to Chapter 50 of the IFC).

5 Effective Strategies for Addressing High Hazard Occupancy Requirements: Managing Control Areas & Hazardous Materials Inventory

When evaluating a new building design, or considering occupying an existing building, it is essential that the existing and realistic future utilization of hazardous materials are considered. While ensuring that building control areas exceeding exempt amounts are properly designed for Group H occupancy, it is also beneficial to take maximum advantage of control areas and quantity expanders like automatic sprinkler systems, approved storage cabinets, and other approved storage containers to expand the storage amounts. Also, consider proper fire resistance of building components and walls to enable the maximum number of control zones per floor.

As a certificate of occupancy is issued for a building, it is essential to think of it as a contract between facility management and the building and fire official that management of hazardous materials will conform to the agreed upon limitations for storage and use throughout the life of the operation. As previously mentioned, the management of hazardous materials to ensure compliance with the IBC and IFC does not stop once the facility is occupied. It requires a continual management commitment to oversee and evaluate conformance. Effective use of management of change (which include evaluation of the building code requirements for chemical changes) and periodic audits of hazardous materials inventories are two key management systems used to ensure compliance with the IBC requirements.

6 Unique Occupancy Classification Issues Involving Combustible Dust

While the majority of situations involving hazardous materials may use the approach detailed in Section 3, the IBC and IFC handle occupancy classification involving combustible dust in a different manner. The entry on Table 307.1(1) for Combustible dust requires the use of H-2 occupancy classification when the conditions referenced in note (q) on the table exist. Rather than use a maximum allowable quantity, the code basically requires the of H-2 occupancy classification when combustible dust is "manufactured, generated or used in such a manner that the concentration and conditions create a fire or explosion hazard based on information prepared in accordance with Section 414.1.3".

Section 414.1.3 requires that an engineering evaluation be prepared to document and characterize the classification and quantity of hazardous materials involved. For the unique situation involving combustible dust, this evaluation requires an analysis as to whether concentrations or conditions create a fire or explosion hazard, which would typically be done through the performance of a Dust Hazard Analysis (DHA). Numerous resources are available for conducting DHAs, including NFPA 652, *Standard on the Fundamentals of Combustible Dust* [7] and the CCPS Text *Combustible Dust Hazard Analysis* [8].

7 Conclusion

The use and storage of hazardous materials in any structure can introduce your operations to a number of regulatory and permitting requirements. One of the most overlooked and misunderstood requirements is how hazardous material utilization affects the building (and fire) code occupancy classification of your facility. The building code establishes fundamental requirements for all buildings, and affects the allowable design and layout features of the structure.

These codes rely on the accurate determination of *high-hazard,* or Group H, occupancy designation to ensure proper building features are provided for occupant and facility safety. The intent of the building codes are not to require the classification of every hazardous material use as a high-hazard occupancy, but rather to apply thresholds that require separation between any traditional occupancy groups from a Group H occupancy.

While appearing complex on the surface, the consistent use of a three-step process will help ensure accurate occupancy classification:

(1) Accurately determining the categories of hazardous materials that are used;
(2) Properly classifying the use category of each hazardous material; and
(3) Properly accounting for the quantity of each category of hazardous material within each control area of the building and comparing it with the maximum allowable amount (exempt amount). It the quantity exceeds the exempt amount, hazardous occupancy classification for the control area must be done.

High hazard occupancy classification has impacts throughout the life cycle of building use. To avoid expensive and disruptive building modifications it is essential to understand the intended use and inventory of hazardous materials to ensure that the building will meet the requirements of the building code both upon initial occupancy and during future growth and changes. Ongoing oversight of operations, routine audits of hazardous materials inventory, and an effective evaluation of building (fire) code implications during management of change reviews are essential for success in this area.

Recent updates to code requirements have affected occupancy classification based on the use of combustible dust. Rather than relying on the maximum allowable quantity per control area for hazardous occupancy classification, the code requires an evaluation about whether the dust is manufactured, generated or used in such a manner that the concentration and conditions create a fire or explosion hazard. With increasing regulatory focus on combustible dust, the methods used for occupancy classification involving this hazard will become more impactful to many manufacturing facilities.

8 References

[1] 2018 IBC: International building code. (2017). Country Club Hills, IL: International Code Council.

[2] 2018 IFC: International fire code. (2017). Country Club Hills, IL: International Code Council.

[3] United Stated Occupational Safety & Health Administration, "Hazard Communication:". Retrieved from: https://www.osha.gov/dsg/hazcom/, Accessed on March 13, 2019

[4] National Fire Protection Association, "NFPA 400: Hazardous Materials Code". 2019, http://www.nfpa.org/400

[5] Atticus, LLC. (2017, November 10). Decimite™ Safety Data Sheet. Retrieved from: https://www.atticusllc.com/wp-content/uploads/2018/01/Decimite-SDS.pdf, Accessed on March 13, 2019

[6] National Fire Protection Association, "NFPA 13: Standard for the Installation of Sprinkler Systems". 2019, http://www.nfpa.org/13

[7] National Fire Protection Association, "NFPA 652: Standard on the Fundamentals of Combustible Dust". 2019, http://www.nfpa.org/652

[8] Guidelines for Combustible Dust Hazard Analysis. (2017). Wiley-AIChE.

Process Safety, The Journey is most important!

James A Martin
Global Subject Matter Expert - PSM
Dow-DuPont
Corteva Ag. Chemicals Div.
974 Centre Road
Wilmington DE 19805-1269
James.A.Martin@DuPont.com

Prepared for Presentation at
American Institute of Chemical Engineers
2019 Spring Meeting and 15th Global Congress on Process Safety
New Orleans, LA
March 31 – April 3, 2019

Process Safety, The Journey is most important!

James A Martin
Global Subject Matter Expert - PSM
Dow-DuPont
Corteva Ag. Chemicals Div.
974 Centre Road
Wilmington DE 19805-1269
James.A.Martin@DuPont.com

Keywords: PSM Culture, Employee Participation

Abstract

This paper will discuss process safety management work and the desired effects of employee participation and culture at all levels within the company.

PSM is viewed at some companies within the chemical process industry as primarily a compliance required costly effort. The goal within these companies is often to generate the required documents for compliance at the least possible cost. This approach is a self-fulfilling prophecy which supports the following **incorrect** assumptions:

- The plant already is as safe as economically feasible
- PSM is a costly government program with little benefits
- That strict adherence to existing operating procedures will solely prevent PSM incidents
- That the "shop floor" employees have little to contribute or gain from PSM participation.
- ETC.

The journey of creating the PSM documents is often a much more effective tool than the document that is produced. The operators, maintenance technicians, engineers and possibly managers (assuming they sit in on a few hours the PHA meetings) that participate on a PHA or a PSI review will gain knowledge that cannot be obtained from normal steady state operation of unit. Further many decades of experience will likely not expose employees to all the potentially hazardous events.

Just the knowledge of how the process will respond to every possible instrument and mechanical failure will provide training beyond what could be economically feasible in a dynamic process simulator. Both young and experienced engineers will gain process knowledge and very useful engineering knowledge through the proper analysis of potential hazardous events. This knowledge and skills can be applied to various engineering efforts. If these examples are not enough, the credibility of management's commitment to a safe operation of the facility will be largely enhanced (or potentially destroyed) by their response findings from throughout the PSM efforts within the company.

1 Introduction

A successful PSM program is a holistic management system that stretches from the CEO to the shop floor. Everyone in this management system has something to learn and something to contribute. Further effective leadership across the organization is enhanced through proper implementation of PSM concepts. A PSM management system consists of 12 to 20 elements. Each of these elements requires input from various levels within the organization.

The purpose of process safety is the prevention of high consequence, low frequency events. To prevent an event that is likely be outside of the experience of the most senior employees, anticipatory methods are necessarily deployed such as PHA (aka HAZOP) were the activities focus on what PSM events are possible within the boundaries of the process under study. Other PSM elements ensure that operation is maintained/operated in a safe manner. The Process Safety Information element supports other PSM activities such as PHAs, MOCs, Mechanical integrity and many others with the information and science necessary to properly analyze the risks associated with the operation and any proposed changes.

Culture and management commitment to the principles of PSM are the key to a successful PSM system. The culture of various organizations with in the chemical processes industry varies widely from "compliance companies" to strong companies committed to the principles of PSM. By aiming for minimum essential efforts towards PSM the compliance-oriented companies often overlook the potential of a PSM management system for improvements across the organization regarding not only process safety performance, and general safety performance but improvements in throughput (from consistent incident free operation… Uptime), and improvements in first pass quality and yield from stable controlled operation of the process. Improvements in capital utilization through proper design, thereby minimizing unnecessary scope changes, last minute changes, and efficient start-up new processes are achieved by correct application of the PSM management system.

Participation of the operations and technical staff on various PSM activities such as: Process Safety Information (PSI) reviews and updates, Process Hazards Analysis (PHA), Layer of Protection Analysis (LOPA), Management of Change (MOC) … contributes greatly to the training and development of those employees. All these activities require creative and critical thinking skills, engineering knowledge, process knowledge as well as many other beneficial skills. Having a workforce with this knowledge and skills will improve decision making during abnormal operation such as upset conditions, emergencies. Process design will improve with less rework as the necessary process safeguards will be built into the initial design of new processes as well as process improvement projects. Troubleshooting skills of the operators and technical staff will be enhanced as many of the potential equipment failures will be examined by the PHA teams using the PSI. As a result, many of the operators and engineers will have discussed most potential equipment failures and the effects of these failures on the process well in advance of the occurrence of these failures in the field. As these failures are often outside of the experience of the operators and engineers involved with the day to day operation of the facility their ability to respond to an upset or emergency will be enhanced.

Participation in the various PSM activities contributes to safer and more profitable operation as well as a much more technically competent workforce at all levels in the organization. This operation and organization created through PSM work processes will have a significant advantage over an organization that considers PSM as a costly program that they have to just "get it over with" so that they can get back to the "real work of running the plant."

2 Problem/Need

PSM is an expensive and resource intensive program that involves all the employees within chemical process organization. As a result, some managers within the chemical process industry encourage their organizations to aim for compliance with the regulations through minimum costs and efforts to control fixed costs and to deploy resources in "more productive" endeavors. Operating under this compliance model is likely to satisfy the regulatory requirements of PSM laws, OSHA 1919.119, Seveso Directive, etc. As result compliance is achieved but many of the potential benefits of a healthy PSM program and resulting culture are left unrealized. This organization will likely reinforce the belief that PSM is a costly government requirement that has little benefit on the safety of the operation. Typically, these organizations have leadership that believes that the operation has always been operated safely and that further expenditures on additional safeguards is an unnecessary cost. By contrast an organization with a healthy PSM culture will understand the need for continuous improvement in all aspects of the operation. By understanding the hazards and risks of the operation a sense of vulnerability regarding process safety events will be created and sustained. This sense of vulnerability is a necessary component at all levels throughout the organization to prevent complacency to PSM incidents. This paper will examine some of the work, attributes, and elements of a healthy PSM system. The benefits of robust PSM management system will be examined relative to the organization as well as the individuals participating in the various PSM activities.

3 Management Commitment

Management commitment to Process Safety Management (PSM) is an essential attribute of a successful PSM program within any organization. Management's commitment to PSM begins at the CEO and extends down to the shop floor. This commitment is not based on motivational slogans and posters, but rather it is based on the actions of management. Throughout the organization and at all levels of management decisions regarding PSM are made nearly every day. The outcome of these decisions is observed by the employees at all levels in the organization. The integration of each of the many PSM related decisions, policies and practices becomes the PSM culture of the organization. Based on the quality, consistency and commitment of management to PSM the resulting PSM Culture is healthy or it could become dysfunctional.

An important component of management's commitment to Process Safety Management (PSM) is a "Felt Presence of Management" across all levels of the organization. Felt presence of management is particularly important on the shop floor. Healthy management presence encourages team effectiveness and mutual respect across the organization, and between all levels. Through discussions regarding the rationale of PSM decisions with shop floor employees, management's commitment to PSM becomes more and more credible.

If your actions inspire others to dream more, learn more, do more and become more, you are a leader.

John Quincy Adams [1]

Management participation in PSM activities provides an essential feedback loop from management to the PSM team regarding the acceptance or rejection of recommendations, findings, and action items demonstrating management's commitment to PSM. The PHA team being present during presentation of recommendations to management with further the credibility of the decisions on the shop floor.

Often PSM specialists will overhear statements regarding the true commitment of management to PSM and it is often not favorable. "Management talks a good game regarding safety, but when it comes to spending money; forget about it!" Many times, these statements represent an incorrect understanding of the commitment to PSM from management. However, often the observations of the shop floor employees regarding management commitment to safety is "Spot on!" Shop floor employees have the best "Bullshit Detectors" known to the human race! Through participation at all levels in the organization the true PSM commitment of the management team become credible to the shop floor employees. This is extremely important as the shop floor employee's decisions are often the last line of defense against a serious or catastrophic PSM incident.

An essential component of Effective PSM leadership is achieved via a thorough understanding of hazards and risks at the shop floor level. This understanding is achieved by management's participation at the shop floor level PSM activities, for example visiting a PHA team and participating as an ad hoc contributor.

4 Sense of Vulnerability

Through participation on various PSM teams and programs employees will understand the hazards that they live with every day. They will also understand how these hazards are not the same as risk. Through this understanding they will learn that while the risks are minimized, the hazards remain in the operation. Thus, instilling a sense of vulnerability to PSM incidents that is essential across the organization from the shop floor to the CEO and everyone in-between.

"A detailed understanding of process hazards must be developed, as a lack of or insufficient understanding is an important factor in many serious process incidents and injuries." [2]

"Sense of Vulnerability" is the knowledge that the worse-case scenario can happen at any time on any day. Employees need to understand that the safeguards are in place to reduce the risk of today being "The Day." In no way does this indicate that the risk is ZERO!

New employees often have an inherent sense of vulnerability. This inherent sense of vulnerability is a result of exposing the new employee to a new and hazardous environment. The new sights, sounds, smells and people naturally create an uneasy feeling that causes the new employee to notice small things about this environment. This is what prompts a new employee to question the practices and conditions within the operation. Our goal within the PSM systems is to maintain this vulnerability as the employee matures within the organization. As knowledge and experienced is gained an employee in a healthy PSM culture will be more likely to follow-up on unusual conditions and respond to upsets with proper thoroughness of thought. As result the employee will make better and better decisions as experienced is gained in the unit. All new members of the organization will go through this process in a healthy organization. Various job functions and levels will experience it differently relative to the specifics of their jobs. For instance:

- A new engineer assigned to the operations group will go through this process much the same as the new operators.

- A new engineer assigned to a capital project group will experience this process from a different prospective. As these employees are typically remote from operations, PSM may seem to be just another project deliverable. While this is true, some/most project deliverables prevent the loss of life and/or protect the environment from catastrophic events. The afore discussed "sense of vulnerability" should be felt by capital project employees. The decisions they make relative to schedule, cost control, technology choice should all have a component of PSM.

- Through participation in PSM programs experienced employees "seed" the organization with the PSM culture prevalent throughout their careers. In a healthy culture these "seeds" will create the required sense of vulnerability throughout the organization. Through participation including training of employees, auditing, PHA, MOC, PSSR and project design teams. The concepts and culture, including a sense of vulnerability, is reinforced within the seasoned employees.

This will prevent a dangerous organizational state of compliancy which is often present prior to a significant or catastrophic PSM event.

- Often managers and business leaders who do not have a PSM background will consider their operations safe because they have been operating safely for decades. By working with PSM teams there is a potential for these employees to understand PSM concepts and the established fact that decades of safe operation does not prove that the plant will continue to operate without a PSM incident. Business focused managers will understand that process safety efforts will ensure a higher level of plant availability, throughput, and quality of products.

- By working with various PSM related teams shop floor employees will learn the importance of compliance to established procedures. While, the operators may suggest an alternative procedure for a plant activity. These suggestions must be reviewed by a team of the technical, management and shop floor employees to ensure that the procedure will work in all phases of operation. Further, everyone involved needs to understand that the process must be operated in a controlled manner to ensure that the PSM work is not inadvertently disabled by variations in operation.

For the various PSM efforts to be effective a trust must exist between all levels of employees. This is especially necessary when reviewing the actual behavior of shop floor employees relative to procedures. It must be established that these proceedings are aimed at creating procedures and work processes are efficient, and that the shop floor employees will follow based on their confidence in the documents and their leadership.

5 Process Hazard Analysis

Process hazards analysis (PHA) teams are multidiscipline teams with typical participation from the operations, maintenance and technical employees. Participation on process hazard analysis (PHA) teams is a great learning opportunity for each of the members. For younger engineers the opportunity is to learn the process, perfect engineering calculation methods and experience the PSM culture of the organization. For more experienced engineers the opportunity is to teach younger employees, provide leadership, and develop professionally. Analogously, younger operators will benefit by analyzing each of the potential hazardous events identified by the PHA team. The analysis will benefit the younger operators by providing detailed knowledge regarding safeguards to minimize the risks associated with hazardous events. Also, they will gain understanding of the safeguards requiring operator intervention and how it applies to their day to day job responsibilities. Experienced operators can provide leadership to both the younger operators and engineers. The result of participation on the PHA team provides knowledge, leadership opportunities and team effectiveness to all the members of the PHA team.

6 Documentation and Review of Process Safety Information (PSI)

Process safety information (PSI) is not just a folder of equipment and interlock specifications. It must contain a through discussion of the inherent hazards of the process. These hazards are inherent to the process as a result of the chemicals being handled and the conditions at which they are handled. No relief device, safety instrument function, equipment design or administrative control can eliminate or change the inherent hazards of the process. By working on the generation and revalidation of PSI the technical staff gains valuable technical information about the hazards of the process, the process chemistry, engineering design methods and the application of RAGAGEPs.

Emergency response has the function of picking up the pieces "when the swiss cheese holes line up." Without a thorough analysis of the inherent hazards of the process and a catalog of the potential hazards events how could they plan a response to these events.

7 Management of Change

Often the first PSM related work assigned to a young technical/engineering employee is minor process improvement projects. These projects often involve the organization's management of change policies. This work can involve editing the unit's standard operating procedures and operating conditions, the installation and start-up of new equipment, changes to a raw material and many other activities that over the course of several of these projects nearly all the elements are PSM are engaged. While management of change activities are necessary and important in and of themselves, the new employee will be essentially training him/herself on the details of many if not all the PSM elements at the unit or site level. More experienced employees are often the approvers and reviewers of this work. This is an excellent opportunity for the more experienced employees to mentor and train the younger/newer employees. This self-training and mentoring are often more beneficial than the simply reviewing if a proposed change meets the process safety requirements of the facility and regulations.

8 Incident Investigation

Investigation of process safety incidents and near misses is an important element of process safety. It is usually performed by a small team typically consisting of a technical representative, a front-line supervisor and a shop floor employee. This investigation and report are legally mandated. The report is also an important piece of the Process Safety Information (PSI) to be used by PHA teams, project teams, and to inform employees of a potential hazard. The investigation and report is a means to communicate to the workforce the PSM culture and sincerity of management's commitment to PSM. For instance: in a healthy PSM culture the incident investigation is not focused on quick answers and directing blame and disciplinary actions. This is an extremely important attribute of a healthy PSM culture, but more importantly an attribute of a healthy culture of leadership within the organization. The healthy organization focuses the incident investigation on the root causes of the incident and reducing the risk of the operation through operational and design changes including consideration of inherently safer designs. This is an excellent venue for management to build trust between management, technical, and shop floor employees. Through communication, investigation of root causes, resisting the temptation of assigning blame, and targeted design changes to reduce risks or inherent hazards real and lasting improvements in process safety are achieved.

The opposite focus if often found in an organization with an unhealthy or nonexistent PSM culture. A "brute force" management culture appears to be a culture of high operating discipline and standards of behavior, however is it truly effective?

"… human error is only inevitable if people are placed in situation that emphasize human weaknesses and do not support human strengths." [3]

"Human error has often been used as an excuse for deficiencies in the overall management of a plant." [3]

Organizations with this style of management will likely crush all healthy communication between management and the shop floor employees, as well as many other levels in the organization. Essentially any bad news, unfavorable observations, and improvement opportunities will be met with such unhealthy feedback that the communication simply stops. Management will become the preverbal ostrich, they will have great incident statistics, because few minor injuries and incidents will be reported. There will be few requests for capital aimed to improve the safety of the operation, and few process safety recommendations that involve more than a superficial procedural change. Management in this culture will likely assume that from a safety and employee communication standpoint everything is just fine. Management will likely have little or no sense of vulnerability regarding the hazards and risks of their operation. Thus the "ostrich manager" may continue believing the shop floor employees respect him. Believing that his operation is as safe as economically possible. Possibly for the remainder of his career, or until a major process safety incident occurs. After this type of incident, the manager may say that he doesn't understand how it could happen after so many years of "safe" operation. This manager will usually make statements regarding their incident and general safety statistics. They make these statements to support the manager's assertion that the incident is a "fluke" with no warning signs. Typically, the warning signs of an incident are present for months if not years.

"To paraphrase Trevor Kletz, a renowned expert on process safety, "Organizations have no memory." Unless positive steps are taken to avoid loss, even serious incident will recur, in simple terms, the things that caused one incident are just as likely to cause another." [1]

9 Operating Procedures and Operational Discipline

Operators and engineers typically write operating procedures. These procedures are typically based operating experience, process safety information (PSI) and documents from the initial research and development teams. Working as a team the operators and engineers will gain a very detailed understanding of the operation of the process and the hazards presented by the process. This information may be applied to various decision making and troubleshooting situations throughout the operation of the process. As in most PSM related activities the operating procedures are very important, but the knowledge gained in developing the operating procedures is priceless for ensuring the continued safe operation of the facility.

10 PSM Auditing

PSM auditing is an essential tool for management to understand the effectiveness of their PSM system. Audits are typically self-audits and second party audits performed by personnel not associated with the operation of the facility. Typically, the auditors are from other operating units and corporate resources. Auditors typically focus on the elements of PSM corresponding to their interests, experience and expertise. As mentioned, auditing is an essential feedback to management. However, auditing presents many opportunities for employee development, and mentoring. When an auditor presents a "finding to management" it is natural for management to "push back" thus requiring the auditor to respond with citations from the regulations or company policies supporting the audit finding. This provides in-depth knowledge to the auditor. If the auditor is from another facility the opportunities for leveraging practices between facilities has the potential for significant improvements to both facilities.

11 PSM Training

A very effective method to create in-depth knowledge of a PSM subject is to be a trainer of that subject. The preparation necessary to respond to questions from 20-30 students all with technical degrees and varying experience can only be achieved by thorough knowledge of the subject matter. In this way the students gain the training and knowledge to perform their functions within the PSM system and the trainers become recognized experts in the subject.

12 Conclusion

The writer has given examples of how management's engagement and support of PSM activities leads to healthy PSM culture and a healthy "sense of vulnerability" relative to process safety incidents. Further examples have been presented to demonstrate how management's responses in an unhealthy or nonexistent PSM culture can drive employees at all level in the corporation to make decisions that put throughput and cost considerations ahead of process safety goals and therefore the lives of the employees and members of the community.

The writer has examined how working on various PSM related activities can be more beneficial than the value of the documents and results created by this work. This is primarily through employee development and ownership of the PSM program. This ownership of the PSM program builds a healthy PSM culture and commitment at all levels in the organization.

13 References (examples given to show format).

[1] "Recognizing Catastrophic Incident Warning Signs in the Process Industries," Center for Chemical Process Safety of the American Institute of Chemical Engineers, American Institute of Chemical Engineers, 3 Park Avenue New York, NY 10016

[2] J. Atherton and F. Gil, "Incidents That Define Process Safety", Center for Chemical Process Safety, Center for Chemical Process Safety, Wiley, 2008.

[3] "Guidelines For Preventing Human Error in Process Safety," Center for Chemical Process Safety of the American Institute of Chemical Engineers, American Institute of Chemical Engineers, 3 Park Avenue New York, NY 10016

Error Propagation and Uncertainty Analysis – Application to Fault Tree Analysis

Raymond "Randy" Freeman
S&PP Consulting
12303 Lake Shore Ridge
Houston, TX 77041
Rafree@yahoo.com

Prepared for Presentation at
American Institute of Chemical Engineers
2019 Spring Meeting and 15th Global Congress on Process Safety
New Orleans, LA
March 31 – April 3, 2019

Error Propagation and Uncertainty Analysis – Application to Fault Tree Analysis

Raymond "Randy" Freeman
S&PP Consulting
12303 Lake Shore Ridge
Houston, TX 77041
Rafree@yahoo.com

Keywords: Fault Tree, Risk Analysis, Uncertainty, SIL

Abstract

Fault tree analysis (FTA) is a widely used method for the evaluation of safety systems. FTA is based on modeling the behavior of a system by a combination of logic gates (AND, OR) in a logic tree that describes the system. Failure data of low level events are used to compute the frequency of the top event of concern. The failure rate data describes the failure of the system components such as pumps, valves, vessels and instruments. All of the failure rate data are uncertain.

This paper presents a methodology for the evaluation of uncertainty in the top event of concern using propagation of error and uncertainty analysis methods. The fundamental theory is presented and a simple fault tree uncertainty analysis is completed to demonstrate the methodology. The uncertainty analysis methodology allows for the identification of the components or assumptions that dominate the risk.

1. Background

Fault tree analysis is a widely used method for representation of failure mechanisms in chemical plants. The methodology is described in the CCPS CPQRA guideline book [Ref. 1] and is extensively discussed in Chapter 9 of Lees [Ref. 2]. Lacking in these discussions is a systematic method to quantify the uncertainty of the results of the fault tree calculations.

Fault tree analysis has been used in the aircraft and aerospace industries, nuclear power plant review and is a recognized by the OSHA Process Safety Management (PSM) standard [Ref. 3] as a method for process hazards analysis.

2. Review of Error Propagation and Variance Contribution Analysis (VCA) Methodology

The study of the propagation of error has been underway for over 100 years. In this section we will review the major assumptions in the development of a general method to estimate the uncertainty in an engineered system. Detailed derivations of the equations used may be found in Ku [Ref. 4]. We will now present a shortened and simplified explanation of the method paying special attention to the assumptions made during the derivation. First we will define an arbitrary function of a set of random variables, x_i, as:

Let:

$$Y = F(x_i) \qquad \text{(Eq 1)}$$

The function $F(X_i)$ should be well behaved. That is, the function is continuous, has a first derivative over the region of interest, and is not subject to deterministic chaos behavior.

We will now create a linear function estimate of $F(X_i)$ by creating a Taylor series expansion about an arbitrary point, a. The Taylor series becomes:

$$Y = F(a) + \sum_{i=1}^{n} (xi - a) \left[\frac{\partial Y}{\partial xi}\right] + R \qquad \text{(Eq 2)}$$

Where:

R = the remainder term in the Taylor series expansion which we will ignore
n = number of variables
a = Taylor series point of expansion of variable x_i

The mean of Y can be estimated by setting the Taylor series point of expansion, a, equal to the expected the value of x_i.

$$a = E(x_i) \qquad \text{(Eq 3)}$$

The first order term will become zero, the remainder terms, R, will drop out, and what is left is:

$$E(Y) = F[\, E(x_i)] \qquad \text{(Eq 4)}$$

Where:
E(Y) = expected value of random variable Y = mean of Y
$E(x_i)$ = expected value of random variables x_i = mean of x_i

The above estimate of the mean of the function assumes:

- Well behaved function
- Existence of the first derivative
- Existence of the Taylor series expansion
- Independence of variables
- Existence of the expected value or mean of all of the uncertain variables

In a similar manner the variance of Y can likewise be estimated as:

$$V(Y) = \sum_{i=1}^{n} \left[\frac{\partial Y}{\partial x_i} \right]^2 V(x_i) \qquad \text{(Eq 5)}$$

Where:

V(Y) = variance of random variable Y as defined above in Equation 1

$V(x_i)$ = variance of random variable x_i as defined above in Equation 1

In the estimate of the variance of the function, we make the additional assumption:

- the variance of the random variable exists

In more precise mathematical terms, we are assuming that the random variable has known probability distribution with existing first and second moment generating functions. Not all probability distribution have first and second moment generating functions. Fortunately, for the probability distributions often encountered in engineering studies, the mean and variance exist along with the first and second moment generating functions.

Note that the variance is simply the square of the standard deviation. Using the variance will simplify the mathematics that is described below. The contribution of each independent random variable to the overall variance in the function is:

$$V(Y \text{ from } x_i) = \left[\frac{\partial Y}{\partial x_i}\right]^2 V(x_i) \qquad \text{(Eq 6)}$$

The relative contribution of each term to the overall variance V(Y) is a measure of the importance in the uncertainty in the particular random variable, x_i. In effect, this is a sensitivity analysis combined with an uncertainty evaluation. The variance contribution combines the sensitivity in the answer to changes in the uncertain random variable, x_i, with a measure of the uncertainty in the random variable, x_i. The overall variance in Y is found by summing the sensitivity weighted variances from each random variable. Note that the relative contribution of each random variable to the total variance does not depend on the probability distribution of Y.

In the above analysis we have found the mean and variance of our arbitrary function $Y = F(x_i)$. To compute confidence limits on Y we need to know the probability distribution of Y. Theoretical work by the Russian mathematician Kolmorgov [Ref. 5] found that the probability distribution for Y is normal if the following conditions are true:

- Each random variable contributes a small amount to the total variance in Y. This implies that there are an infinite number of random variables contributing to the total variance.
- Each random variable contributes essentially an equal amount to the total variance in Y.
- The tails of the probability distribution for each random variable decrease very rapidly (like a parrots tail) as $1/x^2$ where x is the random variable of interest.

In most real world cases we will violate one or more of Kolmogorov's conditions. The following advice is given by the author in the use these methods:

1. Ideally an infinite number of random variables are needed. From statistical theory a sample of 30 variables should be adequate. In my experience, good results can be obtained even when there are less than 10 uncertain variables.
2. Stay out of the tails. Many of the probability distributions used in engineering calculations have fixed limits (see uniform or triangular distributions) and no tails. We will violate Kolmogorov's condition on the random variable distribution tails. The author typically uses a 95% upper band and a 5% lower band to describe the uncertainty in the predicted value.

These methods of uncertainty analysis have been used in many fields of research:

- Civil Engineering and Material Science – Haugen [Ref 6]
- Electrical Circuit Design – Peralta [Ref 7]
- Safety Interlock Analysis – Freeman [Ref 8, 9]
- Corrosion tests – Freeman [Ref 10]

These methods form the basis for the National Institute Science and Technology recommended practices on error propagation and analysis – see section 2.5.5 in Engineering Statistics Handbook [Ref 11].

3. Fault Tree Analysis

Fault trees are often used to analyze potential failures of an engineered system. A fault tree is a failure logic diagram that shows the relationship of device or system failures leading to an accident. Fault tree analysis is covered in detail in the CCPS CPQRA book [Ref. 1]. A fault tree is constructed by starting with a top event or condition and asking the question:

What conditions lead to this outcome?

Each condition in the next layer down is then analyzed by asking the same question. What conditions lead to this outcome? The logic of how these intermediate conditions (termed intermediate events) are related is expressed using AND and OR logic gates. An AND gate output is true if and only if (iff) all of the inputs to the AND gate are all true simultaneously. An OR gate output is true iff one of the inputs to the OR gate is true.

Once the fault tree logic structure is created, numerical data can be used to calculate the frequency or probability of the top event of concern occurring. For fault trees where a basic event is found in multiple branches of the tree, such as power failure, the logic structure of the tree must be simplified to remove the impact of the repeated events on the top event of concern. This simplification is done using Boolean Algebra. Each resulting grouping of basic events can cause the top event to occur. These groupings of basic events are called minimal cut sets. Consider the simple fault tree of Figure 1. The Boolean Algebra representation of a fault tree is:

$$T = BE11 + BE1 \bullet BE2 + BE1 \bullet BE3 + BE2 \bullet BE9 + BE3 \bullet BE7 \quad \text{(Eq 7)}$$

Where:

T = Fault Tree Top Event

BE11 = Basic Event 11

BE1 = Basic Event 1

BE2 = Basic Event 2

BE3 = Basic Event 3

BE7 = Basic Event 7

BE9 = Basic Event 9

+ = Boolean Algebra Addition Symbol

• = Boolean Algebra Multiplication Symbol

For the top event T, the minimal cut sets are:

1. BE11
2. BE1 BE2
3. BE1 BE3
4. BE2 BE9
5. BE3 BE7

For example the existence of basic events BE1 and BE3 guarantees that the top event T will occur. Publically available standard computer software such as SAPHIRE [Ref. 12] may be used to find the minimal cut sets of a large fault tree.

The representation of the top event in equation 7 can now be used to compute the probability or frequency of occurrence of the top event. If the devices describe by the basic events are non-repairable and are tested after a period of time TI, the probability of top event, T, may be calculated as:

$$\text{Prob (T)} = \lambda_{11}\,TI/2 + (\lambda_1\,TI/2)\,(\lambda_2\,TI/2) + (\lambda_1\,TI/2)\,(\lambda_3\,TI/2) + (\lambda_2\,TI/2)\,(\lambda_9\,TI/2) + (\lambda_3\,TI/2)\,(\lambda_7\,TI/2) \qquad \text{(Eq 8)}$$

Where:

λ_{11} = failure rate of device 11 in basic event BE11, failures per unit time

λ_1 = failure rate of device 1 in basic event BE1, failures per unit time

λ_2 = failure rate of device 2 in basic event BE5, failures per unit time

λ_3 = failure rate of device 3 in basic event BE3, failures per unit time

λ_7 = failure rate of device 7 in basic event BE7, failures per unit time

λ_9 = failure rate of device 9 in basic event BE9, failures per unit time

TI = test interval, time

Equation 8 assumes that the device failure rates are small and the product λ TI < 0.1. The other major assumption is that the values of all of the failure rates and test intervals are known. In reality the vital data are not known with a range or probability distribution representing the state of knowledge.

4. Example Problem

The following example is taken from the ISA technical report on fault tree analysis [Ref. 13]. This example should not be considered as a recommendation by either ISA or AIChE. The example is presented to allow for the demonstration of the methods that can be used in a fault tree study to quantify the uncertainty in the resulting calculated failure frequency or failure probability.

Consider the interlock on the intermediate storage tank T101 shown on Piping and Instrument Diagram in Figure 2. The interlock block diagram is shown in Figure 3. The interlock is intended to prevent an abnormal condition in the tank which could lead to a uncontrolled release of the tank contents. The process hazards analysis team has recommended that the interlock be designed to safety integrity level 2 (SIL 2) with a target probability of failure on demand (PFD) of no greater than 1E-2 or a risk reduction factor (RRF) of 100. Does the proposed interlock shown on Figure 1 meet the SIL 2 target? What is the uncertainty in the predicted PFD of the interlock?

A fault tree for the failure of this interlock is shown in Figure 4. A minimal cut set analysis has been completed and the resulting minimal cut sets are presented in Table 1. We can now write the Boolean Algebra equation that represents this fault tree as:

$$T = PE + TS1 \bullet TS2 + LS1 \bullet LS2 + FT1 \bullet FT2 + FT2 \bullet FT3 + FT1 \bullet FT3 + BV1 \bullet BV2 + BV1 \bullet SOL1 + BV2 \bullet SOL2 + SOL1 \bullet SOL2 + PT1 \bullet PT2 \qquad \text{(Eq 9)}$$

Assuming that the interlock is non-repairable until tested at time TI and using lamda-time (λT) approximation of the failure rate (see appendix A). The Boolean representation of the fault tree may now be converted to a failure probability model as:

$$\text{Prob (T)} = \text{Prob(PE)} + (\lambda_{TS1}\ TI/2)(\lambda_{TS2}\ TI/2) + (\lambda_{LS1}\ TI/2)(\lambda_{LS2}\ TI/2) + (\lambda_{FT1}\ TI/2)(\lambda_{FT2}\ TI/2) + (\lambda_{FT2}\ TI/2)(\lambda_{FT3} TI/2) + (\lambda_{FT1}\ TI/2)(\lambda_{FT3}\ TI/2) + (\lambda_{BV1}\ TI/2)(\lambda_{BV2}\ TI/2) + (\lambda_{BV1}\ TI/2)(\lambda_{SOL1} TI/2) + (\lambda_{BV2}\ TI/2)(\lambda_{SOL2}\ TI/2) + (\lambda_{SOL1}\ TI/2)(\lambda_{SOL2}\ TI/2) + (\lambda_{PT1}\ TI/2)(\lambda_{PT2}\ TI/2) \qquad \text{(Eq 10)}$$

Where:

Prob(PE) = failure probability of electronic logic solver, assumed to be constant at 5E-3.

λ_{TS1} = failure rate of temperature switch 1, failures per unit time

λ_{TS2} = failure rate of temperature switch 2, failures per unit time

λ_{LS1} = failure rate of level switch 1, failures per unit time

λ_{LS2} = failure rate of level switch 2, failures per unit time

λ_{FT1} = failure rate of flow transmitter 1, failures per unit time

λ_{FT2} = failure rate of flow transmitter 2, failures per unit time

λ_{FT3} = failure rate of flow transmitter 3, failures per unit time

λ_{BV1} = failure rate of block valve 1, failures per unit time

λ_{BV2} = failure rate of block valve 2, failures per unit time

λ_{SOL1} = failure rate of solenoid valve 1, failures per unit time

λ_{SOL2} = failure rate of solenoid valve 2, failures per unit time

λ_{PT1} = failure rate of pressure transmitter 1, failures per unit time

λ_{PT2} = failure rate of pressure transmitter 2, failures per unit time

TI = test interval, time

The electronic logic solver PE is assumed to have a constant failure probability of 5E-3.

Failure rate data for the other devices are present in Table 2. The mode (most likely value) of the failure rate is set equal to the point value taken from the ISA example [Ref. 13]. The triangular probability distribution was assumed to represent the failure rate data for the equipment items. The upper and lower limits are based on the failure rate data of Appendix 4 of Smith [Ref. 14].

Using the device failure rates represented by the Table 2 column labeled as the mode and the minimal cut sets presented in Table 3, the ISA technical report [Ref.13, page 35] computes an interlock PFD as 7.5E-3 (RRF of 133). This would satisfy the requirement for a SIL 2 interlock {PFD of 1E-2 or a RRF of 100}. Now what is the uncertainty in the interlock PFD?

The first thing to remember is that the PFD should be calculated based on the mean value of the failure rate lambda (λ), not the mode of the failure rate. For a data set, the mean is the best single point representation of the data set. Table 2 presents the PFD of each minimal cut set using the mean of the device failure rate lambda (λ). The total of the PFD becomes 1.47E-2 (RRF of 68). This would not satisfy a SIL 2 interlock requirement. Now we will compute the variance of the predicted probability of top event, T.

To compute the variance of the probability of the top event, T, we will need to compute the sensitivity of the top event to each of the variables. For one, two and three event cut sets the derivations are shown in Appendix A of this paper. For these calculations we will assume that the test interval, TI, is known and has no uncertainty. The equations derived in Appendix A to this paper account for the uncertainty of the test interval. Freeman [Ref. 8] has presented numerical methods to account for the uncertainty in the test interval.

First we re-write the fail rate model for the top event as:

$$\text{Prob (T)} = \text{Prob(PE)} + (\lambda_{TS1}\ TI/2)(\lambda_{TS2}\ TI/2) + (\lambda_{LS1}\ TI/2)(\lambda_{LS2}\ TI/2) + (\lambda_{FT1}\ TI/2)(\lambda_{FT2}\ TI/2) + (\lambda_{FT2}\ TI/2)(\lambda_{FT3} TI/2) + (\lambda_{FT1}\ TI/2)(\lambda_{FT3}\ TI/2) + (\lambda_{BV1}\ TI/2)(\lambda_{BV2}\ TI/2) + (\lambda_{BV1}\ TI/2)(\lambda_{SOL1} TI/2) + (\lambda_{BV2}\ TI/2)(\lambda_{SOL2}\ TI/2) + (\lambda_{SOL1}\ TI/2)(\lambda_{SOL2}\ TI/2) + (\lambda_{PT1}\ TI/2)(\lambda_{PT2}\ TI/2) \quad \text{(Eq 11)}$$

We are interested in obtaining the sensitivity of the Prob(T) with respect to each of 13 uncertain variables. The sensitivities are:

$$\frac{\partial Prob(T)}{\partial \lambda TS1} = (TI/2)(\lambda_{TS2}\ TI/2) = \lambda_{TS2}\ (TI^2/4) \quad \text{(Eq 12)}$$

$$\frac{\partial Prob(T)}{\partial \lambda TS2} = (TI/2)(\lambda_{TS1}\ TI/2) = \lambda_{TS1}\ (TI^2/4) \quad \text{(Eq 13)}$$

$$\frac{\partial Prob(T)}{\partial \lambda LS1} = (TI/2)(\lambda_{LS2}\ TI/2) = \lambda_{LS2}\ (TI^2/4) \quad \text{(Eq 14)}$$

$$\frac{\partial Prob(T)}{\partial \lambda LS2} = (TI/2)(\lambda_{LS1}\ TI/2) = \lambda_{LS1}\ (TI^2/4) \quad \text{(Eq 15)}$$

$$\frac{\partial Prob(T)}{\partial \lambda FT1} = (TI/2)(\lambda_{FT2}\ TI/2) + (TI/2)(\lambda_{FT3} TI/2) = \lambda_{FT2}\ (TI^2/4) + \lambda_{FT3}\ (TI^2/4) \quad \text{(Eq 16)}$$

$$\frac{\partial Prob(T)}{\partial \lambda FT2} = (TI/2)(\lambda_{FT1}\ TI/2) + (TI/2)(\lambda_{FT3} TI/2) = \lambda_{FT1}\ (TI^2/4) + \lambda_{FT3}\ (TI^2/4) \quad \text{(Eq 17)}$$

$$\frac{\partial Prob(T)}{\partial \lambda FT3} = (TI/2)(\lambda_{FT2}\ TI/2) + (TI/2)(\lambda_{FT1} TI/2) = \lambda_{FT2}\ (TI^2/4) + \lambda_{FT1}\ (TI^2/4) \quad \text{(Eq 18)}$$

$$\frac{\partial Prob(T)}{\partial \lambda SOL1} = (TI/2)(\lambda_{BV1}\ TI/2) + (TI/2)(\lambda_{SOL2}\ TI/2) = \lambda_{BV1}\ (TI^2/4) + \lambda_{SOL2}\ (TI^2/4) \quad \text{(Eq 19)}$$

$$\frac{\partial Prob(T)}{\partial \lambda SOL2} = (TI/2)(\lambda_{BV2}\ TI/2) + (TI/2)(\lambda_{SOL1}\ TI/2) = \lambda_{BV2}\ (TI^2/4) + \lambda_{SOL1}\ (TI^2/4) \quad \text{(Eq 20)}$$

$$\frac{\partial Prob(T)}{\partial \lambda BV1} = (TI/2)(\lambda_{BV2}\ TI/2) + (TI/2)(\lambda_{SOL1}\ TI/2) = \lambda_{BV2}\ (TI^2/4) + \lambda_{SOL1}\ (TI^2/4) \quad \text{(Eq 21)}$$

$$\frac{\partial Prob(T)}{\partial \lambda BV2} = (TI/2)\ (\lambda_{BV1}\ TI/2) + (TI/2)\ (\lambda_{SOL2}\ TI/2) = \lambda_{BV1}\ (TI^2/4) + \lambda_{SOL2}\ (TI^2/4) \qquad \text{(Eq 22)}$$

$$\frac{\partial Prob(T)}{\partial \lambda PT1} = (TI/2)\ (\lambda_{PT2}\ TI/2) = \lambda_{PT2}\ (TI^2/4) \qquad \text{(Eq 23)}$$

$$\frac{\partial Prob(T)}{\partial \lambda PT2} = (TI/2)\ (\lambda_{PT1}\ TI/2) = \lambda_{PT1}\ (TI^2/4) \qquad \text{(Eq 24)}$$

We evaluate each of the sensitivities at the expected value (mean) of the random variables contained in their equation. The mean of the uncertain failure rates are found in Table 2. In all cases the test interval is considered fixed at 1 year (8760 hours).

Temperature Switch – Use TS1 as the example

$$\frac{\partial Prob(T)}{\partial \lambda TS1} = E(\lambda_{TS2})\ (TI^2/4) \qquad \text{(Eq 25)}$$

$$E(\lambda_{TS2}) = 10.2E\text{-}6 \text{ fail/hr} \qquad \text{(Eq 26)}$$

$$\frac{\partial Prob(T)}{\partial \lambda TS1} = (10.2E\text{-}6 \text{ fail/hr})((876hr)^2/4) \qquad \text{(Eq 27)}$$

$$\frac{\partial Prob(T)}{\partial \lambda TS1} = 195.7 \text{ hr} \qquad \text{(Eq 28)}$$

Flow Transmitter – Use FT1 as the example

$$\frac{\partial Prob(T)}{\partial \lambda FT1} = E(\lambda_{FT2})\ (TI^2/4) + E(\lambda_{FT3})\ (TI^2/4) \qquad \text{(Eq 29)}$$

$$E(\lambda_{FT2}) = E(\lambda_{FT3}) = 8.0\ E\text{-}6 \text{ fail/hr} \qquad \text{(Eq 30)}$$

$$\frac{\partial Prob(T)}{\partial \lambda FT1} = (8.0E\text{-}6 \text{ fail/hr})\ ((8760hr)^2/4) + (8.0E\text{-}6 \text{ fail/hr})\ ((8760hr)^2/4) \qquad \text{(Eq 31)}$$

$$\frac{\partial Prob(T)}{\partial \lambda FT1} = 307.0 \text{ hr} \qquad \text{(Eq 32)}$$

Level Switch – Use LS2 as the example

$$\frac{\partial Prob(T)}{\partial \lambda LS1} = E(\lambda_{LS2})\ (TI^2/4) \qquad \text{(Eq 33)}$$

$$E(\lambda_{LS2}) = 8.9E\text{-}6 \text{ fail/hr} \qquad \text{(Eq 34)}$$

$$\frac{\partial Prob(T)}{\partial \lambda LS1} = (8.9E\text{-}6 \text{ fail/hr})\ ((8760 \text{ hr})^2/4) \qquad \text{(Eq 35)}$$

$$\frac{\partial Prob(T)}{\partial \lambda LS1} = 170.7 \text{ hr} \qquad \text{(Eq 36)}$$

Pressure Transmitter – Use PT2 as the example

$\frac{\partial Prob(T)}{\partial \lambda PT2}$ = E(λ_{PT1}) (TI^2/4) (Eq 37)

E(λ_{PT1}) = 7.8E-6 fail/hr (Eq 38)

$\frac{\partial Prob(T)}{\partial \lambda PT2}$ =(7.8E-6 /hr) ((8760 hr)^2/4) (Eq 39)

$\frac{\partial Prob(T)}{\partial \lambda PT2}$ =149.6 hr (Eq 40)

Block Valve – Use BV1 as the example

$\frac{\partial Prob(T)}{\partial \lambda BV1}$ = E(λ_{BV2}) (TI^2/4) + E(λ_{SOL1}) (TI^2/4) (Eq 41)

E(λ_{BV2})= 4.2E-6 fail/hr (Eq 42)

E(λ_{SOL1})= 3.8E-6 fail/hr (Eq 43)

$\frac{\partial Prob(T)}{\partial \lambda BV1}$ = (4.2E-6 fail/hr) ((8760 hr)^2/4) + (3.8E-6fail/hr) ((8760 hr)^2/4) (Eq 44)

$\frac{\partial Prob(T)}{\partial \lambda BV1}$ = 153.5 hr (Eq 45)

Solenoid Valve – Use SOL1 as the example

$\frac{\partial Prob(T)}{\partial \lambda SOL1}$ = E(λ_{BV1}) (TI^2/4) + E(λ_{SOL2}) (TI^2/4) (Eq 46)

E(λ_{BV1}) = 4.2E-6 fail/hr (Eq 47)

E(λ_{SOL2}) = 3.8E-6 fail/hr (Eq 48)

$\frac{\partial Prob(T)}{\partial \lambda SOL1}$ = (4.2E-6 fail/hr) ((8760 hr)^2/4) + (3.8E-6fail/hr) ((8760 hr)^2/4) (Eq 49)

$\frac{\partial Prob(T)}{\partial \lambda BV1}$ = 153.5 hr (Eq 50)

The sensitivities of the top event probability are summarized in Table 4. We may now compute the variance contribution of each uncertain variable using equations 1 – 6 in the methodology section of this paper. The relative variance contribution is expressed as a percent of the total variance of the top event probability. In this example the variance (measure of uncertainty) is dominated by the flow transmitters. The flow transmitters account for almost 2/3 of the uncertainty in the top event probability. Failures of the flow transmitters appear in 3 of the minimal cut sets. Since the uncertainty in the transmitter failure rates is large, the flow transmitters should contribute a significant amount to the variance of the fault tree top event.

We can use the variance in the top event to define the likelihood of achieving SIL 2 performance of the interlock. The calculations are done using the normal probability distribution. The normal distribution is tabulated as the standard normal distribution using a normalization factor Z.

The standard normal factor, Z, [Ref. 15] is defined as:

$$Z = \left[\frac{xi - E(x)}{\sigma}\right] \qquad \text{(Eq 51)}$$

Where:

σ = standard deviation. Note that the variance of a random variable is the square of the standard deviation of the random variable.

E(x) = Expected value of the random variable x_i

For the example, the expected value of the probability of the top event of concern was calculated as

$$E[Prob(T)] = 1.47E\text{-}2 \qquad \text{(Eq 52)}$$

From Table 4, the standard deviation in the probability of the top event is

$$\sigma_T = 2.78E\text{-}3 \qquad \text{(Eq 53)}$$

The target PFD for a SIL 2 interlock must not be worse than 1E-2 (RRF of 100).

We compute the normal distribution Z factor at the SIL 2 target value as

$$Z = (0.01 - 0.0147)/2.78E\text{-}3 \qquad \text{(Eq 54)}$$

$$Z = -1.69065 \qquad \text{(Eq 55)}$$

Tabulations of standard normal distribution are presented in most statistics books such as Meyer [Ref. 15]. For this value of Z, the corresponding probability that the interlock will achieve SIL 2 performance is 4.55 percent. We can also compute the risk reduction that the proposed interlock design will achieve.

What is the interlock probability of failure on demand (PFD) that we can be 95% certain that the interlock will provide? At the 95% level, the corresponding Z factor is 1.65 [Ref 15]. We now compute the corresponding PFD of the interlock that corresponds to this Z factor.

$$Z = \left[\frac{x95\% - E[Prob(T)]}{\sigma}\right] \qquad \text{(Eq 56)}$$

In this case:

$x_{95\%}$ = the PFD that we can be confident that the interlock will achieve

σ_T =2.78E-3 (Eq 57)

E[Prob(T)] =1.47E-2 (Eq 58)

Z= 1.65 (Eq 59)

Rearranging equation 59 to find $x_{95\%}$we obtain:

$x_{95\%}$ = E[Prob(T)] + Z σ_T (Eq 60)

$x_{95\%}$ = 1.47E-2 + (1.65) (2.78E-3) = 0.0193 (Eq 61)

or a 95% certain risk reduction factor of

RRF = 1/PFD = 1/0.0193 = 52 (Eq 62)

By the same logic there is 5% chance that the risk reduction is a PFD of 1.01 or an RRF of 99. In summary, this interlock is likely to perform as a mid-range SIL 1 interlock, not the SIL 2 interlock desired by process hazards analysis team.

5. Methodology for Fault Tree Uncertainty Analysis

We can now generalize the methods used in the above example to a proposed method for the evaluation of the uncertainty in a fault tree.

Step 1. Create the fault tree using standard methods outline in the CPQRA book [Ref 1].

Step 2. Determine minimal cut sets by hand or using standard computer software such as SAPHIRE [Ref 12].

Step 3. Define the needed failure rate data for each basic event.

Step 4. Define those basic events that are considered to be uncertain.

Step 5. For the basic events that are uncertain, define the probability distribution and associated parameters needed to numerically define the probability distribution. Appendix B of this paper presents the needed description of 4 commonly used probability distributions. In many cases the triangular distribution will be selected. The minimum, maximum and mode (most likely) parameters will be needed for the triangular distribution.

Step 6. Compute the mean of each of the uncertain variable

Step 7. Compute the probability or frequency of the fault tree top event of interest using the mean value for each uncertain variable.

Step 8. Compute the sensitivity of the top event probability or frequency using analytical methods as outlined in this paper or using numerical methods as outline in the paper by Freeman [Ref 8].

Step 9. Compute the variance of each of the uncertain variables

Step 10. Compute the variance contribution of for each of the uncertain variables to the top event probability or frequency using equation 6.

Step 11. Compute the total variance of the top event of interest probability or frequency by summing all of the contributions determined in Step 10.

Step 12. Compute the variance contribution percent of each uncertain variable by dividing the variable contribution (Step 10) by the total variance of the top event of interest (Step 11).

Step 13. Define the level of risk the project is willing to take. What chance will the project management accept for potential failure of the interlock to achieve the desired risk reduction? In the above example, I have used a 5% risk of failure or a 95% certainty that the interlock will achieve a risk reduction factor of 52. In the example, there is a very low probability that the interlock will achieve a SIL 2 PFD of 0.01 or an RRF of 100.

Step 14. Alternately you can report the 90% range for the top event of interest. For the example the 90% range starts at the 5% RRF of 99 and there is a 95% certainity that that an RRF of at least 52 will be achieved. The expected risk reduction for this interlock is an RRF of 68.

6. Conclusions

A general method has been presented to quantify the uncertainty in a calculations of failure frequencies using the fault tree methodology. The uncertainty analysis method is based on the application of propagation of error and variance contribution analysis techniques to the minimal cut sets created during a fault tree study. The method is simple and can be incorporated into standard fault tree analysis programs such as SAPHIRE [Ref 12].

For the example interlock presented in this paper, the uncertainty in the device failure rates degrades the predicted interlock SIL from meeting SIL 2 requirements to becoming a mid range SIL 1 interlock. This is a general result. Uncertainty in design parameters will reduce the likelihood that the equipment will achieve desired results. The design engineer has two choices:

1. Accept the conclusion that the interlock will not perform as SIL 2
2. Re-design the interlock using more reliable components to achieve the SIL 2 target

The variance contribution analysis shown in Table 4 indicates that the major contributor to the uncertainty in the interlock performance is the Flow Transmitters.

7. References

[1] Center for Chemical Process Safety , *Guidelines for Chemical Process Quantitative Risk Analysis – 2rd edition*, American Institute of Chemical Engineers, New York, 2000.

[2] Frank P Lees, Loss Prevention in the Process Industries – 2rd Edition, Butterworth, ISBN 0-7506-1547-8, Oxford, UK, 1996

[3] US Department of Labor, Occupational Safety and Health Administration, CFR 1910.119 - Process Safety Management of Highly Hazardous Chemicals, 1910.119(e)(2)(vi).

[4] H.H. Ku, "Notes on the Use of Propagation of Error Formulas," *Journal of Research of the National Bureaus of Standards – C. Engineering and Instrumentation*, pages 263 – 273, Vol 70C, No 4, October – December 1966

[5] B.V. Gnedenko and A.N. Kolmogorov, *Limit Distributions for sums of Independent Random Variables*, Translated from Russian by K.L. Chung with Appendices by J.L. Doob and P.IL. Hsu, Addison-Wesley, Menlo Park, CA, 1966

[6] Edward B. Haugen, *Probabilistic Approaches to Design*, John Wiley, New York, 1968

[7] Mike Peralta, *Propagation of Errors*, ISBN 13: 9781469985862, 2013

[8] Raymond. Freeman and Angela Summers, "Evaluation of Uncertainty in Safety Integrity Level Calculations," *Process Safety Progress,* Published online in Wiley Online Library (wileyonlinelibrary.com) DOI 10.1002/prs. 11805, 2016

[9] Raymond. Freeman, "General Method for Uncertainty Evaluation of Safety Integrity Level Calculations – Part 2 Analytical Methods," *Process Safety Process*, Published online in Wiley Online Library (wileyonlinelibrary.com) DOI 10.1002/prs 111915, October 2017

[10] Raymond Freeman and David Silverman, "Error Propagation in Coupon Immersion Tests," *Corrosion*, pages 463-466, Vol 48, No 6, June 1992

[11] National Institute Science and Technology (NIST), Engineering Statistics Handbook, Available online at:
http://www.itl.nist.gov/div898/handbook/eda/section1/eda1.htm
accessed on 6/27/2012 at 2:00:24 PM.

[12] US Nuclear Regulatory Commission, "Systems Analysis Programs for Hands-on Integrated Reliability Evaluations (SAPHIRE) – Version 8," NUREG/CR-7039, Vol 3, June 2011.

[13] ISA, "Safety Instrumented Functions (SIF) – Safety Integrity Level (SIL) Evaluation Techniques Part 3: Determining the SIL of a SIF via Fault Tree Analysis, ISA Technical Report ISA-TR-84.00.02-2002 – Part 3, Approved 17 June 2002.

[14] David J. Smith, *Reliability Maintainability and Risk – 8th Edition*, ISBN 978-0-08-096902-2, Elsevier, London, 2011

[15] Paul L. Meyer, Introductory Probability and Statistical Applications, John Wiley, New York, 1972

Table 1. Basic Events in Minimal Cut Sets For Example

Cut Set No	BE 1	BE 2
1	PE	-
2	TS1	TS2
3	LS1	LS2
4*	FT1	FT2
5*	FT2	FT3
6*	FT1	FT3
7*	BV1	BV2
8*	BV1	SOL1
9*	BV2	SOL2
10*	SOL1	SOL2
11	PT1	PT2

* Minimal cut Set with same basic event as another cut set

Table 2. Lamda (λ) Failure Rate Data With Uncertainty Limits

Device	Minimum Failures per 1E6 Hours	Mode Failures per 1E6 Hours	Maximum Failures per 1E6 Hours	Mean Failures per 1E6 Hours	Lamda Variance (Failures per 1E6 Hours)^2
Flow Transmitters FT1, FT2 and FT3	1.0	2.9	20.0	8.0	18.3
Pressure Transmitters PT1 and PT2	1.0	2.3	20.0	7.8	18.8
Temperature Switch TS1 and TS2	3.0	7.6	20.0	10.2	12.9
Level Switch LS1 and LS2	2.0	4.6	20.0	8.9	15.8
Block Valves BV1 and BV2	0.2	2.3	10.0	4.2	4.4
Solenoid Valves SOL1 and SOL2	1.0	2.3	8.0	3.8	2.3

Table 3. Probability of Minimal Cut Sets Based on Mean Failure Rates

Cut Set No	BE 1	BE 2	PFD based on mean λ
1	PE	-	5.00E-03
2	TS1	TS2	2.00E-03
3	LS1	LS2	1.52E-03
4	FT1	FT2	1.23E-03
5	FT2	FT3	1.23E-03
6	FT1	FT3	1.23E-03
7	BV1	BV2	3.38E-04
8	BV1	SOL1	3.38E-04
9	BV2	SOL2	3.38E-04
10	SOL1	SOL2	2.77E-04
11	PT1	PT2	1.17E-03

Table 4. Variance of Top Event Probability Due to Uncertain Device Failure Rates

Device No	Device Type	Label	Sensitivity of Top Event Probability to Device Failure Rate $Hour^{-1}$	Variance of Device Failure Rate (Failures per 1E6 $Hour)^2$	Variance Contribution to Top Event Variance	Variance Contribution to Top Event Variance %
1	Logic Solver	PE	-	0	0.00E+00	0%
2	Temperature Switch	TS1	195.7	12.9	4.94E-07	6%
3	Temperature Switch	TS2	195.7	12.9	4.94E-07	6%
4	Level Switch	LS1	170.7	15.8	4.60E-07	6%
5	Level Switch	LS2	170.7	15.8	4.60E-07	6%
6	Flow Transmitter	FT1	307	18.3	1.72E-06	21%
7	Flow Transmitter	FT2	307	18.3	1.72E-06	21%
8	Flow Transmitter	FT3	307	18.3	1.72E-06	21%
9	Block Valve	BV1	153.5	4.4	1.04E-07	1%
10	Block Valve	BV2	153.5	4.4	1.04E-07	1%
11	Pressure Transmitter	PT1	149.6	18.8	4.21E-07	5%
12	Pressure Transmitter	PT2	149.6	18.8	4.21E-07	5%
13	Solenoid valve	SOL1	153.5	2.3	5.42E-08	1%
14	Solenoid valve	SOL2	153.5	2.3	5.42E-08	1%

Variance of Top Event= 8.24×10^{-6} (Failures perHour)^2

Standard Deviation of Top Event = 2.78×10^{-3} (Failures per Hour)

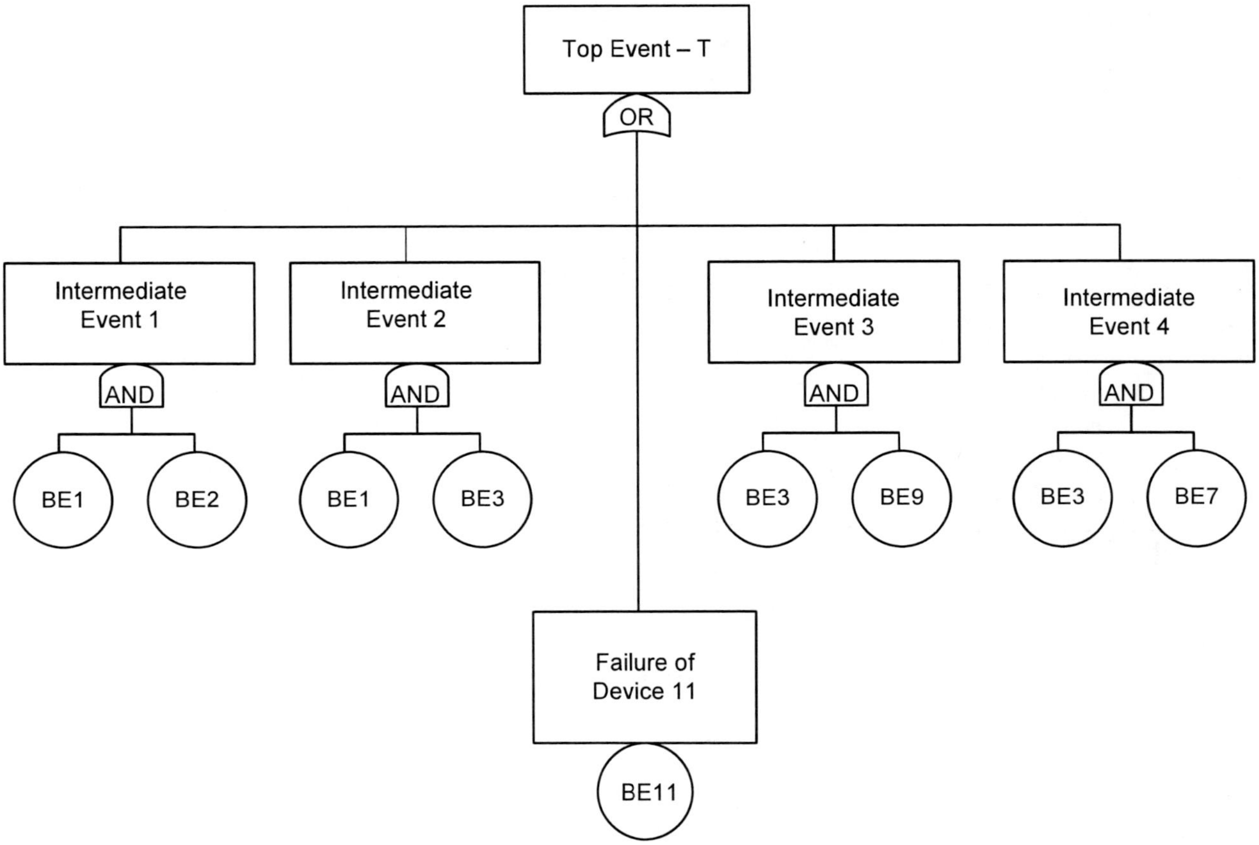

Figure 1. Simple Fault Tree

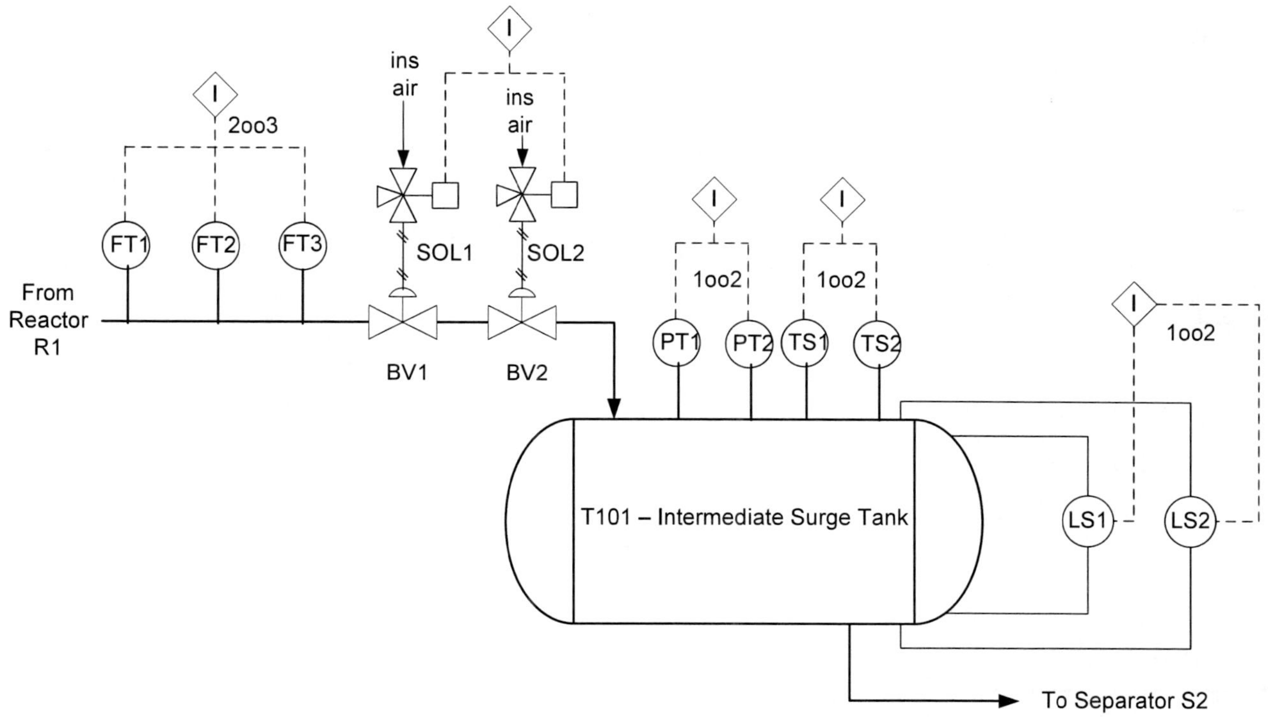

Figure 2. P&ID Diagram for Example

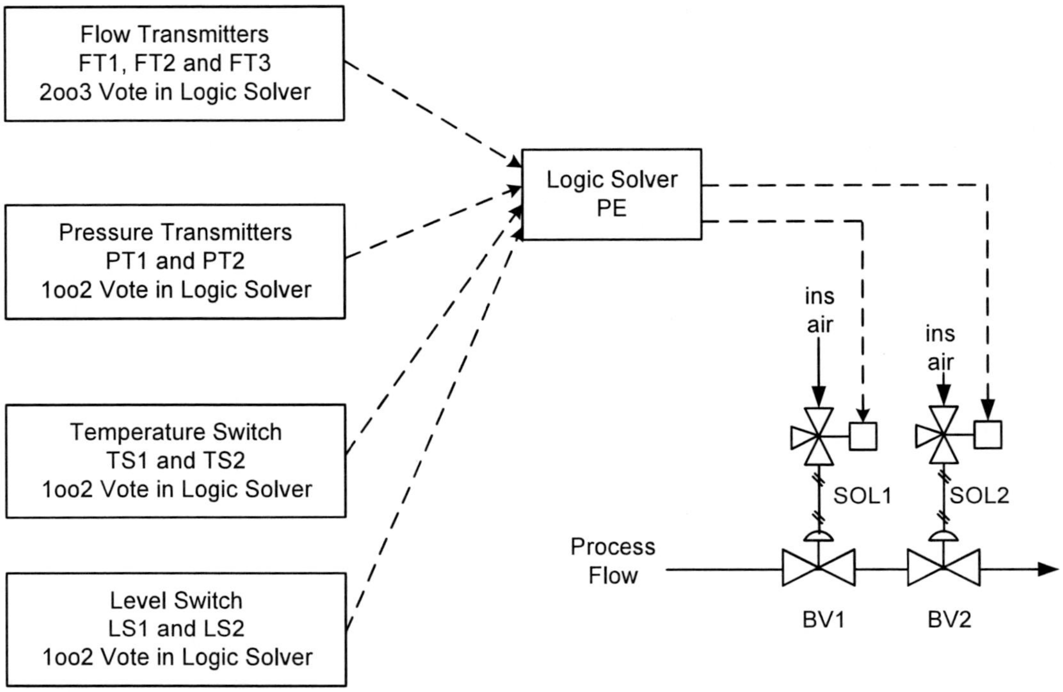

Figure 3 – Interlock Block Diagram

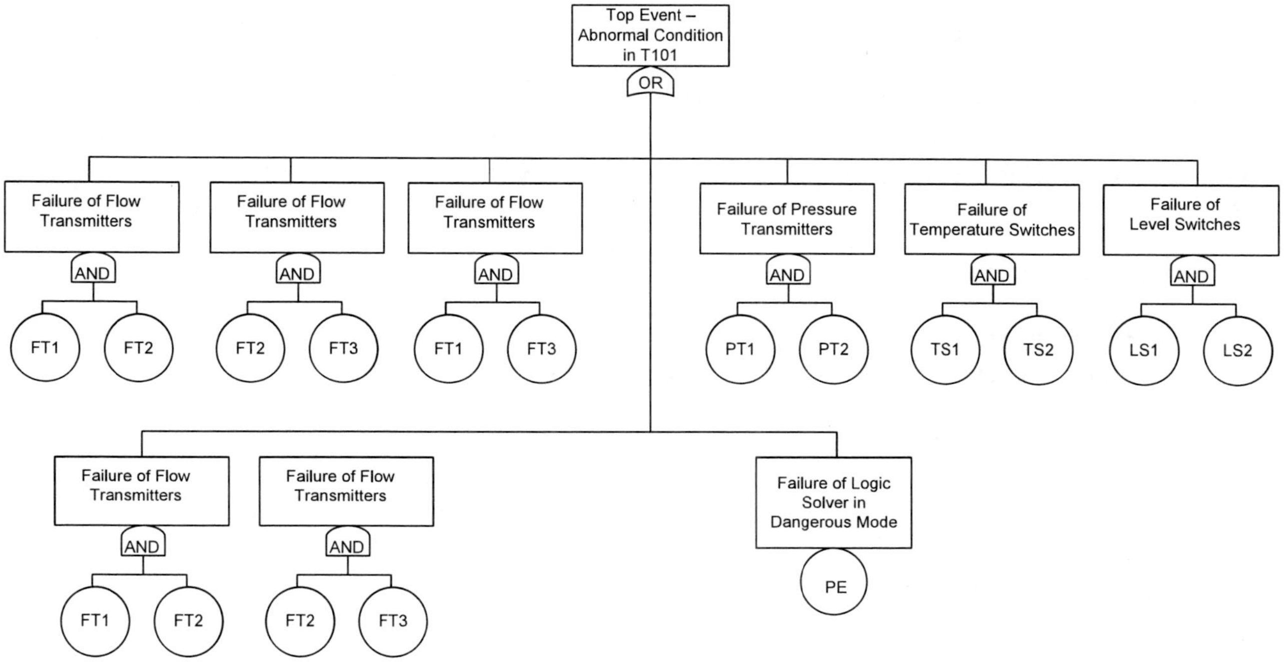

Figure 4. Fault Tree for Example Interlock

Appendix A – Development of Mean and Variance of Fault Tree Cut Sets

This appendix presents the development of equations to find the mean and variance of uncertain variables that may arise when quantifying a fault tree. The development of these equations will use the λ T approximation for the failure analysis of a negative exponential failure rate model. This approximation is very good when $\lambda T < 0.1$. We will calculate the unavailability, $\bar{A}$, to represent the failure estimate. The unavailability may be thought of as the fractional dead time of a safety interlock or the probability of failure on demand (PFD) of a safety instrumented system (SIS). This appendix presents the calculations for the non-repairable device cases.:

The structure of the equation development is the same in all cases. First the failure model is presented as a fault tree cut set. Then the unavailability failure rate model is presented in terms of the device(s) failure rate(s) and a time variable. Next the expected value of the unavailability is computed in terms of the expected values of the random variables. Finally the variance of the unavailability is computed in terms of the variance values of the random variables.

Non-Repairable Devices

<u>A. Single Event Cut Set, Q</u>

Consider a single event minimal cut set Q. The unavailability of Q may be written as

$$\bar{A}_q = \lambda_q T/2 \qquad \text{(Eq A-1)}$$

Where:

$\bar{A}_q$ = unavailability of the device describe by cut set Q

λ_q = failure rate of the device describe by cut set Q

T = test interval for the device describe by cut set Q

The mean or expected value of the unavailability is:

$$E(\bar{A}_q) = E(\lambda_q)\, E(T)\, /2 \qquad \text{(Eq A-2)}$$

Where:

$E(\bar{A}_q)$ = Expected value of the device unavailability

$E(\lambda_q)$ = Expected value of the device failure rate

E(T) = Expected value of the device test interval

The variance of the unavailability of Q is:

$$V(\bar{A}_q) = \left[\frac{\partial \bar{A}q}{\partial \lambda}\right]^2 V(\lambda_q) + \left[\frac{\partial \bar{A}q}{\partial}\right]^2 V(T) \qquad \text{(Eq A-3)}$$

$$V(\bar{A}_q) = \left[\frac{T}{2}\right]^2 V(\lambda_q) + \left[\frac{\lambda_q}{}\right]^2 V(T) \qquad \text{(Eq A-4)}$$

Where:

$V(\bar{A}_q)$ = Variance of the unavailability of the device

$V(\lambda_q)$ = Variance of the failure rate of the device describe by cut set Q

V(T) = Variance of the device test interval

Now we evaluate the variance of unavailability at the expected value of the random variables to yield:

$$V(\bar{A}_q) = \left[\frac{E(T)}{2}\right]^2 V(\lambda_q) + \left[\frac{E(\lambda_q)}{}\right]^2 V(T) \qquad \text{(Eq A-5)}$$

B. Two Event Cut Set, Q W

Consider a single event minimal cut set with two basic events, Q and W. The unavailability of QW may be written as

$$\bar{A}_{qw} = [\lambda_q T/2]\,[\lambda_w T/2] = \lambda_q\ \lambda_w T^2/4 \qquad \text{(Eq A-6)}$$

Where:

$\bar{A}_{qw}$ = unavailability of the cut set Q W

λ_q = failure rate of the device Q describe in cut set QW

λ_w = failure rate of the device W describe in cut set QW

T = test interval for the devices describe in cut set QW

The mean or expected value of the unavailability is:

$$E(\bar{A}_{qw}) = [E(\lambda_q)\ E(T)\ /2]\ [E(\lambda_w)\ E(T)\ /2] \qquad \text{(Eq A-7)}$$

Where:

$E(\bar{A}_{qw})$ = Expected value of the cut set QW unavailability

$E(\lambda_q)$ = Expected value of the device Q failure rate

$E(\lambda_w)$ = Expected value of the device W failure rate

E(T) = Expected value of the devices Q and W test interval

The variance of the unavailability is:

$$V(\bar{A}_{qw}) = \left[\frac{\partial \bar{A}qw}{\partial \lambda_q}\right]^2 V(\lambda_q) + \left[\frac{\partial \bar{A}qw}{\partial \lambda_w}\right]^2 V(\lambda_w) + \left[\frac{\partial \bar{A}qw}{\partial}\right]^2 V(T) \qquad \text{(Eq A-8)}$$

Now we evaluate the partial derivatives at the mean value of the random variables to obtain:

$$V(\bar{A}_{qw}) = [E(\lambda_w) E(T)^2/4]^2 V(\lambda_q) + [E(\lambda_q) E(T)^2/4]^2 V(\lambda_w) + [E(\lambda_q) E(\lambda_w) E(T)/2]^2 V(T) \qquad \text{(Eq A-9)}$$

Where:

$V(\bar{A}_{qw})$ = Variance of the unavailability of the cut set QW

$V(\lambda_q)$ = Variance of the failure rate of the device Q in cut set QW

$V(\lambda_w)$ = Variance of the failure rate of the device Q in cut set QW

$V(T)$ = Variance of the device test interval

Appendix B – Commonly Encounter Probability Distributions

There are four probability distributions used to represent the failure rate of equipment items. These are:

1. Uniform Distribution
2. Triangular Distribution
3. Normal Distribution
4. Log-Normal Distribution

This appendix reviews the properties of these four probability distributions.

Uniform Distribution

The uniform distribution [Ref. B-1, pg 74] is used when a variable is only known within a defined range. Figure B-1 presents a plot of the probability density function (pdf) for the uniform distribution. The distribution is defined using two parameters, A and B.

Where:

A = minimum value of the random variable

B = maximum value of the random variable

The mean or expected value of a random variable, x, that follows the uniform distribution is:

Mean = Expected Value – $E(x) = (A+B)/2$ (Eq B-1)

The variance of a random variable, x, that follows the uniform distribution is:

Variance = $V(x) = (B-A)^2/12$ (Eq B-2)

The standard deviation of a random variable, x, that follows the uniform distribution is:

Standard Deviation = $(\text{Variance})^{1/2} = (B-A)/(12)^{1/2}$ (Eq B-3)

Triangular Distribution

The triangular distribution [Ref. B-2] is used when a variable is known to lie within a defined range and a "best guess" or estimate can be made as to the most likely value to the variable. Figure B-2 presents a plot of the probability density function (pdf) for the triangular distribution. The distribution is defined using three parameters, A. B and C.

Where:

A = minimum value of the random variable

B = maximum value of the random variable

C = most likely (mode) of the random variable

The mean or expected value of a random variable, x, that follows the triangular distribution is:

Mean = Expected Value – $E(x) = (A+B+C)/3$ (Eq B-4)

The variance of a random variable, x, that follows the triangular distribution is:

Variance = V(x) = $[A^2 + B^2 + C^2 - AB - AC - BC]/18$ (Eq B-5)

The standard deviation of a random variable, x, that follows the triangular distribution is:

Standard Deviation = $(\text{Variance})^{1/2} = \{[A^2 + B^2 + C^2 - AB - AC - BC]/18\}^{1/2}$ (Eq B-6)

Normal distribution

The normal distribution is used when a variable when data analysis finds the normal distribution is the best model to describe the spread in the measured variable. Figure B-3 presents a plot of the probability density function (pdf) for the normal distribution. The distribution is defined using two parameters, μ and σ.

Where:

μ = mean of the measured data for variable

σ = standard deviation of the measured data for variable

The expected value of a random variable, x, that follows the normal distribution is:

Expected value = mean = $E(x) = \mu$ (Eq B-7)

The variance of a random variable, x, that follows the normal distribution is:

Variance = square of standard deviation = $V(x) = \sigma^2$ (Eq B-8)

The standard deviation of a random variable, x, that follows the normal distribution is:

Standard deviation = σ (Eq B-9)

Lognormal Distribution

The lognormal distribution [Ref. B- 2] is used when a variable when data analysis finds the lognormal distribution is the best model to describe the spread in the measured variable. Figure B-4 presents a plot of the probability density function (pdf) for the lognormal distribution. Many variables in nature are lognormally distributed (for example the height of people). The log normal distribution is used when the variable of interest has a known physical lower limit of zero. A variable, x, is lognormally distributed if ln(x) is normally distributed. The lognormal distribution is defined using two parameters, μ_y and σ_y.

Let the original data be defined by the variable x with mean and standard deviation as:

μ_x = mean of the measured data for variable x

σ_x = standard deviation of the measured data for variable x

Define a new variable y as:

$y = Ln(x)$ (Eq B-10)

Then variable x is said to be lognormally distributed if y is normally distributed. We can now write the mean and standard deviation of y (μ_y and σ_y.) [Ref. B-3, B-4 and B-5] as

$$\mu_y = \text{Ln}\left[\mu_x^2 /((\sigma_x^2 + \mu_x^2)^{\frac{1}{2}})\right] \quad \text{(Eq B-11)}$$

$$\sigma_y = \left[\text{Ln}(\sigma_x^2 / \mu_x^2 + 1)\right]^{\frac{1}{2}} \quad \text{(Eq B-12)}$$

Where:

Ln is the natural logarithm (base e) of the argument

Appendix B References:

B - 1. P. L. Meyer, *Introductory Probability and Statistical Applications*, 2nd Edition, Addison-Wesley, Reading Mass, Library of Congress Catalog no. 75-104971, 1972, pp. 74

B - 2. D. Vose, *Risk Analysis–A Quantitative Guide*, 3rd edition, Wiley, Chichester, West Sussex, England, ISBN 978-0-470-51284-5, 2008, pp. 686-687

B - 3. *"Lognormal mean and variance - MATLAB lognstat". www.mathworks.com. Retrieved* 27 December *2018*

B - 4. A. M. Mood, F. A. Graybill, and D. C. Boes. *Introduction to the Theory of Statistics.* 3rd ed., New York: McGraw-Hill, ISBN 0-07-042864-6, 1974. pp. 117.

B - 5. E.L. Crow and K. Shimizu, Lognormal Distributions – Theory and Applications, Marcel Dekker, New York, ISBN 0-8247-7803-0, 1988, pp. 9

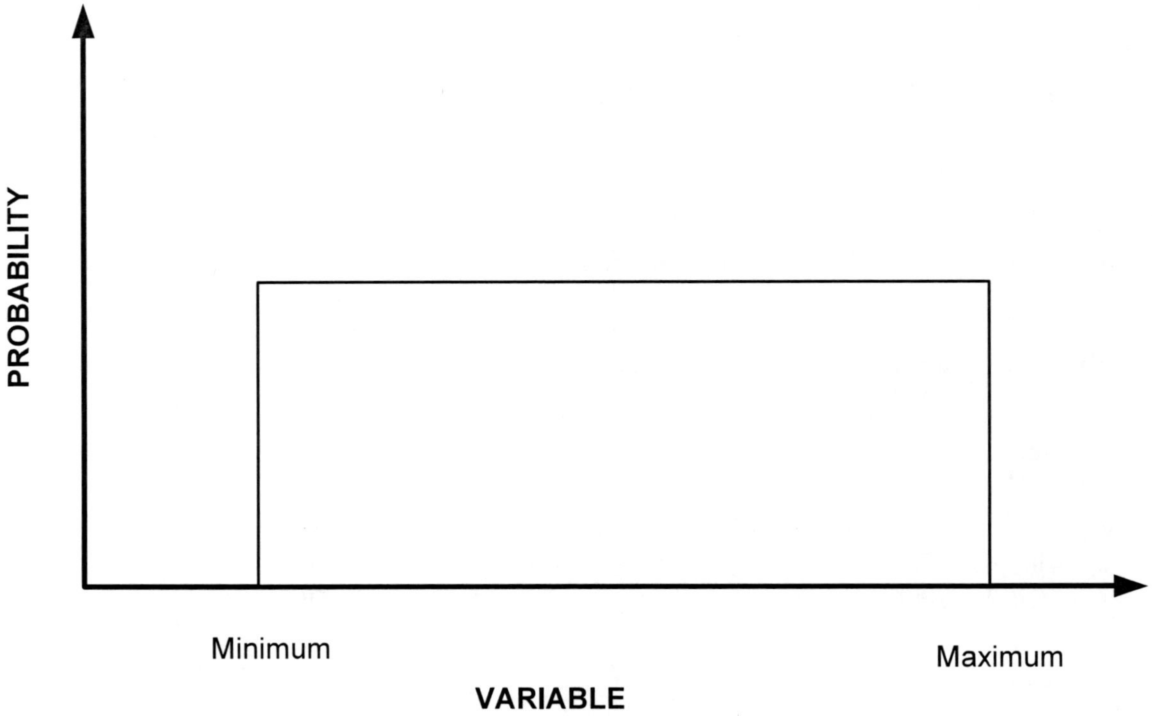

Figure B-1. Uniform Probability Distribution

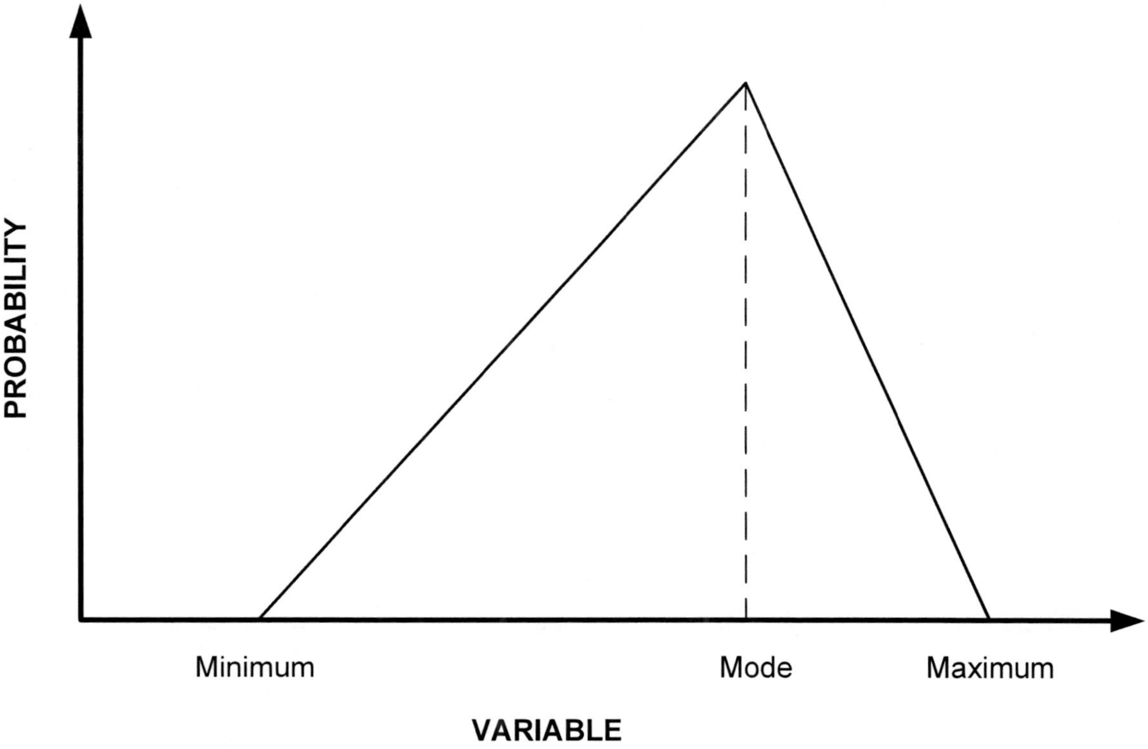

Figure B-2. Triangular Probability Distribution

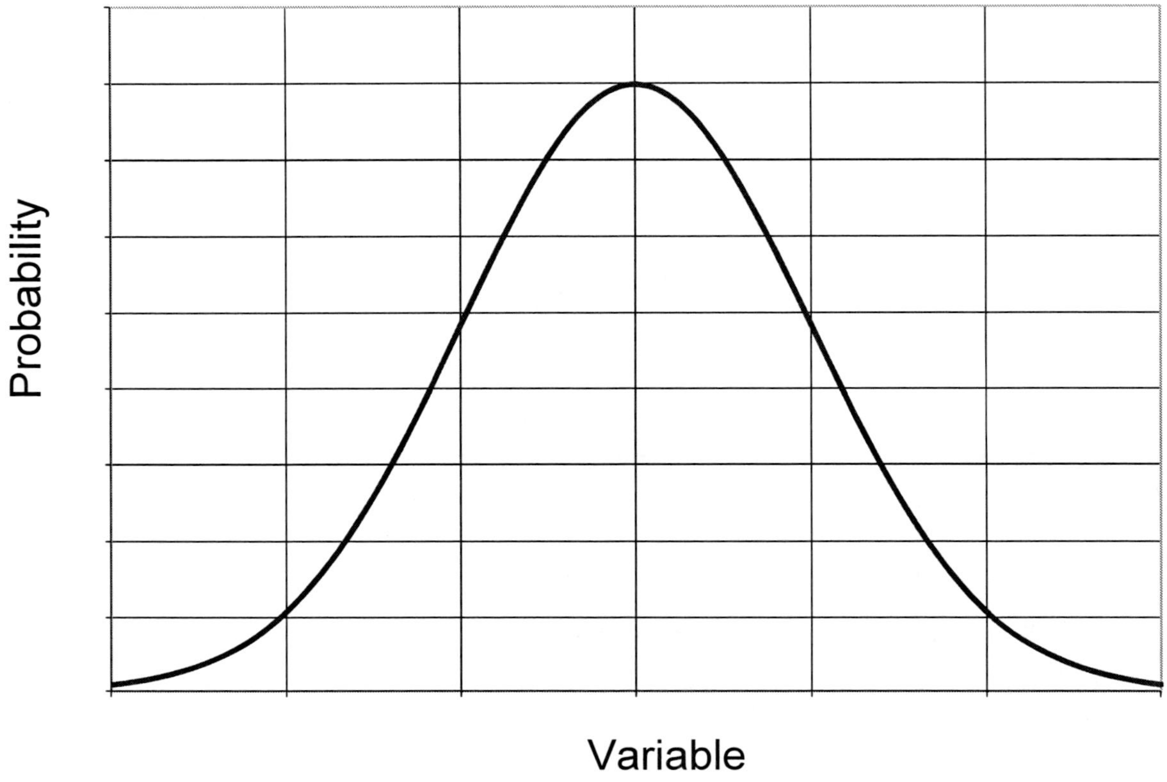

Figure B-3. Normal Probability Distribution

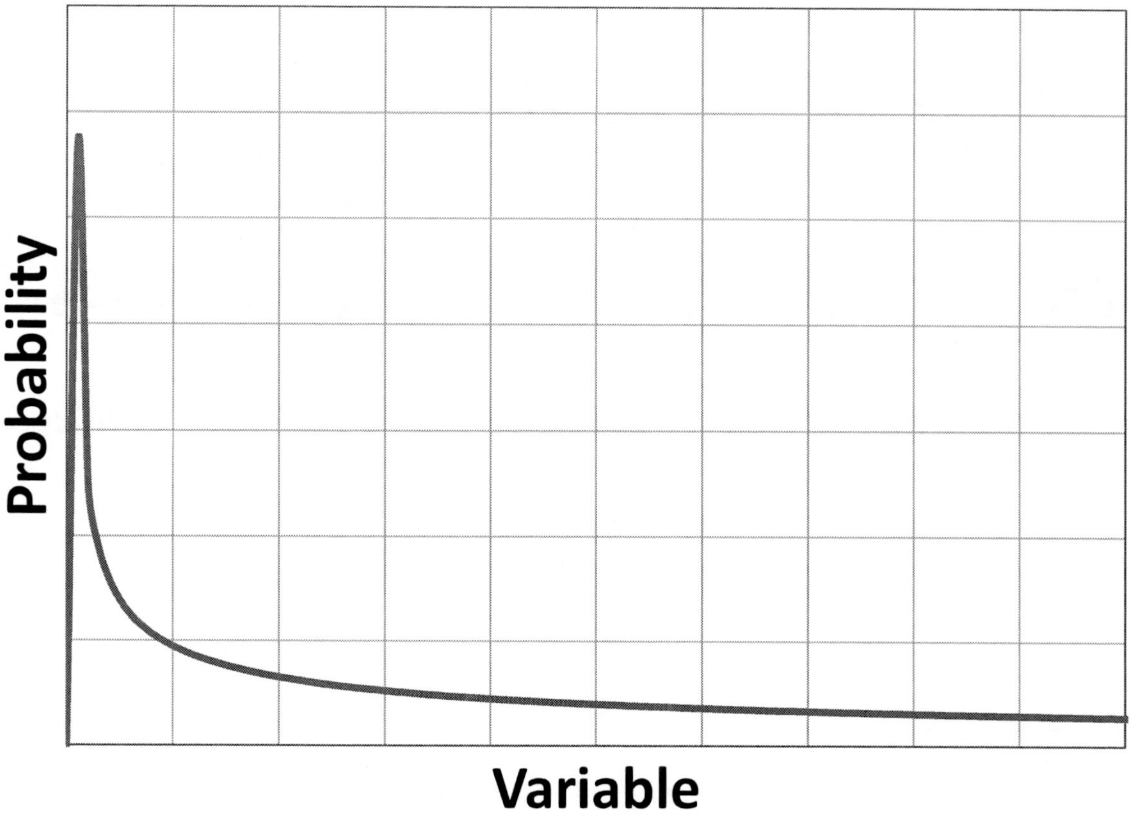

Figure B-4. Lognormal Probability Distribution

Effective Process Safety Management on Small Capital Projects

Kerissa Bowen
Eastman
knbowen@eastman.com

Prepared for Presentation at
American Institute of Chemical Engineers
2019 Spring Meeting and 15th Global Congress on Process Safety
New Orleans, LA
March 31 – April 3, 2019

Effective Process Safety Management on Small Capital Projects

Kerissa Bowen
Eastman
knbowen@eastman.com

Keywords: Process Safety Management (PSM), Process Hazard Analysis (PHA), Small Capital Projects

Abstract

Recognizing that Small Capital Projects are often quick turn-around, limited scope projects, it can be challenging to effectively incorporate PSM while meeting schedule and budget constraints. Process safety professionals must quickly adapt to develop the process safety scope and deliverables needed for each project. This paper will present techniques for providing quality Process Safety Management across all phases of a Small Capital Project – from inception to completion. Examples will show how a generic process can be tailored for each unique project to ensure that risk-based process safety principles and standards are effectively implemented.

1 Introduction

Meeting the targets in a Small Capital Project (SCP) is no easy task. These projects require quick turn-around and a focused scope. Process safety experts then become Safety Project Managers for capital projects and are often forced to confront the Project Management Iron Triangle of "good, fast, or cheap". Unfortunately, when looking at the Project Management Iron Triangle, quality decreases as choices are made in any direction. [1] As Safety Professionals, these choices can lead to ineffective risk management. With the proper approach to risk management on small capital projects, the Iron Triangle can be transformed into a target where all three goals can be met. Incorporating various strategies into the planning, design and execution stages of a project enables the Process Safety bullseye to be hit.

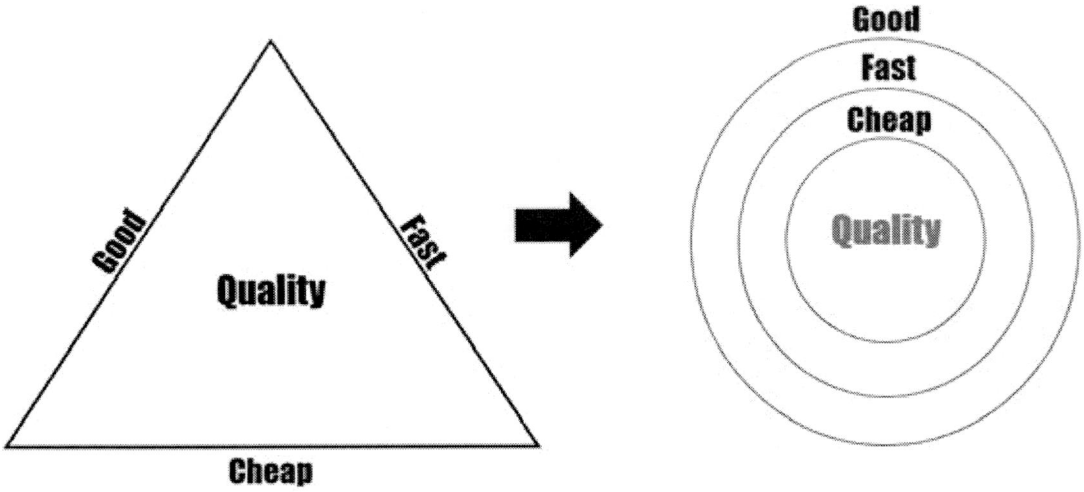

This paper will study several projects and provide examples of how the planning, design and execution of safety scope influenced their success. Techniques used to improve outcomes will be detailed.

2 Planning for Safety Scope

Process safety scope in small capital projects may not be well-defined unless project teams have members with process safety experience. For maximum efficiency, a process safety expert should be assigned to the project as soon as possible, preferably during the conceptual design phase so that a conceptual process safety review can be performed early in the project schedule.

2.1 Conceptual Safety Review

A Conceptual Safety Review is a design review that examines preliminary project information to begin defining Process Safety Scope. This review can give design guidance and help decide project scope when needed. Attendees should include the process safety expert, project lead, team lead and a process expert. Operators, while not required, can provide unique input to the changes being proposed and should be invited to attend if available.

One of the hurdles for the conceptual design phase is the potential lack of project-specific safety information. This preliminary project information can include a basic project scope, redlined piping and instrumentation diagrams (P&IDs), proposed chemicals and chemical hazards, existing equipment ratings, initial equipment layout, current equipment safety interlocks and relief strategy and lessons learned or incident investigations from similar processes.

2.1 Example #1

A recent project had a well-defined scope to change out several heat exchangers and pumps in an existing, well-documented process. The team had quite a bit of information from the existing Process Hazard Analysis (PHA) documentation – including known chemicals and hazards, materials of construction, P&IDs, and electrical classifications. Based on the information from the existing equipment and process, the Conceptual Safety Review highlighted the needs for the boiling liquid/expanding vapor (BLEVE) protection on new pumps, required temperature and pressure ratings for new heat exchangers, potential new relief design requirements and additional instrumentation and interlock protection..

During the Conceptual Safety Review, one part of the project was highlighted as having two scope options. The first was to continue unloading rail cars into the existing storage tank and upgrade the inlet pumps and instrumentation for better unloading efficiency. A major tank issue was that it held approximately half a railcar of liquid. Until the remainder of the railcar was unloaded, the railcar would be held for an extended period with steam supplied to the heating coils. The second option was to change rail unloading to a tank that would hold a full railcar and upgrade pump and equipment for that tank. In both options, level instrumentation and BLEVE controls would need to be added. However, during the safety review, it was noted that the larger of the two tanks would be inherently safer since it could hold a full railcar of material virtually removing a potential tank overfill scenario. Selecting to use the larger tank also eliminated holding railcars of heated chemicals for indeterminate periods of time.

2.2 Example #2

In another project, a compressor installation project team insisted the installation was a copy of a presently operating compressor, which set the initial safety scope. As the project progressed, it became clear that the new compressor skid was not identical to the existing installation. Relief design needed to be evaluated for the new compressor due to piping changes to inlets and outlets. Safety Instrumented Systems needed to be changed due to differences in compressor operation. A further complicating factor in the design process was the timing at which the process safety expert was consulted. Assuming the compressor would be a copy of a current installation, the process safety expert was consulted during detailed design, rather than conceptual design. This presented schedule and budget changes due to the re-design needed to correctly develop the safety scope.

These issues could have been easily avoided by involving the safety expert from the project's inception. This lesson proved to be valuable to the project team. A project team with essentially the same personnel was formed to install a compressor for a different area. This team included a process safety expert from the beginning. A Conceptual Safety Review was scheduled to compare the proposed compressor design to existing compressors and previous projects. This revealed several design differences that affected process safety. The Conceptual Safety Review allowed for better understanding of the project scope and process safety concerns.

The Conceptual Safety Review is intended to find major design issues with a limited amount of process knowledge. An advantage of a Conceptual Safety Review is the amount of time needed for the review. The two Conceptual Safety Review meetings for the projects in these examples took approximately five hours combined. The process safety expert can cover broad topics that need to be considered when entering the design phase of the project. PHA information from the current design can be obtained and examined for any outstanding recommendations. Previous relief analysis can reviewed to check for any deficiencies that need to be addressed during the project. Interlock narratives can be studied for potential changes with the new equipment. This information coupled with the scope and generic safety principles provide a wealth of information for the project team to act on in initial design. The time spent conducting the Conceptual Safety Reviews define the safety scope which in turn enhance the quality of the schedule and budget.

Expert Tip: *Identifying major process safety concerns early will allow the project scope, schedule and budget to be better refined and re-design due to process safety issues minimized.*

⚠ ***Caution:*** *Too little process and project information will not give an effective Conceptual Safety Review. This can cause lack of safety scope definition and potentially lead to multiple design iterations.*

3 Designing with Defined Safety Scope

As projects progress, more information becomes available with respect to Process Knowledge. Along with the refinement of equipment specifications, P&IDs, material balances, and so on, there should be a refinement of safety design. This includes relief and Safety Instrumented Systems (SIS) Design as these are iterative processes. Often, the project needs safety design information to specify and order equipment. At the same time, safety design is requesting information from the project to complete a thorough hazard analysis, relief calculations, and SIS calculations. This request for information from both parties highlights a latent continuous do-loop present in all capital projects. Process safety experts can spend valuable time and resources performing multiple iterations of hazard analyses and calculations based on continually updated design information. One way to circumvent this issue is to identify all design information required to start process safety design when the project schedule is developed. The project team and the process safety expert should establish the approval state for PSI to be used in process safety design. A lack of clear understanding of which information is used for calculations and hazard analysis can lead to the inability to compare final project information to initial project information and ascertain a delta which needs to be reviewed from a process safety standpoint.

3.1 Preliminary relief design

Preliminary relief design can begin once information such as stream compositions and the maximum allowable working pressure (MAWP)/maximum allowable working temperature (MAWT) of vessels have been reasonably well defined by the project. A relief expert should evaluate the equipment for potential relief cases to ensure that the relief scope is well-defined. [2] Once all cases are defined, the worst-case relief scenario can be evaluated for relief inlet and orifice size. Performing the full calculation is not necessary at this stage. The preliminary calculation can be used to verify relief nozzle sizing on vessels and to allow the project to begin quoting relief equipment. This is especially important for long-lead or expensive relief devices. In addition, preliminary sizing can note any unusual relief cases allowing the project team to consider the use of inherently safer design or mitigation of those cases. Final relief calculations will be performed toward the end of detailed design.

3.1 Example #3

During preliminary relief design for a project, the relief designers had the P&IDs, the MAWP/MAWT for the vessels and the process chemistry. The designers had access to relief calculations performed on a similar installation as well. Relief scenarios were evaluated and the worst-case scenario determined. Preliminary relief valve specification sheets were sent for quote and returned to the project. Initial relief design indicated a valve that could be purchased and shipped relatively inexpensively and quickly. During detailed design, the project informed the relief designers that a special type of flange needed to be used for the relief device connections. The project noted that this flange was used in all previous installations. However, a study of previous installation PSI and engineering standards did not call out this specialty flange. The flange type had not been called out in any project meetings. Relief designers updated the specification sheets to reflect the required flange. Once the change was made, the valves went out for quote again. A valve that was relatively inexpensive and quick to procure became a significantly more expensive and long-lead item. The change to the specialty flange was the correct decision to make; however, had the need for the specialty flange been communicated to the relief designers early-on, schedule and budget would not have been affected as much.

3.2 Preliminary Safety Instrumented Systems (SIS) Design

SIS design should also begin early in the project. As project PSI becomes available, the process safety expert can begin performing Layers of Protection Analyses (LOPAs). These can give direction to the appropriate instrumentation and Safety Integrity Levels (SIL) intended to mitigate safety scenarios such as tank overfill, high temperature, and high flow. Preliminary boiling liquid/expanding vapor (BLEVE) mitigation techniques can be designed for pumps. Burner Management System (BMS) actions can be put into place. Like preliminary relief design, early definition of SIS design helps the project team design with SIS instrumentation in mind. It also allows for budget and schedule refinement by allowing for instrumentation to be quoted early on. Specific details surrounding the SIS interlocks such as setpoints can be filled in as the project progresses to the final stages of detailed design.

3.3 Conducting the Preliminary Hazard Analysis (PrHA)

As the preliminary relief and SIS designs are being completed, the project team and safety professional should gather for a Preliminary Hazard Analysis (PrHA). This is a more in-depth study of the project using partially complete design information. The PrHA can best be defined as a Process Hazard Analysis (PHA) rehearsal. In the PrHA, the same technique is used as planned in the PHA whether it be a What-if Analysis, checklist, Hazards and Operability (HAZOP) study or other method. The intent of a PrHA is not to fulfill the regulatory compliance requirements. Rather, the intent of the PrHA is to review design information to date for safety concerns and finalize the safety scope. One challenge of performing the PrHA is the point at which the PrHA is done. With too little process information, the PrHA becomes little more than a Conceptual Safety Review. If the PrHA is performed too far into the project there is risk of potential re-design of processes and equipment to correct any missed safety scenarios.

The Center for Chemical Process Safety (CCPS) calls out quite a bit of information to be used in Process Knowledge Management. [3] These include:

- Chemical Hazards Information

  - Toxicity Information
  - Permissible Exposure Limits
  - Physical Data
  - Reactivity Data
  - Corrosivity Data
  - Thermal and Chemical Stability
  - Hazardous Effects of Inadvertent Mixing
  - Thermodynamic Data
  - Other Special Hazards
- Process Technology Information
  - Block or Simplified Process Diagram
  - Process Chemistry
  - Maximum Inventory
  - Safe Upper and Lower Control Limits
  - Consequences of Deviation from Limits
- Process Equipment Information.
  - Materials of Construction
  - Piping and Instrumentation Diagrams
  - Electrical Classifications
  - Relief Systems Design and Basis
  - Ventilation System Design
  - Design Codes
  - Material and Energy Balances
  - Safety Systems Employed

The above list is not all-inclusive. Furthermore, all items on the above list will not be required for every project. The process safety expert must use judgement as to the PSI required. As a suggestion, those items most helpful to have in a PrHA have blue text. Items in green text should be generated and added to the updated PrHA PSI for the final Process Hazard Analysis (PHA). (This list will be re-visited in Section 4.3 during the discussion of PSI and the final PHA.)

The advantage of the PrHA is identifying hazards early enough in the project to be able to effectively mitigate them during detailed design. It eliminates the potential for scope, budget and schedule increases if the same scenarios were to be found closer to project completion. It also allows for informed process safety decisions to be made during detailed design.

3.3 Example #4

In one project, there was debate about replacing a base heater during the conceptual safety review. The base heater was functional, but the MAWP was a concern. Replacement of the heater could be done but would have been a major undertaking because of the existing process layout. The PrHA team discussed the issue and safety concerns in greater detail. Using the vessel MAWP and updated project process information, a LOPA was performed. The LOPA showed adequate relief and SIS systems for the vessel at the MAWP. After weighing the options and performing a safety analysis, it was decided the existing heater would be re-used in the process, saving time and money without sacrificing safety.

3.4 Getting the most out of the PrHA

In addition to starting the PrHA with effective PSI, another key to an effective PrHA is limiting the number of personnel in attendance. The minimum number of personnel should be the process safety expert, project technical leader, process expert and area operator. This would be same as the minimum personnel required for a PHA as required by OSHA 1910.119 (e)(4). The argument could be made that the small number of personnel may not effectively identify all safety scenarios. One of the most effective ways to combat this is to wisely choose personnel who must attend all portions of the PrHA and have other pertinent project personnel on standby. In the same PrHA discussed above, the main attendees were the process safety expert, project technical leader, process expert and area operators. Because the PrHA was expected to last approximately three weeks, the project technical leader chose to make personnel such as equipment designers, instrumentation engineers, and P&ID developers aware the PrHA was being performed, but not require their attendance. If a question arose during the PrHA that needed expertise from a team member not in attendance, the question was noted and tabled until the team member could attend. At the end of each daily PrHA session, the process safety expert emailed all project team members an update on the items covered in the study along with any recommendations. This allowed project team members the flexibility to work on detailed project design while the PHA team focused on hazard identification and risk assessment. Daily communication kept the entire project team aware of safety scenarios that needed to be considered in their design work.

In instances the project is being performed on an existing process and is schedule-driven, the process safety expert may consider sending the PrHA team the current PHA report for the process that includes the nodes the project is changing. This will allow the PrHA team to begin reviewing safety scenarios for the existing process and contemplate how changes may affect those scenarios. Also, it gives an opportunity to review existing and proposed controls outside of the PrHA.

3.4 Example #5

A project proposed to install a bypass line from an oxidizer to a cat-ox in a different process. The timeline of this project was compressed to allow the mechanical completion during the fall outage. The project was kicked off in the spring. Typically, a week is allowed for the process safety expert to review PSI and prepare for the study. Study length is determined by the magnitude of change. While not a large study, the major project concern were the SIS changes needed for bypassing the existing cat-ox and connecting to the new cat-ox. To expedite the safety instrumented system review, the PHA report for the existing process was given to the PrHA team highlighting the nodes involved in the project. The team reviewed the report for scenarios and updated the SIS information. Reviewing the report also generated a list of safety interlocks to evaluate for applicability to the project. Since all the initial review was outside of the PrHA, the PrHA meeting was able to evaluate and update scenarios and controls more quickly. In addition, project SIS design was expedited since the existing design had been appraised for the interlocks that needed to be modified for the project.

Expert Tip: *Use methods such as limiting PrHA team size and referencing current process PHA reports to enable the PrHA team to more efficiently identify process safety scenarios and recommend mitigation techniques.*

⚠ ***Caution:*** *Lack of consistent and clear communication to the project team can lead to multiple detailed design iterations potentially increasing scope schedule and budget.*

4 Executing Safety Scope

During the PrHA, as potential safety scenarios are identified which need to be mitigated, recommendations should be made to the team as to ways to correct the deficiencies. These should be documented in a format to pass on to the detailed design team. When wording the recommendations, give suggestions to mitigate the scenario described in the scope, but use wording that gives the design team latitude for design and the ability close the recommendation satisfactorily.

The project team and the process safety expert should stay in communication during the full project. When PSI is generated and distributed for approval, the process safety expert should be placed on the distribution list. This is not to say he or she should have approval privileges for the PSI. The process safety expert may be added as an "information only" recipient. This inclusion allows for review of design changes and distribution of the PSI to the process safety designers.

4.1 Final Relief Design

Final relief design can be undertaken while final design and construction is being completed. The obstacle to finishing relief is often a lack of information. While initial calculations should reveal worst-case design scenario and relief sizing, final calculations will ensure sizing is correct considering final process chemistry and flow, vessel specifications, and pipe routing. Furthermore, the final relief calculations should thoroughly discuss all cases considered when sizing the valve. As PSI is generated by the project, it can be distributed to the process safety team to use in final calculations.

4.1 Example #6

A de-bottlenecking project changed the tank inlet flows and heat exchanger sizing for the process. During preliminary design, the relief designers provided the project team with nozzle sizes for the proposed heat exchanger. Designers verified proposed flows to the tanks would not affect those reliefs. As the modification progressed, the project team decided impeller size on the pumps for the tanks would need to be increased. The proposed change was communicated to the process safety team. The communication allowed the process expert to consult the process safety team for pump safety scenario mitigations and re-evaluate the tank reliefs. Despite the larger flow, pool fire remained the worst-case scenario. During final relief calculations, relief size was confirmed, however it was discovered that outlet piping would provide too great a pressure drop. The outlet piping pressure drop could not have been foreseen in preliminary calculations given the piping was field routed by construction forces. Working with the project team, the piping was re-routed to provide allowable pressure drop without a change in relief size.

4.2 Final SIS Design

Safety instrumented design can be the most fluid of safety design during the project process. Simply adding or removing valves can elevate or reduce the SIL rating of a proposed interlock.

As the project progresses, the process safety team should evaluate interlock levels using LOPA and by consulting with the operating area regarding the level of risk the area is willing to accept. While interlocks have basic components that can be assumed, the final components often cannot be specified until the project and the safety expert have discussed the appropriate SIL rating for the interlock. One way to make this process more efficient is to templatize safety interlock design. Once the area and the process expert have agreed on the appropriate SIL rating to mitigate a safety scenario, the SIS designer may apply the template to the scenario, giving a way for the project to quickly order the needed components.

4.3 Final PHA

The final PHA should be performed before start-up but after drawings are Issued for Construction. At this point, all PSI should be generated for the project. The project technical leader can be very helpful in the process of gathering all the PSI generated during detailed design. (The list of recommended PSI is shown in Section 3.3) A deliverables checklist should be given to the technical leader. This will allow him or her to ensure all PSI is available for a thorough study.

A comprehensive study of the final PSI should be conducted by the process safety expert. Comparing the final PSI to that used for the PrHA will allow the expert to look at changes made during detailed design. The process safety expert can make a choice to review all scenarios in the PHA or only those affected by the changes. Reviewing only scenarios involving design changes will allow the PHA to take less time. It should be noted that this should only be done if the safety expert has full understanding of the changes made and the effects on the design. In addition, he or she should understand any affects the changes in one node may have on any other equipment being studied in the process.

During the final PHA, it is most helpful to update the PrHA. Again, the team members present should be those required by OSHA 1910.119 (e)(4). [4] Like the PrHA, additional personnel may be put on standby to ensure they are available if questions need to be answered. Using the PrHA to update makes the PHA more efficient. Most, if not all, safety scenarios have been identified. If a recommendation was made in the PrHA, then the recommendation may be reviewed for completion. The completion details can be entered into the scenario and the recommendation deleted. If the recommendation has not been completed, then the documentation should be kept for follow-up either by the project team or operating area, as dictated by management. Ideally, there will be no recommendations from the PHA as the safety scope addressed in the Conceptual Safety Review and PrHA should have been addressed during the detail design phase.

Changes may be made after the final PHA. The project team should keep a change log so these changes can be evaluated once they are complete. These changes should be small in scope. The project team should be reminded by the process safety expert that all changes need to be assessed with respect to process safety.

4.3 Example #7

A project to install a new tank issued construction P&IDs which were used to complete the final PHA. During the PHA, all recommendations from the initial PrHA were closed because of completion during construction. Relief and SIS design were completed and installed. However, after the PHA was complete, modifications were made in the field resulting in changes to the P&IDs. The project team logged all changes in a spreadsheet format, listing each change made to each P&ID. At project completion, the PHA team met again to review each change on the list and the potential effects to the PHA. During this meeting, the team noted that no changes made affected the PHA results and the project proceeded to start-up on-schedule and under-budget.

Expert Tip: *Celebrate having a quality project that is under-budget, within scope and on-schedule.*

⚠ **Caution:** *Excessive celebration may result in undesirable consequences.*

5 Conclusion

The process safety professional has the opportunity to positively impact project schedule and budge while delivering a quality project by implementing multiple innovative ideas. Consulting with the project team early provides a base for which inherently safer design may be built. The Conceptual Safety Review helps eliminate potential redesign by looking for major process safety issues during initial project stages. Conducting a detailed PrHA using all available PSI allows preliminary relief design and SIS design to be performed, giving the detailed design team information on everything from nozzle sizing to instrumentation requirements. In schedule driven project, using previous PHA, relief and SIS information can help meet time demands. Keeping PrHA and PHA team sized small allow those meetings to be held more efficiently, calling in experts as needed to evaluate various scenarios. The final PHA, relief and SIS design are chances to refine and confirm the safety design using final project PSI. Using these tools gives the process safety professional confidence in a quality product without choosing between good, fast or cheap.

6 References

[1] R. Swisher, Fast, Good or Cheap. Pick Three?, Available at https://www.business.com/articles/fast-good-cheap-pick-three/, Accessed on March 14, 2019.

[2] ANSI/API Standard 521, Pressure Relieving and Depressurizing Systems / ISO 23251, Petroleum, petrochemical, and natural gas industries – Pressure-relieving and Depressurizing Systems

[3] CCPS, Guidelines for Risk Based Process Safety, American Institute of Chemical Engineers, John Wiley & Sons, Inc., Hoboken, New Jersey, 2007.

[4] Occupational Safety & Health Administration, U.S. Department of Labor, Occupational Safety and Health Standards: 1910.119, Process safety management of highly hazardous chemicals

When Emergency Response Goes Fatally Wrong: Lessons Learned from Investigating Two Major Incidents

Scott Davis
Gexcon US
4833 Rugby Avenue Suite 100
Bethesda, MD
sgdavis@gexcon.com

Prepared for Presentation at
American Institute of Chemical Engineers
2019 Spring Meeting and 15th Global Congress on Process Safety
New Orleans, LA
March 31 – April 3, 2019

When Emergency Response Goes Fatally Wrong: Lessons Learned from Investigating Two Major Incidents

Scott Davis
Gexcon US
4833 Rugby Avenue Suite 100
Bethesda, MD
sgdavis@gexcon.com

Keywords: emergency response planning, lessons learned, case studies, major explosion incidents

Abstract

Emergency response plans are an essential, yet oftentimes overlooked, layer of protection for facilities where all other safeguards failed to prevent an incident. While catastrophic accidents such as large releases of chemicals, fires or explosions are devastating for the process industry, experience investigating numerous incidents has shown that a lack of an effective emergency response plan can lead to an unnecessary and tragic escalation of the initial incident. More specifically, investigation of two recent incidents: (1) 2013 West Texas Explosion and (2) 2015 explosion on the FPSO Cidade São Mateus; will demonstrate how fatal consequences resulted from the lack of emergency planning in the first case study and from deficiencies in the emergency response plan or implementation of the plan in the second case study. More specifically, the FPSO explosion resulted in 9 fatalities, eight of which were responding to a leak of a light-end hydrocarbon, and the West Texas explosion that resulted 15 fatalities, most of whom were responding to the fire that preceding the devastating explosion. Lessons will show that the fatal consequences could have been avoided with the proper emergency planning. In addition, while emergency response plans consider "maximum credible" scenarios, past events have shown that low probability high consequences should also be at least considered.

1 Introduction

Emergency response plans are an essential, yet oftentimes overlooked, layer of protection for facilities where all other layers of protection failed to prevent an incident. An emergency plan specifies procedures for handling sudden or unexpected situations, where the ultimate objectives are to be prepared to: (1) prevent fatalities and injuries; (2) reduce damage to buildings, stock, and equipment; (3) protect the environment and the community; and (4) accelerate the resumption of normal operations [1]. Developing the emergency response plan consists of five major steps (see Figure 1). The first step is to assess the vulnerability of the site and determine what hazards pose a threat to the installation. Such an exercise will help determine not only what "credible" events can occur at a given facility, but also what catastrophic or major events can occur. It should be noted that major emergencies are generally rare events.

Next, it is necessary to develop your emergency response plan and determine what means are currently available to execute the plan and what is still needed. At the planning stage, it is important that several groups be asked to participate, such as: subject matter experts and process engineers; the health and safety group; and even appropriate municipal officials should also be consulted since control may be exercised by the local government in major emergencies and additional resources may be required. Once the emergency response plan is developed, the necessary facilities and equipment need to not only be provided but tested and maintained. Communication, training and periodic drills will ensure adequate performance if the plan must be carried out. Lastly, there should be a periodic review of the emergency response plan.

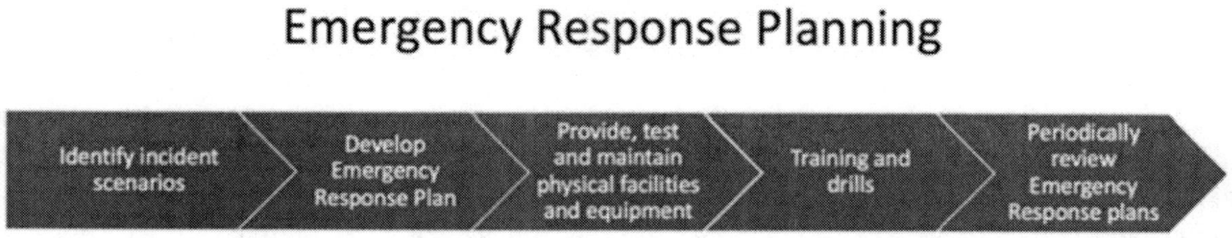

Figure 1: Steps of Emergency Response Planning

While the major benefit is to provide guidance during an emergency, developing the plan has other advantages. For example, previously unrecognized hazardous conditions, which would aggravate an emergency situation, could be discovered during the emergency planning and means put in place to mitigate or eliminate these hazards. The planning process may also identify deficiencies, such as the lack of resources (equipment, trained personnel, supplies), which can be easily corrected before an emergency occurs.

While catastrophic accidents such as large releases of chemicals, fires or explosions are devastating for the process industry, experience investigating numerous incidents has shown that a lack of an effective emergency response plan can lead to an unnecessary and tragic escalation of the incident. Chaos can quickly occur during an emergency if there are lack of resources and trained personnel coupled with an urgent need for rapid decisions and a very short period of time. The stress of the situation can lead to poor judgment resulting in severe losses, and normal channels of authority and communication cannot be relied upon to function routinely. Therefore, it is imperative to eliminate the “it can’t happen here” mentality, as emergencies can occur and preplanning is necessary.

The importance of emergency planning will be discussed within the context of two recent incidents: (1) 2013 West Texas Explosion and (2) 2015 explosion on the FPSO Cidade São Mateus. These two case studies will demonstrate how the lack of emergency planning resulted in devastating consequences in the first case study and from deficiencies in the emergency response plan or implementation of the plan in the second case study.

2 Background of Incidents

2013 West Texas Explosion

On April 17th, 2013, an explosion occurred at the West Fertilizer Company storage and distribution facility in West, Texas. The explosion at West Fertilizer resulted from an intense fire in the seed and fertilizer area of the facility that led to the detonation of approximately 30 tons of ammonium nitrate stored inside a wooden receiving bin. The explosion occurred while emergency services personnel were responding to a fire at the facility. Fifteen people were killed, more than 250 were injured, and numerous buildings were damaged or destroyed.

The seed and fertilizer building (Figure 2) was built in 1958 and stored fertilizer and seed, which were distributed to farmers. Figure 3 shows the layout of the seed and fertilizer building, which included an ammonium nitrate (AN) storage bin neighboring the seed room, as well as an empty bin used for ammonium nitrate storage, and bins containing ammonium nitrate, K-Mag (fertilizer mix of potassium, magnesium and sulfur), Potash, ammonium sulfate, DAP (diammonium phosphate), and a bin for potassium chloride along the west wall. This building was capable of receiving rail shipments directly from the rail car by means of a rail spur junction and an underground conveyor belt and bucket elevator; the bucket elevator dumped directly into the northern AN storage bin, or onto a conveyor belt feeding the bins along the west wall.

Figure 2: West Fertilizer Facility – viewed from the west (upper) and east (lower)

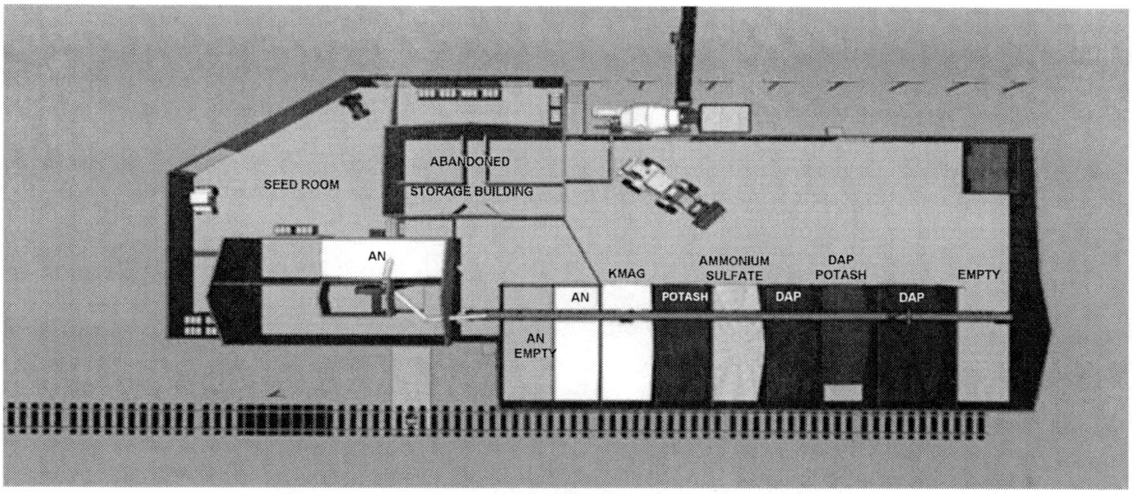

Figure 3: Layout of the Seed and Fertilizer Building.

On April 17th, 2013, at approximately 7:29 pm, a fire was reported, and fire fighters responded to the fertilizer plant [1]. All operations were already stopped for the day and employees had left the plant at about 5:30-6:00 pm. GexCon obtained several videos that recorded the progression of the initial fire [3]. Figure 4 shows the initial fire as viewed from the east and west; note that the highest point on the building represents the location of the bucket elevator by the northern AN storage bin. The image shows that the fire is isolated to the northern section of the building, which includes the seed room and portions of the northern AN storage bin containing about 30 tons of ammonium nitrate.

At approximately 7:50 pm, a massive explosion occurred at the fertilizer plant. Figure 5 shows images from a video taken from the high school parking lot to the southeast and Figure 6 the ensuing blast damage and crater from the detonation. Based on the videos from the day of the incident, there were actually two explosions: one smaller blast followed by a larger blast roughly 300 msec later.

Figure 4: Close up of the Seed and Fertilizer Building during fire as viewed from east (left image) and viewed from the west (right image)

Figure 5 – View of the explosion (viewed from southeast)

Fifteen people were killed as a result of the explosion, which included voluntary firefighters, and over 250 people were injured [3]. Damage was extensive throughout the neighboring community, which included homes, schools, an apartment building and a nursing home.

Figure 6: Blast damage and resulting crater

2015 explosion on the FPSO Cidade São Mateus

Due to the confidentiality of matters in this case, all references are made to two publicly available reports regarding the incident [5], [6]. The Cidade São Mateus was a Floating, Production, Storage and Offloading of Oil and Gas Export (FPSO), with a capacity to produce 4,000m³/day of oil and 10,000,000m³/day of gas as shown in Figure 7.

Figure 7: 2015 FPSO Cidade São Mateus.

On February 11, 2015 there was a leak of condensate (light-end hydrocarbon liquid) discovered in the pump room at a spade blind flange connection. Figure 8 shows that the pump room is situated just forward of the engine room and aft of the storage tanks. The layout and levels of the pump room is given in the left image of Figure 9, while the leaking spade blind flange is in the right image.

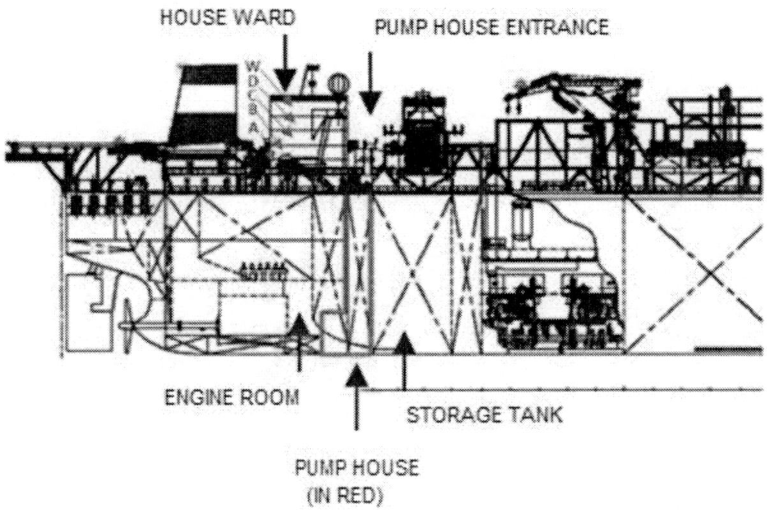

Figure 8: Cutaway side view of the pump room on the FPSO Cidade São Mateus.

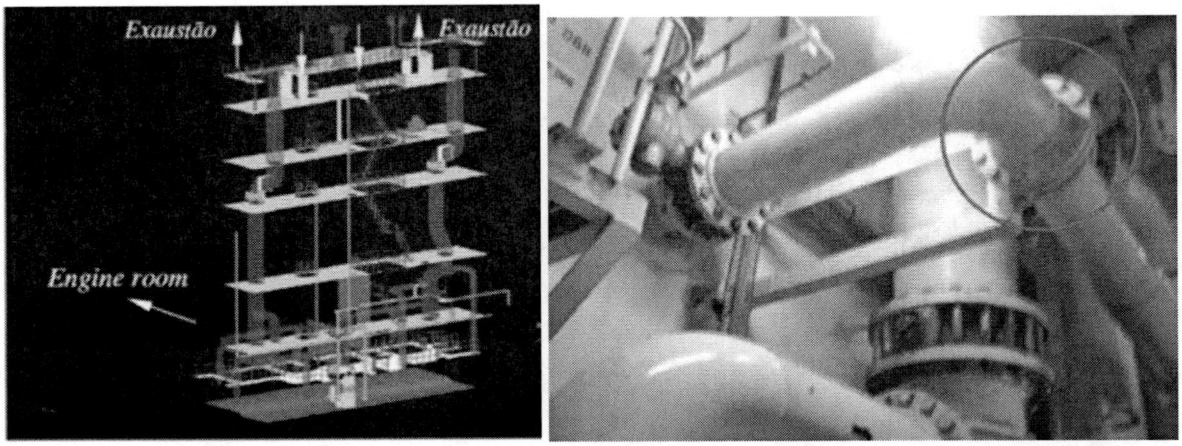

Figure 9: Model of the pump room on the FPSO Cidade São Mateus (left) and spade blind flange (right).

Just prior to the leak, the operations crew was attempting to drain condensate center cargo tank 6 (6C) into an alternate tank and they used an alternate pump alignment (stripping pump). However, the valve on discharge side of stripping pump was closed and they were pumping against a spade blind that was inadequate for such pressures, which ultimately failed the blind and seals. The gas detectors in the pump room began to alarm, and ultimately all three were alarming.

The ventilation was interlocked to shutdown per "confirmed gas" in the pump room. Emergency response teams entered the pump room to attempt clean up and repair the flange. There were three separate entries, and on during the third entry the explosion occurred. Figure 10 through Figure 12 show the level of blast damage to the pump room. Nine people were killed, of which eight were first responders.

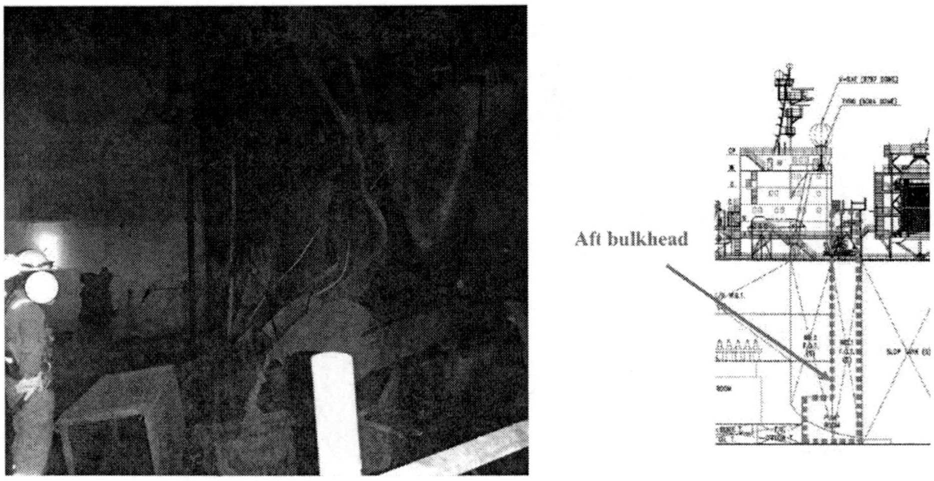

Figure 10: Damage to level above the pumps – complete bulkhead failure.

Figure 11: Damage to the engine room.

Figure 12: Damage to upper level of pump room.

3 Emergency Response

Response during the 2013 West Texas Explosion

As the fire occurred 20 minutes prior to the explosion, fire personnel were deployed to scene. Of the 15 people killed, 12 were first responders (8 volunteer firefighters and emergency services), and the other individuals were good Samaritans who supported the emergency response at the fertilizer plant. One of the volunteer fire fighters, who died in the blast, was employee of West Fertilizer. In fact, he was the plant foreman and knew what was stored in the facility. The emergency response is outlined in Figure 13.

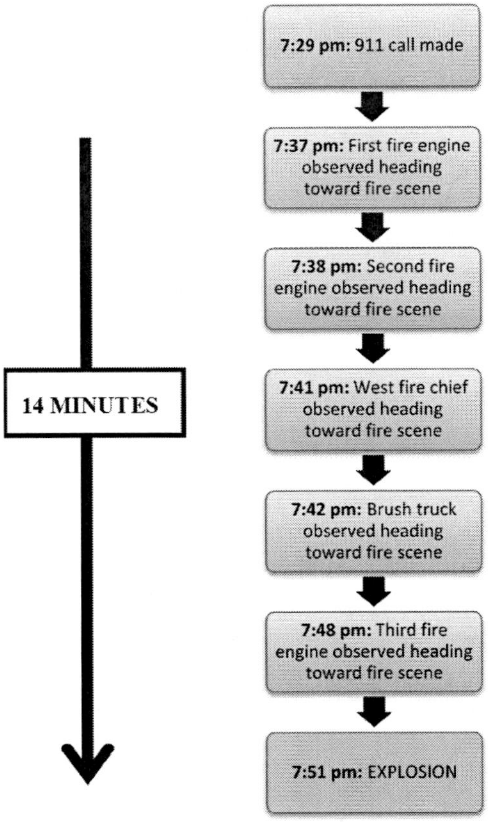

Figure 13: Emergency Response Plan at West Texas

Plant personnel and management did not understand the unique explosive nature of ammonium nitrate. More specifically, while the conditions necessary for detonation are not precisely known, the data suggest that detonations are promoted by severe shock or by heating with confinement, such as the massive fire at the plant. In fact, pools of molten ammonium nitrate may form in the event of fire, and if the molten mass becomes confined (e.g., in the elevator pit) it could explode. These hazards were not known nor understood by plant personnel and emergency responders. Therefore, responders were unaware of the potential of ammonium nitrate to detonate, had no knowledge of how long the fire had been burning, no robust incident pre-planning process in place, and no previous ammonium nitrate-related fire emergency training or drills. Hence, the strategy and tactics utilized by the West Volunteer Fire Department were not appropriate for the rapidly developing situation and they were exposed to unnecessary and extreme risk that proved fatal.

Response during 2015 explosion on the FPSO Cidade São Mateus

The first detector alarmed at 11:31, and by 11:38 all three detectors reached the over limit levels. The 1st team entered the pump room between 11:49-11:54 to assess the gas alarms, and they reported a leak at the flange and a 2-meter diameter pool under the flange. A 2nd team entered the pump room around 12:07 to assess the damage and whether repairs could be made. Despite the confirmed gas on all three detectors, the 2nd team still entered, and a personal gas detector alarmed at 100% LEL. The 2nd team exited around 12:15 and the ship was partially demobilized around 12:20 for lunch. The 3rd team enters around 12:27 to clean and repair, with ladders, wrenches, fire hose and absorbent blankets. The explosion occurred around 12:38 and killed four members in the pump room and four members at the top of the pump room on watch. Due to the failure of the wall into the engine room, the utilities operator was fatally injured in the engine control room.

There was a detailed Emergency Response Plan on the FPSO, along with the muster points for each person involved in the emergency response as shown Figure 14.

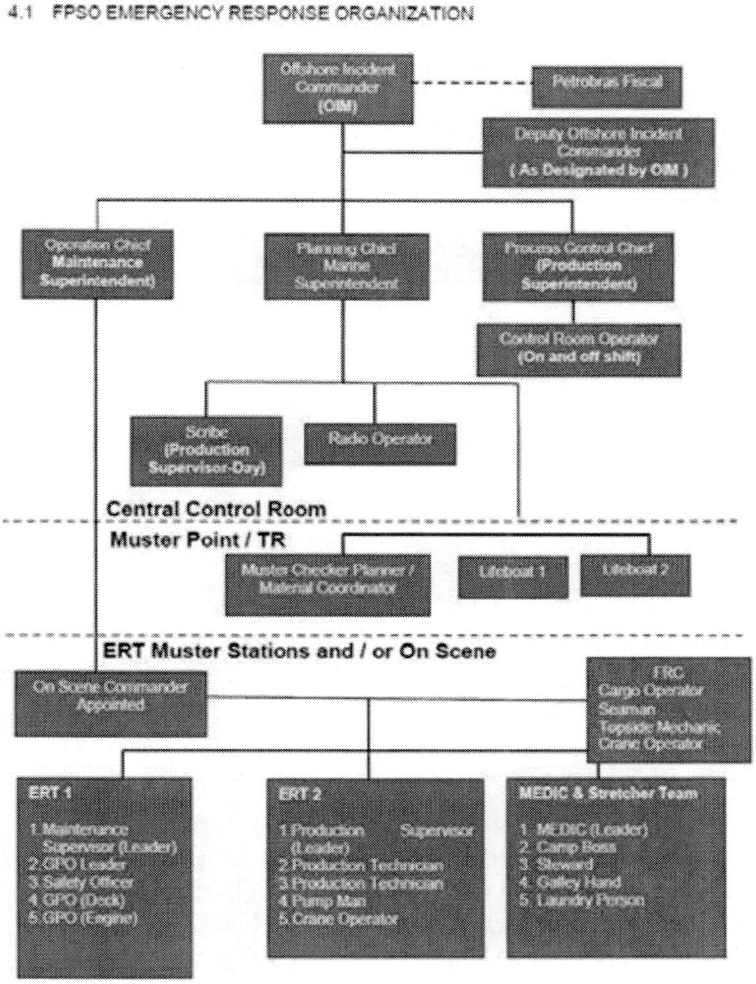

Figure 14: Organization Chart and muster points of the emergency response team

The Offshore Incident Commander is the Offshore Installation Manager (OIM) and would have the roles of implementing a command structure that would satisfy the organizational needs to protect life, environment and the installation, as well as deploying quick and efficiency actions to control incidents, as well as managing response actions to ensure that the escalating potential, risks of injury to personnel and damages to the installation are minimized. In the event of a leak in potentially explosive atmospheres, the emergency response must consider the possibility of the event escalating to a fire or explosion scenario. Thus, systems and procedures should be provided for the crew to control the situation and mitigate the risks of escalation from explosive atmospheres.

Investigation showed that the vessel was not designed for condensate, but for crude and gas. Yet despite the fact that there was confirmed gas detection within the pump room and on personal gas detectors, the crews entered three times. On the third entry, the explosion occurred. It was also reported that an explosive atmosphere in the pump or engine room was not included as an accident scenario in the emergency response plan, and there were no clear instructions of this particular scenario. Despite the knowledge that there was confirmed gas detection on the fixed sensors installed at the bottom of the pump room and also on personal detectors, Emergency Response Teams were sent into the pump room to evaluate the scenario in an explosive atmosphere.

The actions of Command essentially played down the risk of the condensate leak and allowed personnel to enter the pump room on multiple occasions to clean the spill without first performing a risk analysis, and without additional monitoring the room's atmospheric conditions. This noncompliance with the applicable emergency procedure was not only observed during entry operations, but also through the partial or total demobilization of the people irrespective of their muster points with an ongoing confirmed gas detection in the pump room. This demobilization resulted in two members being injured as they were found in a damaged elevator shaft after the blast.

4 Analysis

While deficiencies in the emergency response plans resulted in unnecessary and tragic consequences, the reasons each were lacking are very different and distinct.

For the West Texas Explosion, the Texas State Fire Marshal incident investigation [2] concluded that neither West Fertilizer nor West Fire Department had pre-fire and pre-incident plan at the facility. -Therefore, the recommendations provided were to ensure that local fire departments implement an Emergency Response Plan, with a particular focus on high-hazard occupancies. It was identified that fire departments visit and pre-incident plan responses to commercial structures in their jurisdiction. Further, they concluded that it is imperative that high-risk occupancies containing hazardous materials are identified and appropriate emergency response strategies and tactics are developed through pre-plan documents and SOP directives. Despite the extremely low likelihood of ammonia nitrate detonations, the recommendations state that high-hazard and high-risk occupancies containing heavy fire loads and hazardous materials should be identified by Fire Departments and all members made aware of their potential dangers. This will better prepare them for not only credible hazards, but also those high consequence hazards which may be found in the event of a fire and also provide them with insight as to when it is prudent to attack a fire or evacuate the area. More specific recommendations regarding the emergency response planning included:

- Develop a detailed Emergency Response Plan and Evacuation Plan for the area within 1 mile of existing ammonia nitrate buildings
- Develop standard operating procedures/guidelines to train and educate members on firefighting and emergency scene strategies and tactics for high-hazard occupancies
- Develop pre-incident plans by visiting the commercial and high-risk properties within the response jurisdiction.
- Utilize risk management systems on all emergency responses to determine when to employ alternative tactics.

- Incorporate incident management systems in every emergency incident, training exercise and basic operational process of the fire department to provide appropriate command and control for conducting operations.
- Implement a strategic incident action plan (IAP) for all emergency operations through a coordinated application of appropriate emergency scene tactics for the presented dangers
- Ensure that an incident safety officer (ISO), a person who is independent from the incident commander, is appointed and effectively utilized at every incident that would require an ISO.
- Establish a hazardous materials program and actively train members on the tactics to use when an incident involving hazardous materials occurs

In contrast, for the 2015 FPSO event, there was a detailed Emergency Response Plan for the vessel. Despite this plan, investigations concluded that there were significant deficiencies in the plan that needed to be remedied, and largely included reference to personnel exposure, failure to identify the accident scenarios for the emergency plan, training and mobilization. The execution of the plan failed to properly identify the potential high consequence and catastrophic result of the actions of the emergency responders. More specifically:

- Exposure to personnel
    - In the event of a leak in potentially explosive atmospheres indoors, the plan must consider the possibility of event escalation to a fire and/or explosion. Thus, systems and procedures should be provided once there is the formation of an explosive atmosphere. To ensure the success of mitigation actions, emergency procedures and resources for emergency response should be pre-planned, drawing on the use of all available resources, including the time of scenario assessment.
    - To determine that all functions involved in emergency response, including land-based support, have periodic theoretical training in emergency response procedures.
    - Set in the emergency response plan the available material and human resources to the response of each accidental
- Failure to identify accident scenarios or not included from the risk scenarios
    - Establish criteria and procedures for the migration of risk scenarios identified in the risk analysis and safety studies for emergency response plans. In addition to the selected scenarios, the procedures should contain scenarios that represent challenges to response actions, such as fatalities, and multiple injuries, among others.
    - Establish periodically simulated emergency covering all identified accident scenarios, including scenarios with multiple injuries and fatalities. Such simulations should include the activation of external resources to inform platform that response structure is effectively tested and that corrective measures arising from simulated ensure continuous improvement of the response actions of all entities involved.
- Demobilization of meeting points
    - Periodically coordinate all staff regarding the hazards involved in the assumptions made in the response procedure to the emergency unit.

    - Prohibit the release of people to normal activities before the full control of the emergency.
- Ignition of the explosive environment
    - As the ignition was introduced by the action of the emergency responders within explosive atmospheres, review the procedures related to emergency response and safe work practices considering the possible ignition sources identified in existing rules

5 Conclusion

The lessons from two specific incidents show that critical deficiencies in the emergency response plan or implementation of the plan can result in fatal consequences that could have been avoided. The 2013 West Texas explosion shows how important it is to not only recognize high risk events, but also have the appropriate plans in place to avoid unnecessary escalation of an already tragic incident, where 12 of the 15 people fatally injured were first responders. By preparing first responders for not only credible hazards, but also those high consequence hazards, can provide them with crucial insight as to when it is prudent to attack a fire or evacuate the area. In addition, not properly following or playing down the risk of a high consequence event was observed in the 2015 FPSO event, where eight of the nine crew members killed by the blast were first responders. The failure to recognize the potentially high consequence and catastrophic event, led to an emergency response that unnecessarily exposed responders and ultimately led to an exacerbated event. This study confirms that emergency response plans are an essential layer of protection for facilities where all other safeguards failed to prevent an incident, which could have prevented unnecessary and tragic escalation of the incident.

6 References

[1] Preparedness for Emergency Response - FEMA Training - FEMA.gov

[2] Texas State Fire Marshal, "Firefighter Fatality Investigation FFF FY 13-06," West, TX, 2013

[3] Davis, SG, DeBold, TF, & Marsegan, C (2017). Investigation Findings and Lessons Learned in the West Fertilizer Explosion. Journal of Fire Science. Volume 35, Issue 5, September 5, 2017: pp. 379-395.

[4] "A Public Health Report on Injuries Related to the West (Texas) Fertilizer Plant Explosion," June 24, 2014.

[5] Brazilian Navy Directorate of Ports and Coasts, Maritime Accident Safety Investigation Report, FPSO Cidade São Mateus - Explosion followed by flooding and casualties, February 11th, 2015.

[6] Investigation report of the explosion incident occurred on 11/02/2015 in the FPSO Cidade São Mateus, Brazilian National Agency of Petroleum, Natural Gas and Biofuels, Superintendence of Operational Safety and the Environment, August 2015.

19 AIChE Spring Meeting
15TH GLOBAL CONGRESS ON PROCESS SAFETY
NEW ORLEANS • 2019

Major Explosion at a Chemical Company in China: the Causes and Lessons Learned

Wei Wang
China University of Petroleum (East China)
No. 66, West Changjiang Road, Huangdao District, Qingdao, China
15621097585@163.com

Yi Liu
China University of Petroleum (East China)
No. 66, West Changjiang Road, Huangdao District, Qingdao, China
liuyi@upc.edu.cn

Dongfeng Zhao
China University of Petroleum (East China)
No. 66, West Changjiang Road, Huangdao District, Qingdao, China
zhaodf@vip.sina.com

Yanchun Wang
China University of Petroleum (East China)
No. 66, West Changjiang Road, Huangdao District, Qingdao, China
Wychun25@163.com

Jian Chen
China University of Petroleum (East China)
No. 66, West Changjiang Road, Huangdao District, Qingdao, China
836771528@qq.com

Hanhai Dong
China University of Petroleum (East China)
No. 66, West Changjiang Road, Huangdao District, Qingdao, China
donghanhaiupc@126.com

Prepared for Presentation at
American Institute of Chemical Engineers
2019 Spring Meeting and 15th Global Congress on Process Safety
New Orleans, LA
March 31 – April 3, 2019

Major Explosion at a Chemical Company in China: the Causes and Lessons Learned

Wei Wang
China University of Petroleum (East China)
No. 66, West Changjiang Road, Huangdao District, Qingdao, China
15621097585@163.com

Yi Liu
China University of Petroleum (East China)

Dongfeng Zhao
China University of Petroleum (East China)

Jian Chen
China University of Petroleum (East China)

Yanchun Wang
China University of Petroleum (East China)

Hanhai Dong
China University of Petroleum (East China)

Keywords: Fishbone Diagram; Analytic Hierarchy Process; Process Safety Management; Incident Investigation

Abstract

On July 12, 2018, a devastating explosion incident occurred in Yibin City Hengda Technology Co., Ltd. in Sichuan Province, resulting in 19 deaths and 12 injuries, and impacting an area of about 1,800 square meters. This was one of the most disastrous explosions that occurred in a chemical plant in China in the past five years. The incident reflects the general deficiencies of process safety management in small and medium-scale companies. Therefore, it is significant to analyze the causes of the incident and put forward countermeasures to avoid such kind of incidents.

According to the characteristics of complexity, diversity, and uncertainty about the influencing factors of the explosion, this paper presented a systematic analysis from qualitative and quantitative perspectives. The fishbone diagram model was used to identify influence factors from six aspects of people, design, management, environment, safety prevention, and materials. Furthermore, an assessment index including 26 factors was determined. The analytic hierarchy process method was used to analyze the importance degree of each factor quantitatively, and the key factors leading to the incident were obtained. The results showed that the sum of weights of three main aspects, including people, design, and safety prevention is nearly80%, so they were the dominant causes leading to this incident. Lack of chemical industry experience, non-standard process design, and incomplete automatic control system were the focus of prevention among 26 factors.

Process safety management is the application of management systems to the identification, understanding, and control of process hazards to prevent incidents from occurring. This analysis considered process hazard analysis, pre-startup safety review, operation procedures, training, and emergency plan and response, which were the main elements in process safety management framework. Finally, we proposed safety precautions for the identification, storage, production, and use of hazardous chemicals, as well as recommendations for process equipment management.

1 Introduction to the event

At 18:42:33 on July 12, 2018, an explosion occurred in Yibin Hengda Technology Co., Ltd. in the industrial park of Jiang'an County, Sichuan Province, causing 19 fatalities and injuring 12 people. The explosion caused the second and third workshops to catch fire, and the combustion area is $1800m^2$ [1]. The first workshop, warehouse, drying room, analysis room, office building, and other buildings had been damaged to some extent. The fire department dispatched 42 fire engines and more than 200 firefights. It took more than two hours to extinguish the fire on the spot as shown in Figure 1. It was estimated that the total energy of the explosion in this incident was equivalent to 235 kg of TNT. The direct economic loss was 41.42 million yuan.

At the time of the incident, the company was producing imazamox and triazole. Triazole is a pesticide, and imazamox is an herbicide. The process of the incident was the dehydration of butyramide in the process of synthesis of imazamox. At about 11:30 on July 12, the company collected 2 tons of chemicals labeled as raw materials (actually the explosive chemical sodium chlorate, not the raw material required for the production of butyramide). At around 14:00, the deputy director of the company's second workshop received raw materials from the warehouse. At around 15:30, the workers in the second workshop raised the raw materials for production through the lift to the third floor of the second workshop. At around 16:00, the 2R301 kettle for dehydration of butyramide was completed, and the feed was in an empty state. Before 17:20, the 2R301 kettle was completed. At about 6 pm, the dehydration completed. At 18:42:33, it was due to the on-site handover time. A chemical explosion occurred in the 2R301 kettle on the third floor of the second workshop. The direct cause of the incident was that the operator replaced the unlabeled sodium chlorate with butyramide and put it into the 2R301 kettle for dehydration, which caused the explosion and fire. Sodium chlorate is a kind of explosive chemical, which is easy to form an explosive mixture when mixed with butyramide. The explosion led to the disintegration of 2R301 kettle. The high-temperature toluene vapor rushed out rapidly formed an explosive mixture with the external air and produced a secondary explosion, which also caused the explosion and combustion of the materials such as sodium chlorate, toluene, and methanol stored at the workshop site.

Figure 1. Extinguishing at the scene.

According to the investigation, in this incident, the second workshop was the core area of the explosion, and the overall structure was severely damaged as shown in Figure 2. The platform on the third floor collapsed. The reactor with a height of nearly two meters and a diameter of about one meter fell from the upper level.

Figure 2. Severely damage to the second workshop.

2 Incident evaluation index system based on the fishbone diagram

Fishbone Diagram is also named as cause and branches figure analysis. Kaoru Ishikawa, a Japanese management master, invented it for finding out the root cause of incident [2]. It indicates the relationship between a problem and its underlying causes. It is a qualitative analysis tool to help identify incident causes [3]. With step by step in-depth study to analyze its influencing factors,

Fishbone Diagrams can systematize complicated incident causes. When starting the specific report, we first analyze the factors which affect the safety most and then followed by departure from the overall reason to find the basic reason and to detect and determine the main reason [4,5,6].

After the incident, the government set up a team of experts to identify risk factors from six aspects: human, design, management, environment, safety prevention, and material and establish an incident evaluation index system with 26 elements. The fishbone diagram that induced the explosion is shown in Figure 3.

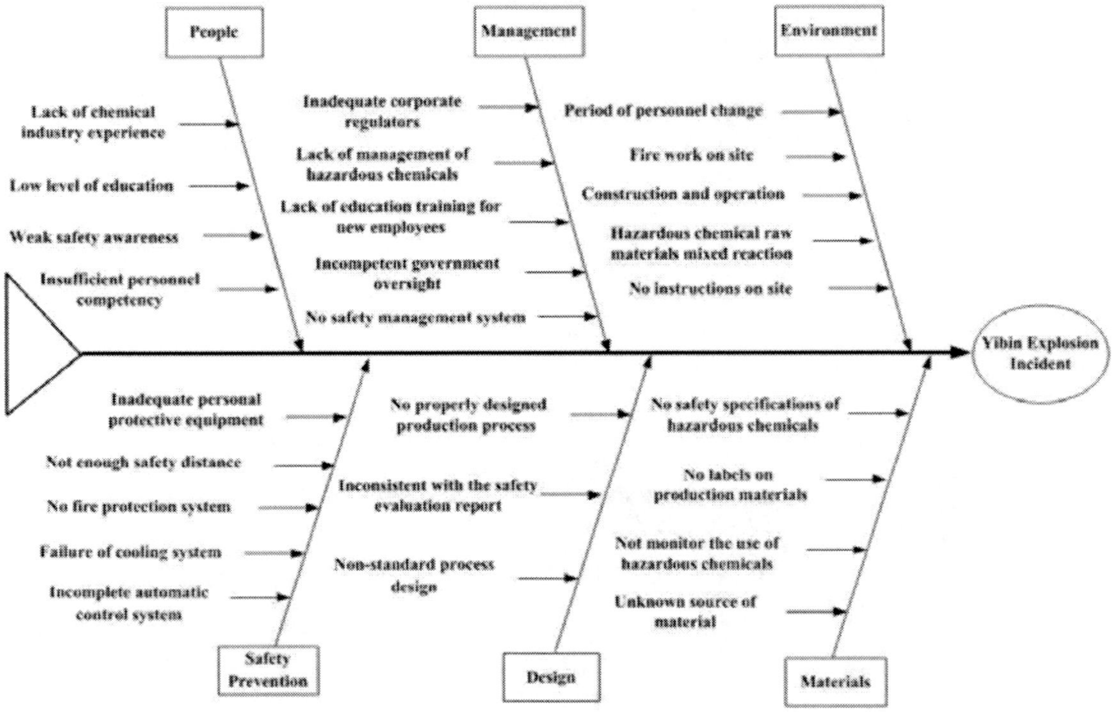

Figure 3. Fishbone diagram of Yibin explosion incident [7].

2.1 People

Neither the safety managers nor the operators demonstrated proficiency in the requirements of process safety competency. There were two main aspects below:

- Corporate management - The general manager of the enterprise had responsibility for the mechanical processing and production contract. However, he did not have appropriate experience in the chemical industry. The technical director did not have experience with management of process safety and production systems. The corporate legal person responsible for finance and logistics also had no experience in chemical engineering.

- Operators - Among the 19 people who died in the incident, 16 were operators, and most of these operators were local farmers. They lacked the basic knowledge of chemical process safety production, and they did not understand the process safety risks in the production process.

2.2 Design

The second workshop where the incident occurred was not constructed and installed strictly by the design drawings, and it was originally planned to have nine sets of 5000L tanks and one set of 3000L tanks and produce 5-nitroisophthalic acid. As built, the workshop had three sets of 5000L tanks, seven sets of 3000L tanks, and eight sets of 2000L tanks were set up. The product was also different from the original, but imazamox and triazole. A similar situation existed in the third workshop. The production technology of imazamox and triazole used by the company had no formal technical source, and the enterprise also did not entrust professional institutions for process calculation and construction drawing design. The technical director of the enterprise only planned the required production equipment and operation process according to his experience, but he knew nothing about the process safety risks of related technologies. It was during the production of imazamox that the distillation kettle used for the dehydration of butyramide by toluene first exploded. According to the *Notice on Further Strengthening the Safety Design Management of Hazardous Chemicals Construction Project*, Hengda company needed to start construction in April 2018, but the company had already started construction in July 2017. Also, the fire separation design of each building did not meet the requirements of the *Petrochemical Enterprise Design Fire Code (GB50160).*

2.3 Management

The lack of process safety culture is the main cause of management failure. Safety culture is developed through the interactions of factors such as safety policies and safety goals, the discipline of the organization, the behavior and attitude of the company staff, and the competence of workers. Improper operating discipline and procedures can lead to a negative process safety culture and an increased likelihood of process safety incidents. The enterprise had not formulated the post safety responsibility system, safety management rules, and regulations. Furthermore, the enterprise manager and the person in charge of each department only had the oral authorization, and there was no formal appointment document to clarify the post responsibility. The safety management organization was not complete. The company's safety and environmental protection department had only one safety officer who did not have safety management capabilities. There were also no full-time or part-time safety officers in each workshop. There was a lack of management of hazardous chemicals in the process safety management. The technology provider concealed some chemical name marks for confidentiality, and there was no safety specification. Required knowledge and competence connected with safety for workers were the pivotal content of safety training. New employees did not have security education directly work or internship. The government had not strictly regulated the safety departments of local key hazardous chemicals companies.

2.4 Environment

The factory built the workshop while producing, and used the polypropylene pipe which was easy to produce static electricity to transport flammable and explosive liquid materials. The building materials, hardware, and chemicals in the warehouse were mixed and piled with hidden dangers. The incident occurred during shift time, resulting in a large number of casualties. In the on-site operation workshop, there was neither operation instructions nor warning signs for hazardous chemicals. At the time of the incident, the second workshop produced two products at the same time. This practice of producing multiple produced at the same time in the same workshop has excellent security risks.

2.5 Safety prevention

Compared with the 20 elements of CCPS, China drew on the process safety management experience of the Occupational Safety and Health Administration (OSHA) and further optimized the formation of 12 elements for the production of Chinese chemical companies. After that, a series of process safety technologies were developed and promoted for the key elements of risk management and equipment integrity. In 2013, the SAWS issued *Guidance for enhancing chemical process safety management*. The Ministry of Emergency Management, which was integrated by the State Administration of Work Safety(SAWS) in 2018, attaches great importance to the supervision of hazardous chemicals, hazardous chemical processes, and major hazardous sources, referred to as "Two Priorities and One Major." If an enterprise violates The relevant provisions, it shall be punished according to the upper limit of economic punishment according to law.

In the production process of imazamox and triazole, the nitrification, oxidation, diazotization, hydrazine hydrate, hydrogen peroxide, methanol, and other dangerous chemicals and processes involved were all "Two Priorities and One Major" projects, which needed strict supervision. However, the company's automatic control system, flammable and toxic gas alarm system and fire water system and other safety facilities were not installed. Once there was a danger, the enterprise itself could not deal with an emergency. The lack of an automated control system led to more than ten operators performing manual operations around the reactor, which was an important cause of major casualties.

2.6 Materials

According to the investigation, the safety evaluation report made by the enterprise before the incident did not fully consider the process safety information, such as the lack of description of dangerous chemicals and the risk identification of the process, so the enterprise had great potential risks. Workers didn't know the source of raw materials, and many chemical production raw materials were not marked. Workers knew nothing about the risks of these dangerous chemicals, and company officials failed to communicate information about the hazards of the process to employees. The workers compiled codes to replace the names of products and raw materials, and these numbers were not abbreviated as specifications. The raw materials used by Hengda Company, such as hydrazine hydrate, nitric acid, toluene, hydrochloric acid were dangerous chemicals or precursor chemicals, but the company still purchased them illegally without relevant licenses. In the production process, regulators did not supervise the management of hazardous chemicals.

3 Determination of the weights of the incident incentive index based on the analytic hierarchy process

The Analytical Hierarchy Process (AHP) is a decision-making method proposed and used by American operations researcher T.L.Saaty in the mid-1970s [8]. AHP is a process that breaks down a problem in a hierarchical matrix with increasing level of detail to determine the most significant causes of an incident. According to the obtained fishbone diagram of the Yibin explosion incident, an incremental layer structure for analytic hierarchy analysis was constructed, as shown in Fig. 4, where the A, B, and C layers illustrated the primary effect and hierarchy of causes.

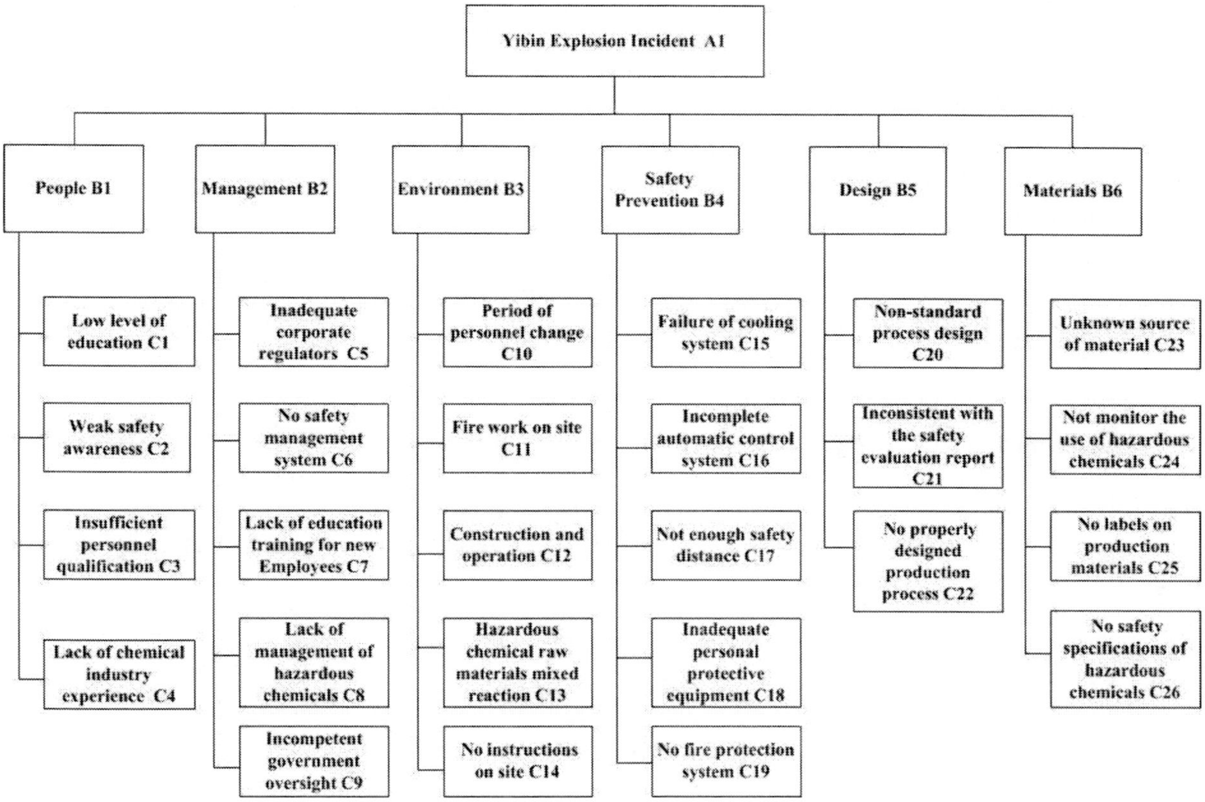

Figure 4. Hierarchical structures of safety evaluation system for Yibin explosion incident.

3.1 The single-layer index weights of the influencing factors

According to the actual data collected and expert opinions, we used 1-9 scale to construct the judgment matrix (A) of the first-level evaluation index, as shown in table 1, and obtained the influence of various factors in layer B on layer A. λ_{max} is the maximum eigenvalue, CI is the consistency index, RI is the average random consistency index, and CR is the random consistency ratio; When CR<0.1, it passed the consistency test [9]. According to table 1, the weights of design is the largest, followed by people and safety prevention, with the sum of the weights of the three as high as 0.8. It can be seen that the leading cause of this incident is due to the low level of process design, operators' insufficient qualification of chemical safety production and non-compliant protection measures. Therefore, enterprises should pay attention to intrinsic safety design, improve the safety awareness of staff by popularizing safety knowledge and operational guidelines, and finally achieve the goal of safe production.

Table 1. Judgment matrix of primary evaluation indicators and weights results

A	B1	B2	B3	B4	B5	B6	Weights	Consistency	Ranking
B1	1	2	3	1/2	1/3	7	0.1645	λ_{max}=6.4698 CI=0.094 RI=1.26 CR=0.0746<0.10 Satisfy consistency	3
B2	1/2	1	2	1/3	1/4	6	0.1060		4
B3	1/3	1/2	1	1/4	1/5	5	0.0700		5
B4	2	3	4	1	1/2	8	0.2545		2
B5	3	4	5	2	1	9	0.3810		1
B6	1/7	1/6	1/5	1/8	1/9	1	0.0239		6

3.2 The comprehensive weights of the influencing factors

We built each judgment matrix of layer C to layer B and calculated the single layer weights of layer C, and finally, we obtained the comprehensive weights of incidents caused by bottom influencing factors. We considered the product of these two layers as the comprehensive weights of the sub-factor. For example, the comprehensive weights of C1 is equal to the single-layer weights of C1 to B1 multiplied by the single-layer weights of B1 to A. The comprehensive weights result and ranking of each sub-factor were shown in table 2.

Table 2. Single and synthetic weights of C layer indexes

		Single weights	Consistency	Synthetic	Ranking
B1	C1	0.0843	λ_{max}=4.1375 CI=0.0458 RI=0.9 CR=0.0515<0.1 Satisfy consistency	0.0139	16
	C2	0.2120		0.0349	8
	C3	0.1544		0.0254	12
	C4	0.5492		0.0904	3
B2	C5	0.1556	λ_{max}=5.2550 CI=0.0638 RI=1.12 CR=0.0569<0.1	0.0165	15
	C6	0.0566		0.0060	21

	C7	0.4735	Satisfy consistency	0.0502	5
	C8	0.2578		0.0273	11
	C9	0.0566		0.0060	22
B3	C10	0.6253	λmax=5.3598 CI=0.08995 RI=1.12 CR=0.0803<0.10 Satisfy consistency	0.0438	7
	C11	0.0901		0.0063	19
	C12	0.0435		0.0030	23
	C13	0.1976		0.0138	17
	C14	0.0435		0.0030	24
B4	C15	0.0781	λ_{max}=5.1717 CI=0.042925 RI=1.12 CR=0.0383<0.1 Satisfy consistency	0.0199	13
	C16	0.5895		0.1501	2
	C17	0.1271		0.0324	9
	C18	0.0781		0.0199	14
	C19	0.1271		0.0324	10
B5	C20	0.6833	λ_{max}=3.0246 CI=0.0123 RI=0.58 CR=0.0236<0.1 Satisfy consistency	0.2604	1
	C21	0.1168		0.0445	6
	C22	0.1998		0.0761	4
B6	C23	0.1242	λ_{max}=4.1891 CI=0.0630 RI=0.9 CR=0.0708<0.1 Satisfy consistency	0.0030	25
	C24	0.2626		0.0063	20
	C25	0.5465		0.0131	18
	C26	0.0667		0.0016	26

As can be seen from table 2, the full weights of C16 (Incomplete automatic control system) and C20 (Non-standard process design) was greater than 0.1. The full weights of C4 (Lack of chemical industry experience) and C22 (No properly designed production process) were very close to 0.1. The combined weights of these four factors reached 0.577, which was more than half of the total. Therefore, these four influencing factors were the primary causes of the explosion incident in Yibin, Sichuan, as strict control and prevention were given priority.

Among them, both C20 and C22 belong to the design aspect, and enterprises should entrust qualified formal companies to design the process and production plan. In the method of safety design, the operation and design should be analyzed in real time to check the status of dangerous materials, effectively reduce the operating temperature and pressure, and control the influence of hazards factors in the process on chemical equipment. The equipment used in production shall meet the strength and anti-corrosion performance required by the process, and shall be designed and manufactured by qualified factories.C4 is a human problem. Therefore, it is necessary to conduct chemical safety education and operation training for operators before entering the plant. Many of Hengda's workers were the relatives and friends of management personnel. They were generally poorly educated and skilled in operation, and had not received any training in chemical safety. The workers did not go through systematic learning, but the workshop director directly taught them operating skills. C16 is a problem of safety prevention. The automatic control system can not only improve the operation efficiency of production equipment but more importantly, it can protect the safety of chemical personnel and reduce the loss caused by incidents. Enterprises should follow the safety production regulations, install automatic control systems, toxic and harmful gas alarm systems, and fire protection systems to effectively prevent incidents.

The weights of C7 (Lack of education training for new employees), C10 (Period of personnel change) and C21 (Inconsistent with the safety evaluation report) were all around 0.05. The weights of C2 (Weak safety awareness), C19 (No fire protection system) and C17 (Not enough safety distance) also exceeded 0.03. These five factors were important causes of incidents and should be effectively controlled.

C7 is a new employee training management problem. Enterprises should pay attention to the educational training of employees before entering the plant and implement the three-level safety training education of the plant, workshop, and team. C10 involves change management. At the time of the incident, it was during the shift time of employees, the change of personnel and the change of materials and equipment, which had great potential risks. Enterprises should establish a change management system, conduct comprehensive training for the staff who play a vital role in the change management process and clarify the position and responsibility of each person in the change system.

4 Lessons learned from the causes of the incident

The Yibin explosion was a pharmaceutical chemical incident. Through the investigation and analysis of historical disasters in the chemical industry in China, it was found that the proportion of pharmaceutical chemical incidents in the chemical industry was huge [10]. It was characterized by a random change of chemical process, small scale, low level of process control, involving dangerous materials and complex chemical reaction. Especially in small and medium-sized chemical enterprises, the lack of process safety management is an essential cause of such incidents. From the perspective of process safety management, this paper summarized and put forward suggestions for preventing such incidents.

- Process Hazard Analysis - In the design stage, the risk identification and assessment tool system should be applied to identify the risks of materials in the process of technology. In the review stage, HAZOP, SIL are used to review the process, layout and material flow. Companies should use scientific models to calculate the explosion-proof grade and safety distance of buildings. By adding the explosion-proof wall and the explosion-proof film, the building close to the installation area can reach the required safety factor. After the device is put into operation, it needs to be compared with the hazard analysis in the design stage, and the hazard that is inconsistent with the design should be analyzed again. Hengda's actual operating equipment, raw materials, and products were inconsistent with the original plan, and the company had not discovered and controlled the real hazards [11].

- Pre-startup Safety Review - The company shall organize a safety inspection team, whose task is to confirm whether the improvement measures and safety guarantee measures in the process hazard analysis report have been implemented as required; Whether adequate training of personnel in operation, maintenance, and management of process equipment has been completed; Whether necessary safety information of process equipment is completed and updated; Whether necessary equipment testing and checking have been completed and confirmed; Whether the process equipment is constructed and installed according to the design requirements [12].

- Operation Procedures - The incident occurred during the change management phase, and there was no review or other risk assessment during the shift. Staff should review the operating procedures as often as possible to ensure that the operating procedures reflect current operating conditions, including changes in chemicals, process equipment, and facilities. The enterprise shall confirm the adaptability and effectiveness of the operating procedures and ensure that the operators understand and master the operating systems [13].

- Training - A complete process safety management is inseparable from the participation of operators, and the human factor is the most unstable factor in process safety management. Therefore, Enterprises should establish strict production operation disciplines and require operators to master process safety information, equipment integrity management points, and emergency response policies and finally improve workers' ability to control equipment safety operations [14].

- Emergency Plan and Response - Enterprises need to establish a set of overall emergency plans, which usually stipulate in written form how the factory should deal with abnormal or urgent situations. Also, enterprises should establish emergency response teams, traditionally composed of enterprise personnel, but also include external staff; The responsibilities of each team member should be clear, ensuring that members have no questions about accountability and delegation. In case of an emergency, the person in charge may determine the safe area according to the emergency response manual and direct the person to evacuate to a safe place [15].

5 Conclusion

The circumstances of the incident described in this article are typical in China, and the incident reflects the common safety problems in small and medium-sized chemical enterprises. Managers should understand these lessons we learned from this incident. By using fish-bone diagram method, the causes of the explosion incident were obtained through qualitative analysis, including six aspects, such as people, design, management, environment, safety prevention and materials, and 26 factors influencing the incident were further analyzed. Then the analytic hierarchy process was used for quantitative analysis, and the results showed that lack of chemical industry experience, non-standard process design, and incomplete automatic control system were the main factors affecting the incident, so the enterprise should strictly control and prevent these factors. Finally, we summarized the causes of such incidents and put forward the measures from five aspects of process safety management, including process hazard analysis, pre-production safety inspection, operation procedures, training, and emergency plan and response plan. An incident brings not an only economic loss but also bad social influence to the factory. It is our hope that this paper will help prevent future accidents.

6 References

[1] Sichuan Provincial Emergency Management Office, Yibin Hengda Technology Co., LTD. "7.12" Major Explosion Fire Accident Investigation Report, Available at http://www.chinasafety.gov.cn/gk/jsts/201902/t20190214_224397.shtml, Accessed on February 14, 2019.

[2] Tongyuan Luo, Chao Wu and Lixiang Duan, Fishbone diagram and risk matrix analysis method and its application in safety assessment of natural gas spherical tank, Ind. Eng. J. 25(02), 2007, 138-141.

[3] Xing, W., QC Team Based Materials, China Society Press, Beijing, China, 2004.

[4] T Deng, X., Chen, Q. and Jiang, D., Application of fishbone chart analysis method in prevention for steam turbine overspeed incident, Guangdong Electr. Power, 20(02), 2013, 73-77.

[5] Changlong Zhu, Jiayong Zhu, Li Wang and M. Sam Mannan. Lessons learned from analyzing a VCE incident at a chemical plant, Journal of Loss Prevention in the Process Industries,50,2017, 397-402.

[6] Dennis P. Nolan, Chapter 5: Specialized Reviews—CHAZOP, EHAZOP, Bow-Tie Analysis, Layers of Protection Analysis, Safety Integrity Level, FishboneDiagram, and Cyber Security Vulnerability Analysis, Safety and Security Review for the Process Industries (Fourth Edition), Gulf Publishing Company,2015,Pages 17-27.

[7] Xinhua News Agency, Emergency management department: 7·12 incident enterprises have no law and discipline, and government supervision is lax, Available at http://www.gov.cn/xinwen/2018-07/17/content_5307196.htm?_zbs_baidu_bk, Accessed on July 17, 2018.

[8] Seong KonLee, GentoMogi and Jong WookKim. Decision support for prioritizing energy technologies against high oil prices: A fuzzy analytic hierarchy process approach, Journal of Loss Prevention in the Process Industries,22(6), 2009, 915-920.

[9] Saaty T.L., A scaling method for priorities in hierarchical structures, Journal of Mathematical Psychology, 15 (3),1977, 234-281.

[10] Heda Zhang and Xiaoping Zheng, Characteristics of hazardous chemical incidents in China: A statistical investigation, Journal of Loss Prevention in the Process Industries, 25(4), 2012, 686-693.

[11] Hao WU, Igor PEÑARRUBIA, Lin CUI and Jinsong ZHAO, Process safety management considerations for biofuel production, Higher Education Press, 4(3), 2017, 357-367.

[12] Yuan Gao, Yunxiao Fan, Jing Wang, Xi Li and Jingjing Pei, The mediating role of safety management practices in process safety culture in the Chinese oil industry, Journal of Loss Prevention in the Process Industries, 57, 2019, 223-230.

[13] Paul R.Amyotte and Cathleen S.Lupien, Chapter Three - Elements of Process Safety Management, Methods in Chemical Process Safety, 1, 2017,87-148.

[14] T. Kontogiannis, M.C. Leva and N. Balfe, Total Safety Management: Principles, processes and methods, Safety Science, 100(B), 2017, 128-142.

[15] Hao WU, Igor PEÑARRUBIA, Lin CUI and Jinsong ZHAO, Process safety management considerations for biofuel production, Front. Eng. Manag, 4(3), 2017, 357- 367.

Taming the Wild River Rapids – How Process Safety Can Apply Outside the Process Industries

Trish Kerin
IChemE Safety Centre
Level 7, 455 Bourke St, Melbourne, Victoria, Australia
tkerin@icheme.org

Prepared for Presentation at
American Institute of Chemical Engineers
2019 Spring Meeting and 15th Global Congress on Process Safety
New Orleans, LA
March 31 – April 3, 2019

Taming the Wild River Rapids – How Process Safety Can Apply Outside the Process Industries

Trish Kerin
IChemE Safety Centre
Level 7, 455 Bourke St, Melbourne, Victoria, Australia
tkerin@icheme.org

Keywords: lessons learned, process safety, design, maintenance

Abstract

On the 25th October 2016 the Thunder River Rapids ride at Dreamworld on Australia's Gold Coast suffered a pump failure, resulting in the tragic deaths of four riders. This incident was investigated by the police and later a coronial inquest to determine how a tragedy like this could occur. As a process safety professional, it was interesting to follow the coronial inquest findings and then think about how process safety management may have helped to prevent this incident. Assuming that a process safety event can be defined as a high consequence or potential consequence low likelihood loss of control event [1], this incident had all the hallmarks. Therefore, it is possible to apply the IChemE Safety Centre (ISC) framework [2] for process safety to identify where the system failed and understand how it could be averted in future. This paper will explore various elements of the incident and how process safety management techniques may have helped, including that the ride had been modified to be different from all other similar rides, that the staff did not know how to stop the ride in an emergency and that all means to stop the ride required manual detection and intervention, with no alarms or shut down systems installed.

1 The Incident

The Thunder River Rapids ride at Dreamworld in Queensland, Australia, was a water ride, simulating white water rafting. Up to six riders sit in a circular raft, which flows along a turbulent river, including travelling up a conveyor to ride down a waterfall. The water flow is driven by pumps and made turbulent by the installation of underwater obstacles. It was in full operation on 25th October 2016, when the pump managing the water flow tripped out. This resulted in the water level on the ride dropping. This drop in water level caused a raft to become stuck on the rails at the conveyor leading to the disembarkation platform. This raft remained stationary for some 57 seconds while the ride continued to run before it was impacted by the next raft in the sequence. This following raft contained 6 passengers, four adults and two children. When this raft struck the stationary one, it flipped over, killing the four adult passengers. The two children managed to escape the raft [3]. This incident is currently under a Coronial Inquest. As such this paper is written from several media outlets that have been reporting the proceedings and based on the evidence reported, not the findings of the Coroner. The Coronial Report is due to be published in mid 2019.

2 History of the Ride

The Thunder River Rapids ride was completed in 1986 and was likely an in-house design and construction [4]. Sometime after construction, the ride was modified to remove some of the timber slats that form the base of the conveyor [4]. Two in every three slats were removed, though is unclear why this was done. This did create significant gaps in the base of the conveyor belt. The ride was also refurbished some months prior to the incident. [5].

Over the history of the ride there had been a number of incidents that had occurred with the rafts and the pumps. This in itself is not unusual, in that pumps can trip out from time to time, however there had been an incident in 2001 where the pump had tripped, and a raft became stuck [6]. This was then struck by a second raft, which flipped over. In this incident there were no passengers in the raft, and therefore no injuries. The investigation found that the ride operator had failed to adequately monitor the ride and he was dismissed. Details of this incident were not shared broadly and the operators and maintainers of the ride in 2017 had no knowledge of it. At the time an email was written by a park engineer discussing the incident and stated "I shudder when I think if there had been guests on the ride" [3]. This demonstrates that someone at some time recognized the significance of this incident, but it was then lost in history.

3 Technical and Organizational Causes of the Incident

There is one key immediate cause of this incident, however there are many different and wide-ranging causal factors. The immediate cause has been determined to be the tripping of the water pump, which in turn allowed the water level to drop, causing the first raft to become stuck. This provided the obstacle that the second raft, with six passengers, struck, causing it to then flip over. But let us examine the broader aspects of this incident, from the pump reliability to the safety systems and organizational capability. The ISC framework [2] provides a useful template to examine the causes. This framework in represented by Figure 1. Fundamentally leadership underpins all aspects of process safety. There are six functional elements that must be robust in an organization to manage major hazard safety. These are; knowledge & competence; engineering & design; systems & procedures; assurance; human factors and culture. In this incident there were clear failures in each of these elements.

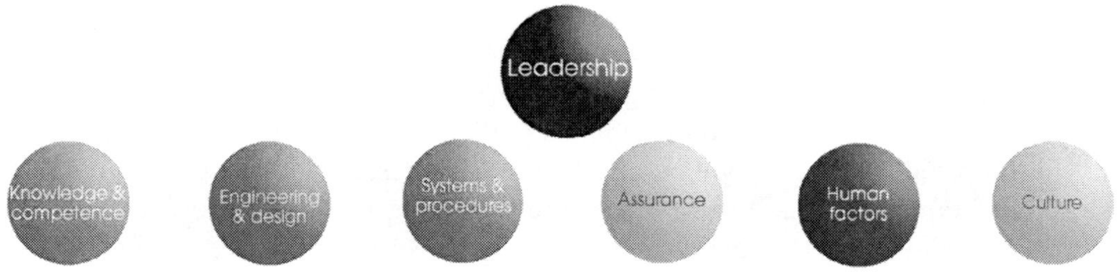

Figure 1. ISC Framework for Process Safety

3.1 Knowledge & Competence

The training of the ride operators took only 90 minutes and was unstructured with no evidence of competence consistently recorded [7]. There was a procedure manual however it was thought to be confusing by the ride operators [8]. The operator running the ride on 25th October 2016 had been trained for 90 minutes that morning, as it was her first day running the Thunder River Rapids. The junior operator stated that the trainer "showed me the front page of the manual and just flipped through it. I had to go through it myself. She said just sign off on it when you are happy." [9] The trainer stated that she sometimes had to "invent" training techniques because of the complexity of the control panel [8]. There was no clear guidance on how and when to use the emergency stop for the ride. The new operator testified that she was told "don't worry about it, you don't need to use it" with regard to the emergency stop button [10]. This is disputed by the person who delivered the training. The new operator believed it was the role of the senior operator to shut the ride down. It should be noted that the emergency stop would shut down the moving equipment with in two seconds, where the normal shutdown took around eight seconds. The ride operator that day stated that she "would have waited for the senior operator to tell her to hit the e stop" [10]. The senior operator that day was "surprised" to learn there was an emergency stop that would have shut the ride down within two seconds [7].

3.2 Engineering & Design

Intamin Amusement Rides designed and built the first river rapids rides in the world, and following this tragedy released a statement stating that the ride was "neither built by Intamin nor supplied by Intamin". The ride at Dreamworld was a custom build, and this was said to be common for rides built in the 1980's in the Gold Coast[4].

3.2.1 The Timber Slats

A few years after the ride was opened, a modification was made to the timber slats in the conveyor system [4]. The slats appear to serve at least two safety purposes. The first was to provide a solid platform for a passenger if they fell from the raft, essentially isolating them for the moving machinery and providing a platform to stand on [11]. The second was so that the rafts did not get caught in gaps and become stuck or indeed flip over. The Thunder River Rapids ride had two of every three slats removed. This provided large gaps between the remaining slats. There does not appear to be a theme park anywhere else that has made this change. There also appears to be no evidence as to why the change was made, nor was there a considered Management of Change (MoC) review undertaken. It is not clear if the organization had a structured and systematic MoC process. The safety regulator did not approve the modification and believed it added in additional hazards [4, 12]. In this incident, the raft became lodged in the gap, and as the conveyor continued to move forward, the raft was driven up vertically. When the passengers then fell from the raft, there was no solid base and they landed in the mechanism of the conveyor. There had also been conveyor framework modifications made to the ride in early 2016 in an attempt to stop rafts flipping over [13]. It is unclear what this modification was or indeed why it was made.

Figure 2. Aerial view of Thunder River Rapids showing the large gaps in the timber slat conveyor after the incident [10].

Figure 3. Aerial view of Infinity Falls, Orlando, showing close spacing of the boards on the conveyor [14].

3.2.2 The Pump

The water pumps had failed three times in the days leading up to the incident, and then twice on 25th October 2016, before the final failure that lead to the tragedy [15]. The pump's purpose was to keep the water flowing through the ride to keep the rafts moving. If the pumps stopped the water not only stopped flowing, it also drained back, and the water level dropped such that rafts could become stuck on the underwater obstacles that produced the rapids [16]. The reason for the pumps tripping out was not determined, the protocol seemed to be to just reset them to get the ride operating again [3]. There also is some confusion as to whether an alarm was sounded when the pumps tripped out, with an old procedure having the comment about an alarm crossed out [17]. At the time of the incident the pumps were approximately 10 years old, and the manufacturer stated that they had a life expectancy of about a decade. There did not appear to be any plans to replace the pumps, however there was a maintenance request into the pump supplier in the week prior, but this visit did not eventuate [18].

3.2.3 Low Water

Low water in the ride was a critical hazard, as the rafts could become stuck or indeed flip. There had been previous incidents where low water level hazard led to rafts flipping, notably in 2001. There was no low water level switch that would shut down the ride if the water level became too low [3]. The only indication the operators had was to notice the water stain level on the wall of the ride canal to determine if the water was too low [19]. When the water level dropped on 25th October, one empty raft became stuck on the rails near the base of the conveyor. As the second raft containing four adults and two children engaged with the conveyor it was then driven into the stuck raft. A safety review conducted after the event stated that the primary cause was a lack of a safety rated water level detection system. It was believed that such a device could have been fitted for as little as AUD 3,000. The ride had been refurbished some seven months prior to the incident, but a device was not fitted. The nearby log ride at the park had such a safety device fitted [18].

3.2.4 Emergency Stop

Around two years prior to the incident an external safety report was written suggesting the emergency stop system should be simplified, however this was rejected by the company as the emergency system in place was seen to be an "acceptable risk". The stopping procedure for the ride contained several individual steps. There was a single button which would stop all components of the ride, however this was located at the unloading area and not the control panel where the operators worked. [20]

There was a great amount of confusion regarding the emergency stop button and what it actually did. The senior operator stated that they were "surprised" to learn there was an emergency stop that would stop all components and the junior operator on her first day was unsure what the button did [7]. She stated she was told to not worry about the button because she did not need to use it and testified, she "would have waited for my senior operation to tell me to press it". She also did not know the button would stop the conveyor from moving. [10] The maintenance workers who performed the daily checks on the ride also stated they did not know what the emergency stop button did as they never tested it [21].

3.2.5 Maintenance

Seven months prior to the incident a directive to cut all repairs and maintenance spending was ordered by the owners. At the time of the incident the facility was approximately AUD125,000 over maintenance budget for the year to date. Capital expenditure could continue to be spent during this time [16]. One maintainer stated that he performed comprehensive daily, monthly and annual checks on the ride, which include consumable replacements such as nuts and bolts [22]. When the regulator inspected the ride after the incident, he noted it was in "generally poor condition" with excessive corrosion and rust on the electrical cables [23].

The maintenance planner noted that if spending could be deferred it would be. He also noted that the ride was shut annually for review, but on the nine occasions he witnessed a shutdown, he never saw an engineer inspect the ride nor were any risk assessment carried out [24].

3.2.6 Seatbelts

The Thunder River Rapids rafts had seat belts fitted for the safety of the riders. It should be noted that in May 2017, some seven months after the Dreamworld incident, that an 11 year old girl died on a similar ride, Splash Canyon, at Drayton Manor in the United Kingdom. The Splash Canyon ride rafts had no seatbelts installed at all, and as the child stood to swap seats with a fellow rider she fell overboard and struck her head. She died days later in hospital. This was not the first time a rider had fallen from the Splash Canyon ride [25]. This incident highlights the importance of proper restraints on these types of rides.

The seatbelts provided on the Thunder River Rapids rafts consisted of Velcro straps. It was vital that these were engaged correctly to provide the necessary restraint. Unfortunately, the Velcro was worn, and this made correct engagement more difficult. It is thought that at least one of the four riders may have survived the incident if the rafts had a better quality seatbelt, such as a mechanical buckle type that are on aircraft. This is because as the raft was shaken by the mechanism the belt failed to hold and the rider dropped to the conveyor system below [23, 26].

3.3 Systems & Procedures

Dreamworld has a series of operating procedures for the Thunder River Rapids ride however they were thought to be confusing [8]. During the investigation into the 2001 incident where empty rafts flipped, it was determined that the operator failed to follow the startup and emergency procedures. The investigation also reviewed the operating procedures and determined that the chance of a similar incident occurring with passengers on the rafts "is nil" [6]. A senior engineer believed that the cause of the 2016 incident was the operator not following procedures and that if procedures were followed the "system is a safe system" [27, 28].

There was also a lack of adequate emergency response procedures. Operators had discussed wanting better procedures and undertaking emergency drills, however, even two years after the incident, these have not been conducted at all [29].

3.4 Assurance

Over the history of the ride there had been a number of third-party reviews, inspections or audits carried out. Three years prior the incident a safety audit was conducted, and it recommended replacing the complicated emergency shut down process with a single emergency stop button, but this recommendation was never actioned [30].

There was also a legal requirement for the rides to have a "certificate to operate" and the Thunder River Rapids ride certificate was due to expire in January 2016. The company sought at extension as they were unable to find a "competent person" as defined under the legislation to inspect the ride [16]. Eight day before the incident the ride was inspected by a third-party engineer. He declared the ride as "structurally sound and properly functioning" even though he did not complete the full safety audit and merely did a visual inspection. He also did not inspect the log book as was required by law in a push to have the ride certified before it expired, though he admitted that he could have refused to certify the ride because of a "lack of documentation" [31, 32].

3.5 Human Factors

The control panel for the ride was confusing [3] and the operators needed to perform 36 different tasks in a one minute period, consistently [10]. This made for a system that encouraged human error, as it was not possible for the operators to be effective under these conditions. In addition to this, the engineering team believed the operators needed to simply follow procedures for the ride to be safe, not taking into account how difficult the procedures were or the actual workload. A senior engineer stated "In my view if procedures are followed the system is a safe system, you may as well not open the park. Procedures are there to be followed, if we don't follow procedures in everyday life, things happen." This blind focus on procedures does not take into account how the human interfaces with the machine and how the system advantages or disadvantages human performance.

3.6 Culture

A maintenance worker stated that he believed that the "safety culture at Dreamworld is very sound" [22] however this is in contrast to other workers who testified that familiarity had bred contempt with respect to the Thunder River Rapids ride [16], where new attractions were given more focus that existing ones [17]. Junior operators believed there was a culture of fear and had engaged with their union over their working conditions and the demands put on them regarding lack of rest breaks [33].

The culture also appeared to not learn from incidents, with the 2001 incident of flipping rafts said to be a one off and having "nil" chance of occurring with riders on board [6]. even though the consequence appeared to be understood at the time with an engineer writing in an email "I shudder when I think if there had been guests on the ride." [3].

The park relied on its apparent good safety history rather than challenging how and when an incident could occur [11]. They did not have an effective incident tracking system to help with the learning process. They introduced a database but there was "general reluctance to use what was a fairly complex database" by the operators [34].

Safety did not appear to be a value in the company culture. The former safety manager stated that "everyone thought it was someone else's responsibility". He believed that deficiencies in the corporate structure led to a lot of problems. In 2018 his corporate safety role was made redundant.

Around five months prior to the incident, a memo was issued to staff to 'remain vigilant when monitoring conveyor movements" for items that might interfere with the rafts. This suggests there was some understanding of hazards and consequences, but it does to appear to have driven any improvement activities or modifications to prevent an incident [24].

4 Ramifications

The full ramifications of this incident are not yet clear. The Queensland government has initiated some legislative improvements to manage theme parks, but there is likely to be more recommendations coming from the Coronial Inquest when it releases its findings. Aside from the tragic loss of four lives, from the Dreamworld perspective, their guest numbers have remained low since the incident occurred, and it is struggling to attract patrons. This has led to their parent company suffering financial losses for the past two years and 50% decreases in the share price over the two year period [35].

The incident also provided an interesting lesson in crisis management. The company was scheduled to hold their annual general meeting in the days following the incident. The meeting went ahead and in it the CEO was awarded a performance bonus of $840,000 including both shares and a cash component despite the four deaths days before [36]. From a public relations perspective this was not a successful event. The CEO also stated they had been in contact with the victim's families, but this was publicly disputed by the families [37]. This further damaged the company reputation at a difficult time.

This incident will not only have ramifications for the families of the deceased, but for the survivors and the staff who responded due to the nature of injuries sustained by the victims. Four former staff members are suing Dreamworld for psychological damage [38].

These ramifications show how a process safety type incident can have significant impact on a business and its ability to continue operations in a positive way.

5 Could the Principles of Process Safety Have Helped?

As had been shown, there were clear and systematic failures across the whole organization that led to this tragedy. Starting with knowledge and competence, which relates to organizational capability, had the operators been trained better in the safe operation of the ride as well as in emergency response, their performance that day may have been improved. From an engineering and design perspective, the design and installation of the ride was inadequate, it underwent undocumented changes which introduced hazards, and it was not adequately maintained. The systems and procedures for the operation and maintenance the ride was confusing and did not appear to contain necessary information. The assurance processes conducted did not seem to be thorough, rather they focused on keeping the ride operating and recommendations were not actioned. The ride was not designed with human factors in mind, from the high workload to the confusing operating panel. The organization culture did not seem to place a value on safety, it only focused on cost reduction. It also did not appear to be a learning culture, where lessons from the past are understood and embedded. Lastly at the time of crisis, the overall leadership displayed by the parent company was publicly found to be lacking.

There are transferrable learnings for the process industries from this incident, with some clear questions to ask. How are you managing your competency program to ensure you have the right capability? Are your designs adequate, maintained and any changes managed? Are your procedures thorough, identifying hazards, consequences and controls and easy to comprehend? How are you monitoring the operation of your facility? Does the design and operation of your facility enhance human performance or make it more difficult? Does your culture support safe operations?

6 Conclusion

A focus on the elements of process safety can help identify what is important in a business. In this instance there had been recommendations for a low-cost instrumented safety system to be installed that would have prevented the incident from occurring. The importance of this was not understood by the organization, and the opportunity to upgrade the ride during its recent refurbishment was ignored. It is clear that the hazard and potential consequences and controls were not understood. As is often the case, when the hazard is not understood and therefore not managed, people suffer tragic consequences.

7 References

[1] Kerin, T. *Managing Process Safety. In the Core Body of Knowledge for Generalist OHS Professionals*. Safety Institute of Australia, Tullamarine, VIC. 2017

[2] ISC (IChemE Safety Centre). Safety Centre Framework, Available at https://www.icheme.org/knowledge/safety-centre/framework/, Accessed on January 2, 2019

[3] ABC News, Available at https://www.abc.net.au/news/2018-06-18/dreamworld-fatal-accident-ride-had-broken-down-inquest-told/9871898, Accessed on January 2, 2019

[4] Australian Financial Review, Available at https://www.afr.com/news/dreamworlds-rapids-ride-different-from-others-around-the-world-20161101-gsfd07, Accessed on January 2, 2019

[5] 9 News, Video 11, Available at https://www.9news.com.au/2018/12/06/03/33/dreamworld-inquest-enters-final-days, Accessed on January 2, 2019

[6] news.com.au, Available at https://www.news.com.au/national/queensland/courts-law/coroner-james-mcdougall-requests-proof-all-dreamworld-rides-were-certified/news-story/e77275b97b0bbee9c1c541bda6df0035, Accessed on January 2, 2019

[7] ABC News, Available at https://www.abc.net.au/news/2018-06-19/dreamworld-operator-on-first-day-when-four-people-died-on-ride/9885214, Accessed on January 2, 2019

[8] ABC News, Available at https://www.abc.net.au/news/2018-06-27/dreamworld-trainer-promoted-since-four-killed-on-ride/9914188, Accessed on January 2, 2019

[9] 9 News, Video 5, Available at https://www.9news.com.au/2018/12/06/03/33/dreamworld-inquest-enters-final-days, Accessed on January 2, 2019

[10] ABC News, Available at https://www.abc.net.au/news/2018-06-20/dreamworld-operator-impossible-load-responsibilities-manage-ride/9889776, Accessed on January 2, 2019

[11] 9 News, Video 1, Available at https://www.9news.com.au/2018/12/06/03/33/dreamworld-inquest-enters-final-days, Accessed on January 2, 2019

[12] 9News, Available at https://www.9news.com.au/2018/11/20/03/32/dreamworld-ride-passed-wphs-safety-audit, Accessed on January 2, 2019

[13] ABC News, Available at https://www.abc.net.au/news/2018-10-08/dreamworld-inquest-hears-total-failure-to-ensure-safety/10350034, Accessed on January 2, 2019

[14] Intamin Worldwide, Available at https://www.intaminworldwide.com/infinity-falls-orlando/ , Accessed on December 12, 2018

[15] 9 News, Video 8, Available at https://www.9news.com.au/2018/12/06/03/33/dreamworld-inquest-enters-final-days, Accessed on January 2, 2019

[16] ABC News, Available at https://www.abc.net.au/news/2018-06-25/dreamworld-inquest-told-of-ride-maintenance-cutbacks/9906214, Accessed on January 2, 2019

[17] Australian Financial Review, Available at https://www.afr.com/real-estate/commercial/hotels-and-leisure/dreamworld-prioritised-new-attractions-over-maintenance-of-old-rides-inquest-20180626-h11urb, Accessed on January 2, 2019

[18] 9 News, Video 10, Available at https://www.9news.com.au/2018/12/06/03/33/dreamworld-inquest-enters-final-days, Accessed on January 2, 2019

[19] ABC News, Available at https://www.abc.net.au/news/2018-06-21/dreamworld-ride-operator-highly-distressed-give-evidence-inquest/9893308, Accessed on January 2, 2019

[20] news.com.au, Available at https://www.news.com.au/national/queensland/courts-law/dreamworld-engineer-tells-families-i-apologise-for-not-being-to-make-eye-contact-with-you/news-story/40545dc747c95eb96b0293c888d77d96, Accessed on January 2, 2019

[21] Sunshine Coast Daily, Available at https://www.sunshinecoastdaily.com.au/news/dreamworld-thunder-river-rapids-button-wasnt-regul/3543997/, Accessed on January 2, 2019

[22] ABC News, Available at https://www.abc.net.au/news/2018-06-26/dreamworld-ride-operator-says-conveyor-was-never-stopped/9910192, Accessed on January 2, 2019

[23] news.com.au, Available at https://www.news.com.au/national/queensland/courts-law/safety-report-shows-simple-change-on-dreamworld-ride-could-have-saved-lives/news-story/9c1fa2c195efd12f86a3e8ad516ea1b8, Accessed on January 2, 2019

[24] 9 News, Video 20, Available at https://www.9news.com.au/2018/12/06/03/33/dreamworld-inquest-enters-final-days, Accessed on January 2, 2019

[25] news.com.au, Available at https://www.news.com.au/travel/travel-updates/incidents/girl-11-dies-after-falling-into-the-water-on-splash-canyon-ride-during-school-trip-to-uk-theme-park/news-story/f83330b0ce12ea11df2d089ece676519, Accessed on January 2, 2019

[26] 9News, Available at https://www.9news.com.au/2018/12/06/03/33/dreamworld-inquest-enters-final-days, Accessed on January 2, 2019

[27] 9News, Available at https://www.9news.com.au/2018/10/16/14/43/dreamworld-inquest-thunder-river-rapids-ride-engineers-give-evidence, Accessed on January 2, 2019

[28] 9News, Available at https://www.9news.com.au/2018/10/17/10/49/dreamworld-inquest-engineer-to-appear-again, Accessed on January 2, 2019

[29] Australian Financial Review, Available at https://www.afr.com/news/dreamworld-ride-trainer-promoted-after-tragedy-staff-wanted-emergency-training-20180628-h11yke, Accessed on January 2, 2019

[30] 9 News, Video 12, Available at https://www.9news.com.au/2018/12/06/03/33/dreamworld-inquest-enters-final-days, Accessed on January 2, 2019

[31] ABC news, Available at https://www.abc.net.au/news/2018-10-09/dreamworld-ride-declared-safe-days-before-tragedy/10352164, Accessed on January 2, 2019

[32] ABC News, Available at https://www.abc.net.au/news/2018-10-17/dreamworld-supervisor-sympathy-has-families-in-tears-inquest/10387186, Accessed on January 2, 2019

[33] 9 News, Video 9, Available at https://www.9news.com.au/2018/12/06/03/33/dreamworld-inquest-enters-final-days, Accessed on January 2, 2019

[34] 9 News, Video 16, Available at https://www.9news.com.au/2018/12/06/03/33/dreamworld-inquest-enters-final-days, Accessed on January 2, 2019

[35] Australian Financial Review, Available at https://www.afr.com/research-tools/AAD/share-prices?days=1095, Accessed on November 26, 2018

[36] The Guardian, Available at https://www.theguardian.com/australia-news/2016/oct/27/deborah-thomas-set-to-receive-840000-in-bonuses-despite-dreamworld-deaths, Accessed on December 19, 2018

[37] ABC News, Available at https://www.abc.net.au/news/2016-10-27/dreamworlds-shareholders-ask-parent-company-to-support-victims/7970104, Accessed on January 2, 2019

[38] news.com.au, Available at https://www.news.com.au/lifestyle/real-life/news-life/dreamworld-first-responders-to-sue-theme-park-for-psychological-damage/news-story/e8fd7399eec9d165c657dbdb3ac35968, Accessed on January 2, 2019

19 AIChE Spring Meeting
15TH GLOBAL CONGRESS ON PROCESS SAFETY
NEW ORLEANS • 2019

Fatal Toxic Chemical Release at DuPont

Tamara Qureshi
U.S. Chemical Safety and Hazard Investigation Board
1750 Pennsylvania Ave, NW Suite 910 Washington, DC 20006
tamara.qureshi@csb.gov

Dan Tillema, Steve Cutchen, Lauren Grim, and Chris Lyon
U.S. Chemical Safety and Hazard Investigation Board
1750 Pennsylvania Ave, NW Suite 910 Washington, DC 20006

Prepared for Presentation at
American Institute of Chemical Engineers
2019 Spring Meeting and 15th Global Congress on Process Safety
New Orleans, LA
March 31 – April 3, 2019

Fatal Toxic Chemical Release at DuPont

Tamara Qureshi
U.S. Chemical Safety and Hazard Investigation Board
1750 Pennsylvania Ave, NW Suite 910 Washington, DC 20006
tamara.qureshi@csb.gov

Dan Tillema, Steve Cutchen, Lauren Grim, and Chris Lyon
U.S. Chemical Safety and Hazard Investigation Board

Keywords: DuPont, toxic release, gas detection, alarm

Abstract

On November 15, 2014, four workers were killed after approximately 24,000 pounds of highly toxic methyl mercaptan released from an insecticide production unit (Lannate® Unit) at the E.I. du Pont de Nemours chemical manufacturing facility in La Porte, Texas. This paper will discuss the events leading to the methyl mercaptan release, the deficiencies in DuPont La Porte's toxic gas detection system, and offer lessons on how to prevent some of the deficiencies in the toxic gas detection system from occurring. An effective toxic gas detection system can protect workers by alerting them of a potentially dangerous situation and preventing them from unknowingly entering a potential toxic chemical release.

1 Introduction

On November 15, 2014, approximately 24,000 pounds of highly toxic methyl mercaptan released from an insecticide production[1] unit (Lannate® Unit) at the E.I. du Pont de Nemours chemical manufacturing facility in La Porte, Texas. Four workers at the DuPont La Porte facility entered the Lannate® Unit's manufacturing building (manufacturing building) and the highly toxic methyl mercaptan release. Three of these workers were responding to a coworker's distress call from inside the manufacturing building. The fourth worker entered after all the workers inside the manufacturing building stopped communicating with workers inside the control room. The Lannate® Unit's methyl mercaptan toxic gas detection system did not effectively warn workers of the large chemical release occurring within the manufacturing building. Additionally, this toxic gas detection system did not provide emergency responders with sufficient information to effectively evaluate the potential harm to the public.

---

[1] The Insecticide Business Unit (IBU) was divided into two primary insecticide product lines: Lannate®, which is DuPont's trade name for methomyl; and Agricultural Products Intermediates (API), which produced insecticides and nematicides known as Vydate®, the trade name for oxamyl [3, p. 11].

The DuPont La Porte methyl mercaptan toxic gas detection system had several deficiencies. First, the methyl mercaptan detector system did not meet its design intent. The alarm set point for methyl mercaptan concentration and the method of conveying alarms did not forewarn workers of the immediately dangerous to life and health (IDLH) conditions they would encounter by entering the manufacturing building. The methyl mercaptan gas detectors only alarmed on the Lannate® control room operator's computer screen. There was no audible or visible alarm provided in the field for outside personnel. Second, the placement of the methyl mercaptan toxic gas detectors did not fully portray the extent of the November 15, 2014 methyl mercaptan release. Third, the DuPont La Porte site normalized the hazards associated with methyl mercaptan. Finally, the alarms did not prompt protective action or investigations into the sources of alarms. During the hours prior to the November 15, 2014 incident, multiple highly toxic chemical gas detectors alarmed (sounded). However, the DuPont emergency response team was not notified nor was the area cleared of personnel. In addition, methyl mercaptan releases on November 13 and 14, 2014, were identified by methyl mercaptan detectors, but were never reported as releases nor investigated as serious process safety incidents. Personnel believed they knew the source of the releases and normalized their responses to these alarms. Consequently, there was a delayed awareness of the release by operations, and workers who entered the manufacturing building after the release began were not alerted of nor prepared for the lethal concentrations of methyl mercaptan being released, resulting in the death of two of these workers.

The DuPont La Porte incident is the CSB's third investigation of a fatal DuPont incident in five years.[2] On September 30, 2015, the CSB issued a presentation, safety video, and an Interim Recommendations report [1],[3] which detailed needed safety improvements at the facility and issued recommendations to the DuPont La Porte facility to be implemented prior to resuming operations of the Lannate® Unit.[4] Due to the amount of methyl mercaptan stored at the DuPont La Porte site, the Lannate® process was covered by Occupational Safety Health Administration's (OSHA) Process Safety Management (PSM) program and the Environmental Protection Agency's (EPA) Risk Management Plan (RMP) Rule [2].

An effective toxic gas detection system can make the difference between personnel being prepared for a potentially dangerous situation or entering a potential toxic chemical release. To be effective, a toxic gas detection system needs to consider the chemical's hazards, the gas detector's

[2] The CSB also investigated an incident at a DuPont plant in Belle, West Virginia (three incidents that occurred in a 33-hour period over January 22-23, 2010, resulting in one fatality), and another in Buffalo, New York (a single incident that occurred on November 9, 2010, resulting in one fatality and one serious injury) [16], [17].

[3] In the Interim Recommendations, the CSB identified significant process safety management deficiencies at the DuPont La Porte facility, including delayed maintenance of safety-critical equipment, lack of written procedures, poor hazard analysis practices, lack of implementation of important inherently safer design concepts, and a lack of hazard recognition.

[4] In response to the CSB findings and recommendations, DuPont committed to making changes at the DuPont La Porte facility. DuPont La Porte, however, was not able to address all of these findings and recommendations, because in the spring of 2016 DuPont announced that it would not reopen the Lannate® Unit. The company had determined that "significant changes in market conditions during the period of the shutdown [would] persist over the long term [14]" and the cost required for the restart "[was] not a long-term viable and cost-efficient option for the DuPont Crop Protection business [15]." In 2017, DuPont dismantled and removed the buildings and equipment from the DuPont La Porte facility that were associated with its insecticide and herbicide business units, which included the Lannate® Unit.

capabilities, the location of the gas detectors, and the alarms associated with the gas detectors. This paper will discuss some of the events leading to the methyl mercaptan release, the deficiencies in DuPont La Porte's toxic gas detection system and offer lessons on how to prevent some of the deficiencies in the toxic gas detection system from occurring.

2 Events Leading to the Incident

On Monday, November 10, 2014, an inadvertent chemical dilution caused operating difficulties that forced a shutdown of DuPont La Porte's Lannate® Unit [3, pp. 20-21]. On Wednesday, November 12, 2014, operators tried to restart the Lannate® process. During the shutdown, the methyl mercaptan piping to the reaction section of the process plugged due to the formation of a clathrate hydrate (hydrate),[5] halting restart. At that time, operators did not know the source of the plugging and began troubleshooting the process to try to clear the piping, unaware that the solid hydrate had formed.

Two days later, on Friday, November 14, 2014, the troubleshooting efforts to clear the plugging were still ongoing. That morning, a technical team composed of engineers and other employees experienced in the Lannate® process, met with operations personnel to discuss troubleshooting options to clear the plugging. This meeting identified the likely scenario that water had entered the methyl mercaptan system and a solid hydrate formed as a result. Based on their understanding of the hydrate,[6] the technical team asked operators to put hot water on the outside of the methyl mercaptan piping, under the insulation, to warm the piping and its contents, to break-up (dissociate[7]) the plugging. The team also realized that when heated, methyl mercaptan would expand, requiring a safe place to vent to avoid over-pressuring the piping. To address this concern, operations personnel opened valves between the methyl mercaptan piping and a waste gas vent header, not recognizing that this alignment created a pathway for liquid methyl mercaptan to eventually release from the piping inside the manufacturing building when drain valves were opened. In addition, when operations personnel opened other valves while troubleshooting, such as drain valves, methyl mercaptan released to the atmosphere causing a strong methyl mercaptan odor[8] that could be smelled by site personnel.

---

[5] At low temperatures (≤ 52 °F) water and methyl mercaptan form a solid, ice-like material called methyl mercaptan clathrate hydrate [10]. Temperatures in the Houston area for the 24 hours preceding the incident averaged approximately 40 °F and had been consistently below 55 °F since Tuesday, November 11, 2014. Although the potential for methyl mercaptan hydrate formation had been identified in DuPont's methyl mercaptan technical standard and in a process hazard analysis years earlier, DuPont never implemented safeguards, such as heat tracing, or developed a procedure to dissociate the hydrate safely [3, p. 22].

[6] DuPont's methyl mercaptan technical standard identifies the potential to form a hydrate at low temperatures. The technical standard states, "[methyl mercaptan] will form a hydrate with water, which is a solid below 40 deg F [degrees Fahrenheit] per information provided by a [methyl mercaptan] supplier."

[7] Methyl mercaptan clathrate hydrates dissociate back to the liquid phases of methyl mercaptan and water at approximately 52 °F [10].

[8] Methyl mercaptan has a low odor threshold. It gives off a smell like rotten cabbage. It can be smelled at only two parts per billion (ppb) [9, p. 1]. OSHA's permissible exposure limit (PEL), however, is 10 parts per million (ppm), which is 5,000 times the minimum concentration it can be smelled. The low odor threshold of mercaptans makes them useful as fuel gas odorants [9, p. 2].

During this troubleshooting process, methyl mercaptan released both outside and inside the manufacturing building, triggering 32 methyl mercaptan gas alarms on the control panel throughout the 17 hours preceding the incident. While methyl mercaptan is a toxic and flammable chemical, and at high enough concentrations lethal, operations personnel did not respond in accordance with the nature of the emergency circumstances, because they associated the alarms with ongoing troubleshooting efforts. They did not perceive the methyl mercaptan alarms as signifying a serious hazard because they normalized[9] both the methyl mercaptan odor within the Lannate® Unit as well as the detector alarms associated with them.

Early in the morning on Saturday, November 15, 2014, the hot water from the hoses warmed the hydrate plugging, causing it to dissociate, clearing the plugging. Due to the valve alignment, liquid methyl mercaptan flowed into the waste gas vent header, located inside the manufacturing building [3, pp. 25-26]. This vent header, however, was not intended for liquid methyl mercaptan.

At 2:51 am alarms began to sound on the control system indicating high pressure in equipment located inside the manufacturing building. Operations personnel did not realize that liquid methyl mercaptan was causing the high pressure. Instead, they attributed the high pressure to a common, long-standing problem with process condensate (that DuPont La Porte personnel believed to be mostly water) accumulating in the vent header piping [3, p. 28]. The Lannate® control room operator (Board Operator) contacted the night shift supervisor (Shift Supervisor) and another operator (Operator 1) separately, asking them to help troubleshoot the high-pressure situation. The Shift Supervisor and Operator 1 each went separately to the manufacturing building, likely with the intent to drain liquid from the vent header piping manually.[10] Sometime between 3:01 am and 3:13 am, two sets of drain valves on the vent header piping, located on the third floor of the manufacturing building were opened (Figure 1). But instead of being comprised mostly of water as workers believed, liquid methyl mercaptan that had filled the piping escaped out of the valves, vaporized, and killed the Shift Supervisor.

Between 3:24 am and 3:26 am, three methyl mercaptan detectors inside the manufacturing building sensed at least 25 parts per million (ppm) methyl mercaptan[11] and triggered alarms at the control panel. The Board Operator—still focused on reducing the high pressure—did not realize a major chemical release inside the manufacturing building was occurring.

---

[9] Normalization of deviance is a phenomenon in which individuals or work teams gradually accept a lower standard of performance until the lower standard becomes the norm. It is typically the result of conditions slowly changing and eroding over time [11, p. 4].

[10] In 2011, DuPont installed a NOx (nitrogen oxides) reducing scrubbed (NRS) incinerator system. The waste gas vent header piping to the NRS Incinerator was designed without sufficient consideration of liquid accumulation. It contained low points where liquid could accumulate, and there was no engineered equipment provided, such as a knock out drum, to remove liquid safely from the waste gas vent header piping system. The liquid accumulation in the piping would create pressure in process equipment. Pressure buildup within the equipment was resolved by manually draining liquid from the waste gas vent header. See DuPont La Porte, Texas Chemical Facility Toxic Chemical Release: Interim Recommendations [3, p. 18].

[11] The methyl mercaptan detectors alarm at 25 ppm–well above the National Institute of Occupational Safety and Health (NIOSH) recommended exposure limit (REL) of 0.5 ppm and the OSHA enforceable (PEL) ceiling limit of 10 ppm. OSHA recommends that employers consider using the alternative occupational exposure limits because the agency believes that exposures above some of these alternative occupational exposure limits may be hazardous to workers, even when the exposure levels are in compliance with the relevant PELs [5]. The Cal-OSHA permissible exposure limit (PEL) is an 8-hour time-weighted average (TWA) of 0.5 ppm and NIOSH recommended exposure limit (REL) is a 15- minute ceiling limit of 0.5 ppm [7], [6].

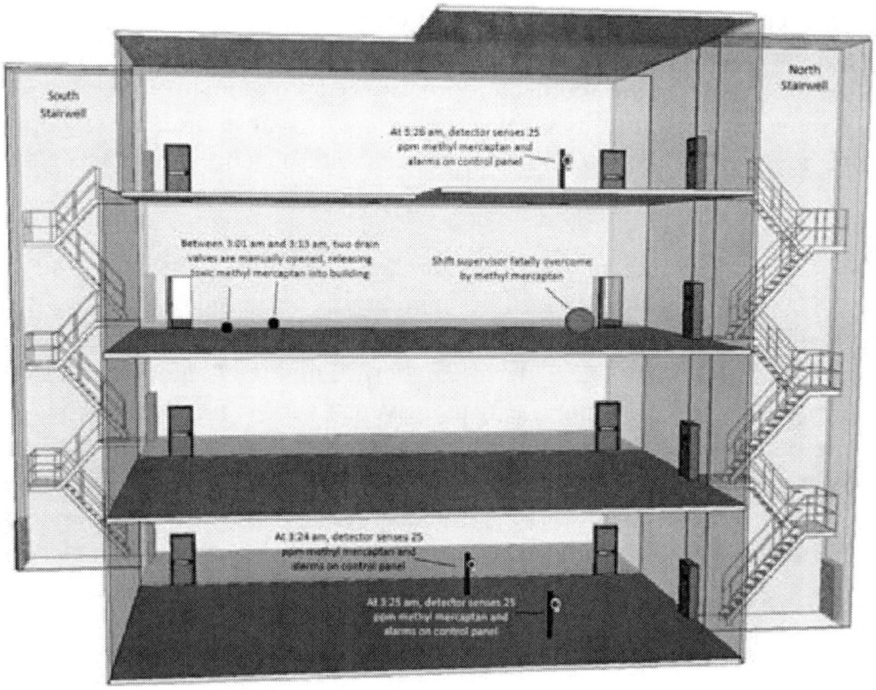

Figure 1. Depiction Showing Location Where Drain Valves Were Opened. These drain valves released toxic methyl mercaptan into the manufacturing building. Methyl mercaptan detectors on the first and fourth floors detected high concentrations of methyl mercaptan shortly after the release began. This depiction is a simplified graphic of the manufacturing building and does not show the location of Operator 1. Source: CSB

At about 3:30 am, Operator 1 made an urgent call for help over the radio. Various personnel interpreted her communication in different ways, generally as either "We need help!" or "I need help on the fourth floor!" The Board Operator tried to get more information through radio communication, but neither the Shift Supervisor nor Operator 1 ever responded [3, p. 29].
Two operators (Operator 2 and Operator 3) who were in the control room and heard the distress call ran to the manufacturing building to help. Another operator (Operator 4) saw Operator 2 and Operator 3 running to the manufacturing building, and he followed them. None of these operators knew there was a major release of toxic methyl mercaptan inside of the manufacturing building, and therefore they did not wear any respiratory protection when they ran into the manufacturing building. The manufacturing building lacked automatic visual or audible alarms to alert field workers or prevent them from entering a potentially toxic atmosphere.

Operator 2, Operator 3, and Operator 4 each entered the south stairway and took different routes inside the manufacturing building to try to find Operator 1. Operator 2 went to the third floor, where he was fatally overcome by methyl mercaptan [3, p. 29]. Operator 4 went to the second floor. He walked about ten feet and hit what he later described as a "wall" of methyl mercaptan, but he managed to retreat to the stairwell [3, p. 29]. Operator 3 went to the fourth floor but did not find anyone. He then announced on the Lannate® Unit's public-address system that he did not see anyone on the fourth floor. The Board Operator responded that the Shift Supervisor and Operator 1 may be on the third floor. Operator 3 then began to feel light headed. He made his way to the stairwell and lost consciousness while descending from the fourth floor [3, p. 29].

When Board Operator could not reach people inside the manufacturing building, an operator in the control room, Operator 6[12] (the brother of Operator 2), then grabbed three 5-minute escape respirators. Other operators in the control room warned him not to enter the manufacturing building because they did not know what was going on or where the non-responsive operators were. Operator 6, however, took the escape respirators and rushed to the manufacturing building [3, p. 29].

At about 3:40 am, Operator 6 encountered Operator 4 in the south stairway of the manufacturing building and put an escape respirator on him. The breathing air helped Operator 4 recover, and he exited the manufacturing building safely. A worker—likely Operator 6—then manually activated the manufacturing building fume release alarm, an alarm intended to alert area workers of a toxic chemical release in the building. Operator 6 then went to the third floor of the manufacturing building. At some point inside the manufacturing building after helping Operator 4, Operator 6 retrieved a self-contained breathing apparatus (SCBA). He tried to rescue his brother (Operator 2) by putting one of the escape respirators on his face. Operator 6 was then fatally overcome by methyl mercaptan before he could connect his own respirator mask to his SCBA tank.

3 Lessons from the DuPont La Porte Incident

3.1 Ensure Toxic Gas Detector Systems Operate as Intended

DuPont La Porte's methyl mercaptan technical standard stated, "[a]utomatic or continuous specific chemical detection systems *must* be installed where a [methyl mercaptan] release could have any off-site impact [...] [Emphasis in the original]." DuPont La Porte's methyl mercaptan toxic gas detection system consists of methyl mercaptan detectors, alarms, distributed control system (DCS), control panel, and a control room operator. The manufacturing building is equipped with gas detectors for methyl mercaptan in various locations. The third floor, where the release occurred, however, did not have any methyl mercaptan detectors. The detectors inside the manufacturing building were also set to alarm on the control panel when they detected a 25 ppm concentration of methyl mercaptan and were intended to protect workers from exposure to methyl mercaptan.[13]

DuPont used hydrogen sulfide (H_2S) detectors to identify methyl mercaptan. DuPont records indicate that the manufacturer has stated the detector should be capable of detecting methyl mercaptan within 1 ppm.[14] The alarms activated when the detectors sensed the hydrogen sulfide equivalent to 25 ppm methyl mercaptan.[15] The implementation of the methyl mercaptan toxic gas

[12] Operator 6 was a board operator for another process in the Lannate® Unit.

[13] The 25 ppm alarm set point is 50 times the NIOSH's REL of 0.5 ppm and more than double OSHA's ceiling limit of 10 ppm [7]. Therefore, alarming at this detection level could not protect workers from exposure to methyl mercaptan.

[14] DuPont La Porte's detectors for methyl mercaptan are calibrated for 0-50 ppm using a hydrogen sulfide (H_2S) calibration gas.

[15] The methyl mercaptan detectors were evaluated post-incident. Although the detectors inside the building have the ability to span 0-50 ppm, these detectors were set to span 0-25 ppm. Post-incident, the span was increased to 50 ppm. This configuration of the alarms limited the amount of information communicated to DuPont La Porte personnel.

detection system inside the manufacturing building, however, did not meet the design intent, federal safety regulations, or DuPont corporate standards.

The methyl mercaptan detectors outside and inside the manufacturing building were installed for two different purposes—the outside detectors were geared towards an offsite release and the inside detectors were meant to protect workers from a hazardous environment. DuPont La Porte documentation stated that the basis for the alarms outside the manufacturing building was to provide an early warning mechanism for a potential offsite methyl mercaptan release. The alarm setpoint was based on the ERPG (Emergency Response Planning Guidelines)-2[16] concentration[17] for methyl mercaptan which is 25 ppm. DuPont La Porte installed the methyl mercaptan detectors inside the manufacturing building after the methyl mercaptan detectors outside the building. Unlike the outside detectors, there was no documentation for the basis for setting the alarm setpoint to 25 ppm for the methyl mercaptan detectors inside the manufacturing building nor was there evidence that these setpoints were set at 25 ppm to prevent methyl mercaptan alarms from becoming a nuisance. It appears that the setpoints were copied from the outside methyl mercaptan detectors even though the inside detectors served a different purpose—worker protection. The methyl mercaptan detectors alarmed at 25 ppm—well above the NIOSH recommended exposure limit (REL) of 0.5 ppm and the OSHA enforceable PEL ceiling limit of 10 ppm.

Depending on the purpose of a toxic gas detection system alarm, there are different sets of values for facilities to consider. For instance, ERPG values are designed for the protection of the public during an emergency and not intended for the everyday routine protection of workers [4].[18] For worker protection, OSHA recommends that employers consider using the lower alternative occupational exposure limits because the agency believes that exposures above some of these alternative occupational exposure limits may be hazardous to workers, even when the exposure levels are in compliance with the relevant PELs [5]. For example, the Cal-OSHA permissible exposure limit (PEL) is an 8-hour time-weighted average (TWA) of 0.5 ppm and NIOSH's REL is a 15- minute ceiling limit of 0.5 ppm [6], [7].

Additionally, information from the methyl mercaptan detectors inside the manufacturing building only displayed on the DCS monitor and was not recorded by the site's DCS and data historian [3, p. 37]. CSB investigators, however, were able to obtain information from alarm logs about the detectors.

DuPont La Porte's management of change documentation for installation of methyl mercaptan detectors inside the manufacturing building stated the condition prompting the change was the

---

[16] ERPG-2 is the maximum airborne concentration below which nearly all individuals could be exposed for up to 1 hour without experiencing or developing irreversible or other serious health effects or symptoms which could impair an individual's ability to take protective action. At concentrations above ERPG-2, some members of the community may experience significant health effects which could impair them from taking protective action. Emergency responders often alert the community to take protective action in areas susceptible to the ERPG-2 concentration and above [19].

[17] The ERPG values for methyl mercaptan in parts per million (ppm) are: ERPG-1, 0.005; ERPG-2, 25; ERPG-3, 100 [18, p. 6].

[18] Emergency Response Planning Guidelines (ERPGs) are chemical concentration levels that are used during emergency response and when preparing for emergencies. They are designed to anticipate public health effects not occupational health effects from exposure to airborne chemicals [4].

possibility that personnel could enter a potentially hazardous location without knowing methyl mercaptan had been released:

> *Present Conditions and Description of Proposed Change:*
>
> *Personnel can enter a potentially hazardous location without knowing MeSH [methyl mercaptan] has been released to the environment.*

The methyl mercaptan gas detectors, however, only alarmed on the Lannate® control room operator's computer screen.

DuPont La Porte site did not have audible or visual alarms on or in the manufacturing building to alert field personnel of a methyl mercaptan release automatically. DuPont relied on verbal communication of alarms that automatically displayed on a continuously manned control board. Without being made aware of the major chemical release inside the building, Operator 2, Operator 3, and Operator 4 entered without respiratory protection. As a result, Operator 2 was killed from methyl mercaptan exposure, and Operator 3 and Operator 4 were exposed to dangerous concentrations of methyl mercaptan.

Facilities should make strategic decisions when choosing a toxic gas detector to ensure that a toxic gas detection system operates as intended. Considerations should include regulatory requirements, NIOSH recommendations, chemical interferences, and utility (worker protection, release modeling, etc.). The methyl mercaptan detectors at the DuPont La Porte facility had a span of 0-50 ppm. This range may be appropriate for early detection of a toxic leak when coupled with alarms and clear instructions to evacuate the premises at lower concentrations. If the goal is to understand other aspects of a release, like potential lethal concentration (150 ppm is methyl mercaptan's IDLH concentration) or modeling an offsite release, then a wider concentration range should be considered.

Additionally, facilities should also consider the use of automatic alarms designed to alert personnel of hazardous conditions—such as a release of a toxic chemical or explosive atmosphere—which can relay critical safety information immediately to personnel, without potential delays inherent in a system that solely relies on a control room operator to verbally communicate safety alarms to operators in the field.

3.2 Ensure Toxic Gas Detector Placement is Strategic

Although the DuPont La Porte site had 13[19] fixed methyl mercaptan detectors inside and outside the manufacturing building, there were no fixed methyl mercaptan detectors located on the third floor inside the manufacturing building where a majority of the November 15, 2014 methyl mercaptan release occurred. Additionally, no workers were provided with either portable or personal methyl mercaptan monitors. Part of the Lannate® process occurred inside the closed manufacturing building, a four-story building containing a large amount of piping and process equipment, dividing walls, multiple stairwells, and an elevator. The Lannate® Unit had methyl

[19] The site had a total of 15 methyl mercaptan detectors—there were two additional methyl mercaptan detectors installed in another process building adjacent to the manufacturing building.

mercaptan process piping both inside and outside the manufacturing building. A facility siting study and a process hazards analysis had identified the necessity of having methyl mercaptan detectors in the Lannate® manufacturing building. These studies identified the need for protecting workers from toxicity and explosivity hazards related to potential methyl mercaptan releases.

DuPont La Porte installed 10 methyl mercaptan detectors near process piping outside the manufacturing building. The design basis for the outside methyl mercaptan detectors was to provide an early warning of a methyl mercaptan release. DuPont La Porte placed these detectors around its methyl mercaptan storage tank and the area where methyl mercaptan could be unloaded. There were, however, no fixed methyl mercaptan detectors along the fence line that could relay whether there was an offsite release.

DuPont La Porte installed three methyl mercaptan detectors inside the manufacturing building. The stated design purpose for the methyl mercaptan detectors inside the building was worker protection. The inside detectors were located within the building on the first and fourth floors. These detector locations were determined based on potential release scenarios from a 2001 process hazard analysis (PHA).

The methyl mercaptan release occurred on the third floor where there were no methyl mercaptan detectors. Although the 2001 PHA did not identify methyl mercaptan detector locations on the third floor, DuPont La Porte routed the waste gas vent header piping along the third floor and this piping was directly connected to the top of the methyl mercaptan storage tank. During abnormal or upset conditions, the waste gas vent header could have been a potential source for a methyl mercaptan release. For example, ineffective hazard management while troubleshooting the plugged methyl mercaptan feed piping during the November 15th incident allowed liquid methyl mercaptan to flow into the waste gas vent header piping towards the NRS incinerator—a location where it was never intended to go.

Additionally, the routine draining of liquid from the waste gas vent header should have triggered a management of change review and considered whether additional safeguards like methyl mercaptan detectors were necessary. As discussed earlier, DuPont La Porte never fully resolved liquid accumulation in the waste gas vent header piping to the NRS incinerator with an engineered solution. Instead, operators were directed to drain the system by area management through written instructions in a daily instruction log book to operations staff [8, p. 18].[20] This routine draining presented potential dangers because DuPont La Porte personnel never sampled or analyzed the liquid workers routinely drained from this system [8, p. 18]. Daily instructions before the incident requested the operators to sample the uncharacterized liquid. DuPont La Porte management wanted samples of the liquid for a subsequent laboratory analysis to identify its chemical composition. Workers reported that the liquid had a strong, noxious odor of sulfur compounds [8, p. 18]. Even though DuPont La Porte was aware that the liquid likely contained more than just water and was unaware of the actual chemical composition [8, p. 18], instructions to workers did

[20] Beginning January 14, 2014, and continuing until the incident, the daily instructions directed operators to drain the methyl mercaptan waste gas vent header's "low point" on the 3rd floor once per shift. After the installation of a flexible hose from the waste gas vent header drain valve to the floor, the instructions changed to include directing the liquid through the hose under a running safety shower until the liquid stops flowing from the hose [8, p. 18].

not specify that additional precautions—like wearing additional PPE—were needed when draining the liquid from the waste gas vent header to the floor inside the manufacturing building.

Consequently, DuPont La Porte workers dealt with the common problem of liquid accumulation in the waste gas vent header on a routine basis by draining the liquid (line breaking) without engineered solutions or ensuring the use of safety procedures or personal protective equipment. On November 15, 2014, when the liquid drain valves were opened this time, flammable and highly toxic methyl mercaptan flowed onto the floor and filled the manufacturing building with toxic vapor and there no methyl mercaptan detectors on the third floor to alert personnel of the hazardous conditions that existed there.

As with toxic detector selection, the placement of toxic gas detectors must be strategic in order to capture the potential areas where a release could occur and the extent of the release. The methyl mercaptan detector placement both externally and internally at the DuPont La Porte facility did not operate as intended in part because the detectors did not fully cover the areas where the majority of the methyl mercaptan was released. For example, there was no real time data for emergency responders to use to evaluate the potential for offsite concentrations during the methyl mercaptan release. Although emergency responders conducted offsite monitoring hours after the initial release, the information obtained from air monitoring could not give an accurate portrayal of the methyl mercaptan release. Furthermore, the methyl mercaptan monitors inside the building did not provide accurate information about the release because they were not located near the open drain valves.

Facilities can adopt a variety of methods to select the location of toxic gas monitors with the understanding that they should be prepared for scenarios where the monitoring location may not be effective. One technique that a facility could use is to develop a series of release scenarios and based on those scenarios place monitors in the path of those releases (including by the potential sources of a leak). This technique, however, may not cover all potential release areas. Additionally, facilities should evaluate additional locations for toxic gas detectors if there is a change in process that creates new potential release scenarios. Another technique, if focusing on worker protection, is to give workers personal monitors when they enter toxic gas hazardous areas, especially if these areas are not normally occupied or congested with equipment. At DuPont La Porte, commercially available personal hydrogen sulfide monitors could have been issued to workers. Additionally, if using fixed toxic gas monitors for personnel protection, they should be installed near the breathing zone.

3.3 Ensure Personnel Understand Toxic Gas Hazards

DuPont La Porte personnel did not fully appreciate the effects of methyl mercaptan's physical characteristics. Although methyl mercaptan[21] is a highly toxic, flammable chemical, its low odor

[21] Personnel at DuPont La Porte would refer to methyl mercaptan as "MeSH." MeSH is shorthand based upon methyl mercaptan's chemical formula: CH_3SH. "Me" represents the methyl group (CH_3) and "SH" represents the mercaptan group (SH)—and when combined, these initialisms forms MeSH.

threshold[22] and ability to cause olfactory fatigue in personnel can generate a false sense of security when working with the chemical.

Although DuPont's methyl mercaptan technical standard states, "[a]ll DuPont sites *will* protect the site and neighboring businesses and communities from the escape of a highly toxic material [,]" methyl mercaptan detectors provided essentially no value to DuPont personnel [Emphasis in the original]. DuPont operations staff became accustomed to the odor of methyl mercaptan being present in their work environment. The low odor threshold of methyl mercaptan resulted in a false perception of safety. Consequently, the workers did not associate detecting the odor of methyl mercaptan as representing an immediate hazard.

> *"And I smell MeSH [methyl mercaptan] all the time ... but it never ... I smelled MeSH but I never felt I couldn't breathe. Like whether it's, there's -- whether it's a small leak or anything, it just never -- I didn't think MeSH was that dangerous. I thought it, you know, but it is."*

As seen by this comment from a DuPont La Porte worker, smelling methyl mercaptan at low concentrations without having any observable, adverse effects can potentially normalize the danger of a highly toxic chemical.

Furthermore, extended exposure to methyl mercaptan could potentially eliminate the perception of a methyl mercaptan release because olfactory fatigue will prevent workers from being able to smell the chemical. The odor threshold for methyl mercaptan is approximately two parts per billion (2 ppb). However, olfactory fatigue is known to occur with methyl mercaptan, resulting in the exposed person to no longer being able to detect the chemical by smell. As a result, using odor as an indicator of a release may not provide adequate warning of hazardous concentrations [9].

In the events leading to the major methyl mercaptan release, there were many smaller releases, filling the area with methyl mercaptan odor. In part because personnel believed they understood the source of releases and normalized the risk associated with these smaller releases, personnel did not apply additional safeguards like clearing the area or wearing the required personal protective equipment required by the site when dealing with a methyl mercaptan release.

Facilities should ensure that employees understand the hazards of chemicals that they are working with and may be potentially exposed to if there is a release. In cases where there are multiple hazardous chemicals, such as methyl isocyanate[23] (MIC) and methyl mercaptan (as was the case at DuPont La Porte), it is important that companies effectively train workers on the hazards of each chemical at their facility. Furthermore, companies should train workers that handle multiple highly toxic chemicals, such as methyl mercaptan and MIC, to treat these chemicals with equal

[22] Odor threshold, in general, is the lowest concentration of a gas or a material's vapor that can be detected by odor. Odor threshold values are not fixed physiological parameters or physical constants but are statistical points representing the best estimate value from a group of individual responses [12, p. 2].

[23] Methyl isocyanate (MIC) is a raw material used in the production of insecticides at La Porte including the Lannate® process. In December 3, 1984, a methyl isocyanate release at the Union Carbide insecticide plant in Bhopal, India resulted in an estimated 3,800 people that died within days, and tens of thousands that were injured. Eventually, the release killed tens of thousands of people. See CSB Safety Message on 30th anniversary of fatal chemical release that killed thousands in Bhopal, India [13].

importance. This training is important for workers to help ensure they have a working knowledge of those chemical hazards.

3.4 Ensure Response to a Toxic Gas Detector Alarm Is Appropriate

The toxic gas detection system at DuPont La Porte had a history of false readings. Because of these false readings, personnel supplanted their ability to observe and detect a leak over the toxic gas detection system. Personnel adapted their responses to a potential toxic gas release to include determining whether the toxic gas detectors were functioning properly by using their perception of smell before implementing the site's leak response procedures.

For example, the Lannate® Unit used chlorine in addition to methyl mercaptan. Workers could observe from a distance a chlorine leak by the green color of the gas or the immediate adverse effects (i.e., difficulty breathing) of the gas. Workers, similarly, would also use their sense of smell to detect methyl mercaptan. Methyl mercaptan gas, however, is colorless and the odor is not a reliable indicator of a release (for the reasons noted above). In both situations where workers used their physical observations to determine where there was a leak, workers were at risk for being overcome by a toxic chemical release because they were not using additional safeguards like personal protective equipment (PPE).

DuPont's highly toxic material standards require that toxic gas detection systems be treated the same as safety interlocks. DuPont defines a safety interlock as a system or function that detects an out-of-limits (abnormal) condition or improper sequence and brings it to a safe condition. In this case, an alarm should trigger a predetermined response from operators to bring the system to a safe condition. In practice, the methyl mercaptan detectors were not treated as a safety interlock.

Multiple DuPont employees informed the CSB that odor was the primary method used to locate a methyl mercaptan leak. DuPont personnel routinely responded to methyl mercaptan detector alarms without any additional respiratory protection. When they responded, if a significant leak was identified, these personnel consistently stated they would leave to get a SCBA and then return.

On the night of the incident, multiple methyl mercaptan detector alarms activated, but because the workers were aware that methyl mercaptan was being released from their activities associated with clearing the hydrate, these alarms were normalized and not treated as a safety interlock. The alarms were not treated as a near miss and there was no actual investigation into the source or how to prevent these releases. No additional personal protective equipment, such as air-purifying respirators or SCBAs was used. In addition, DuPont did not order a unit evacuation and did not notify the emergency response team until hours later, after Operator 1 made the distress call.

When a toxic detection system has a history of false readings it creates an opportunity for workers to normalize toxic gas detector readings and their response, making the toxic gas detection system ineffective. Facilities should take steps to address the potential for false readings and reduce opportunities for workers to normalize false readings and their alarms. One method is to treat toxic gas detector alarms as a near miss and investigate them even if there are no injuries or the readings are determined to be false. This method should trigger an effort to understand alarm management and safety management system issues and address them. Additionally, facilities should ensure that workers are using leak response procedures to prevent personnel from entering into potential hazardous situations without proper safeguards, like PPE.

4 Conclusions

The Lannate® Unit's toxic gas detection system did not effectively warn workers of the November 15, 2014 methyl mercaptan release occurring inside the manufacturing building. The manufacturing building, including the building's overall construction, detectors, and alarms, was not designed such that personnel could be readily notified of a major chemical release in the manufacturing building. Additionally, the methyl mercaptan toxic gas system did not provide real-time, accurate information to warn of potential offsite consequences.

There are several lessons that facilities can learn from the incident to help improve their toxic gas detection systems. Gas detectors must operate as intended to prevent workers from normalizing the hazard the detectors are protecting workers from. For toxic gas detection system to be effective, the detector selection and placement have to be strategic to identify and quantify releases. Finally, personnel have to understand and appreciate the hazards associated with the toxic gases they work with to help ensure they respond appropriately to potential toxic gas releases.

5 References

[1] U.S. Chemical Safety and Hazard Investigation Board (USCSB), "DuPont La Porte Facility Toxic Chemical Release," [Online]. Available: https://www.csb.gov/dupont-la-porte-facility-toxic-chemical-release-/. [Accessed 1 March 2019].

[2] U.S. Environmental Protection Agency (USEPA), "List of Regulated Substances under the Risk Management Plan (RMP) Program," [Online]. Available: https://www.epa.gov/rmp/list-regulated-substances-under-risk-management-plan-rmp-program. [Accessed 2017 August 28].

[3] U.S. Chemical Safety and Hazard Investigation Board (USCSB), "DuPont La Porte, Texas Chemical Facility Toxic Chemical Release: Interim Recommendations," 30 September 2015. [Online]. Available: https://www.csb.gov/assets/1/20/dupont_la_porte_interim_recommendations_2015-09-30_final1.pdf?15526. [Accessed 28 February 2019].

[4] National Oceanic and Atmospheric Administration (NOAA) Office of Response and Restoration, "Emergency Response Planning Guidelines (ERPGs)," [Online]. Available: https://response.restoration.noaa.gov/oil-and-chemical-spills/chemical-spills/resources/emergency-response-planning-guidelines-erpgs.html. [Accessed 10 March 2019].

[5] United States Department of Labor (USDOL) Occupational Safety and Health Administration (OSHA), "Permissible Exposure Limits - Annotated Tables," [Online]. Available: https://www.osha.gov/dsg/annotated-pels/. [Accessed 28 February 2019].

[6] United States Department of Labor (USDOL) Occupational Safety and Health Administration (OSHA), "Annotated Table Z-1," [Online]. Available: https://www.osha.gov/dsg/annotated-pels/tablez-1.html. [Accessed 28 February 2019].

[7] Centers for Diseases Control and Prevention National Institute of Occupational Safety and Health (NIOSH), "NIOSH Pocket Guide to Chemical Hazards - Methyl Mercaptan," 11 April 2016. [Online]. Available: https://www.cdc.gov/niosh/npg/npgd0425.html. [Accessed 28 February 2019].

[8] U.S. Chemical Safety and Hazard Investigation Board (USCSB), "DuPont La Porte, Texas Chemical Facility Toxic Chemical Release: Interim Recommendations," 30 September 2015. [Online]. Available: https://www.csb.gov/assets/1/20/dupont_la_porte_interim_recommendations_2015-09-30_final1.pdf?15526.

[9] Agency for Toxic Substances and Disease Registry (ATSDR), "Methyl Mercaptan," [Online]. Available: https://www.atsdr.cdc.gov/MHMI/mmg139.pdf. [Accessed 28 February 2019].

[10] A. H. Mohammadi and D. Richon, "Equilibrium Data of Sulfur Dioxide and Methyl Mercaptan Clathrate Hydrates," *Journal of Chemical & Engineering Data,* vol. 56, pp. 1666-1668, 2011.

[11] Center for Chemical Process Safety (CCPS), Recognizing Catastrophic Incident Warning Signs in the Process Industries, Hoboken: Wiley, 2012.

[12] A. H. Lehocky, S. S. Murnane and D. Patrick, Odor Thresholds for Chemicals with Established Occupational Health Standards, Falls Church: AIHA, 2013.

[13] USCSB, "On 30th Anniversary of Fatal Chemical Release that Killed Thousands in Bhopal, India, CSB Safety Message Warns it Could Happen Again," [Online]. Available: http://www.csb.gov/on-30th-anniversary-of-fatal-chemical-release-that-killed-thousands-in-bhopal-india-csb-safety-message-warns-it-could-happen-again-/?pg=4.

[14] K. Baker, "DuPont Closing Texas Plant Where 4 Died," The News Journal, 1 April 2016. [Online]. Available: https://www.usatoday.com/story/money/2016/04/01/dupont-closing-texas-plant-where-4-died/82512270/. [Accessed 29 June 2017].

[15] DuPont, "Form 8-K," 31 March 2016. [Online]. Available: https://www.sec.gov/Archives/edgar/data/30554/000003055416000115/dd33116form8-k.htm.

[16] U.S. Chemical Safety and Hazard Investigation Board (USCSB), "E.I. DuPont de Nemours & Co., Inc. Belle, West Virgina Investigative Report," September 2011. [Online]. Available: https://www.csb.gov/assets/1/20/csb%20final%20report.pdf?13966. [Accessed 1 March 2019].

[17] U.S. Chemical Safety and Hazard Investigation Board (USCSB), "E. I. DuPont De Nemours Co. Fatal Hotwork Explosion," [Online]. Available: https://www.csb.gov/e-i-dupont-de-nemours-co-fatal-hotwork-explosion/. [Accessed 1 March 2019].

[18] American Industrial Hygiene Association (AIHA), "Current ERPG Values," 2014. [Online]. Available: https://www.aiha.org/get-involved/AIHAGuidelineFoundation/EmergencyResponsePlanningGuidelines/Documents/2014%20ERPG%20Values.pdf.

[19] F. Cavender, S. Phillips and M. Holland, "Development of Emergency Response Planning Guidelines (ERPGs)," *Journal of Medical Toxicology,* vol. 4, no. 2, pp. 127-131, 2008.

553063 Hurricane Harvey 20 Months Later - Insights on the Arkema Crosby Flooding

On August 24, 2017, Hurricane Harvey, a Category 4 hurricane, made landfall in southeast Texas producing an unprecedented amount of rainfall. Hurricane Harvey killed 68 people and flooded over 300,000 structures, forcing roughly 40,000 people to evacuate their homes. Hurricane Harvey required a large emergency response effort supported by local, state, and Federal officials.

At Arkema's Crosby site, Harvey delivered an unprecedented amount of rainfall. The National Oceanic and Atmospheric Administration estimates show that the rainfall during Harvey had less than a one-in-one-thousand-year likelihood of occurring. Harris County has estimated that at Crosby, Harvey was a 5,000 year to 20,000 year event. The plant's hurricane preparedness procedure was followed and resulted in a safe shutdown and hurricane preparedness of the manufacturing units. However, finished goods inventory remaining in the warehouses was overwhelmed by high water and overcame multiple layers of protection used to safely store organic peroxide finished goods.

This presentation will explain the sites preparedness ahead of time, the events leading up to the flooding, the incident itself, and the emergency response.

The presentation will focus on the impacts, the main lessons learned and the key recommendations from the CSB report.

AUTHOR INDEX

AUTHOR INDEX

AIChE
120 Wall Street, FL 23
New York, NY 10005-4020

ISBN 978-1-5108-8689-6

CURRAN ASSOCIATES INC.
proceedings.com